CHEMICAL EQUILIBRIA AND KINETICS IN SOILS

R

CHEMICAL EQUILIBRIA AND KINETICS IN SOILS

Garrison Sposito
University of California
at Berkeley

New York Oxford
OXFORD UNIVERSITY PRESS
1994

Oxford University Press

Oxford New York
Athens Auckland Bangkok Bombay
Calcutta Cape Town Dar es Salaam Delhi
Florence Hong Kong Istanbul Karachi
Kuala Lumpur Madras Madrid Melbourne
Mexico City Nairobi Paris Singapore
Taipei Tokyo Toronto

and associated companies in
Berlin Ibadan

Published by Oxford University Press, Inc.,
198 Madison Avenue, New York, New York 10016

Library of Congress Cataloging-in-Publication Data
Sposito, Garrison, 1939–
Chemical equilibria and kinetics in soils /
Garrison Sposito.
p. cm. Rev. and expanded ed. of:
Thermodynamics of soil solutions. 1981.
Includes bibliographical references and index.
ISBN 0-19-507564-1
1.Soil solutions. 2. Thermodynamics.
3. Chemical kinetics.
I. Sposito, Garrison, 1939–
Thermodynamics of soil solutions.
II. Title
S592.5.S66 1994 631.4'1—dc20 93-46714

9 8 7 6 5 4 3 2
Printed in the United States of America
on acid-free paper

FOR WERNER STUMM

lux mentis lux orbis

PREFACE

Chemical thermodynamics is the theoretical structure on which the description of macroscopic assemblies of matter at equilibrium is based. This branch of physical chemistry was created 120 years ago by Josiah Willard Gibbs and was perfected by the 1930's through the work of G. N. Lewis and E. A. Guggenheim. The fundamental principles of the discipline thus have long been established, and its scope as one of the five great subdivisions of physical science includes all the chemical phenomena that material systems can exhibit in stable states. It is indeed powerful enough to provide unifying principles for organizing and interpreting compositional data on natural waters and soils. Although these data are known to represent only transitory states of matter, characteristic of open systems in nature, they can be analyzed in a thermodynamic framework so long as the time scale of experimental observation is typically incommensurable with the time scales of transformation among states of differing stability, a point stressed admirably some 70 years ago by Gilbert Newton Lewis and Merle Randall.[1] The practitioners of chemical thermodynamics applied to soil and water phenomena thereby have drawn success from an acute appreciation of the natural time scales over which these phenomena take place and from a perceptive intuition of how to make the "free-body cut": the choice of a closed model system whose behavior is to mimic an investigated open system in nature.

Given the firm status of chemical thermodynamics, its application to describe chemical phenomena in soils would seem to be a straightforward exercise, but experience has proven different. An obvious reason for the difficulties that have been encountered is the preponderant complexity of soils. These multicomponent chemical systems comprise solid, liquid, and gaseous phases that are continually modified by the actions of biological, hydrological, and geological agents. In particular, the labile aqueous phase in soil, the soil solution, is a dynamic, open, natural water system whose composition reflects especially the many reactions that can proceed simultaneously between an aqueous solution and a mixture of mineral and organic solids that itself varies both temporally and spatially. The net result of these reactions may be conceived as a dense web of chemical interrelations mediated by variable fluxes of matter and energy from the atmosphere and biosphere. It is to this very complicated milieu that chemical thermodynamics must be applied.

An attempt to make this application prompted the appearance of *The Thermodynamics of Soil Solutions* (Oxford University Press, 1981). Besides its evident purpose, to demonstrate the use of chemical thermodynamics, this book carried a leitmotif on the fundamental limitations of chemical thermodynamics for describing natural soils. These limitations referred especially to the influence of kinetics on stability, to the accuracy of thermodynamic data, and to the impossibility of deducing molecular mechanisms. The problem of mechanisms vis-à-vis thermodynamics cannot be expressed better than in the words of M. L. McGlashan:[2] "what can we learn from thermodynamic equations about the microscopic or molecular explanation of macroscopic changes? Nothing whatever. What is a 'thermodynamic theory'? (The phrase is used in the titles of many papers published in reputable chemical journals.) There is no such thing. What then is the use of thermodynamic equations to the chemist? They are indeed useful, but only by virtue of their use for the calculation of some desired quantity which has not been measured, or which is difficult to measure, from others which have been measured, or which are easier to measure." This point cannot be stated often enough.

The intervening years have brought the limitations as to kinetics and mechanism into sharper focus, necessitating the present volume, which is a revised and expanded textbook version of *The Thermodynamics of Soil Solutions*. The need for revision was based especially on a growing awareness that the quantitative description of soils in terms of the behavior of their chemical species cannot be considered complete without adequate characterization of the *rates* of the chemical reactions they sustain. Full recognition must be given and full account taken of the fact that few chemical transformations of importance in natural soils go to completion exclusively outside the time domain of their observation at laboratory or field scales. A critical implication of this fact is that one must distinguish carefully between *thermodynamic* chemical species, sufficient in number and variety to represent the stoichiometry of a chemical transformation between stable states, and *kinetic* chemical species, required to depict completely the mechanisms of the transformation. The difficulty in bringing to fulfillment the study of rate processes in natural systems derives from the fact that no general laws of overall reaction kinetics exist in parallel with the general laws of thermodynamics, and no necessary genetic relationship with which to connect kinetic species to thermodynamic species is known. The result of these conceptual lacunae is a largely empirical science of chemical rate processes, at times still rife with inadequate theory and confusing data.

This textbook is intended primarily as a critical introduction to the use of chemical thermodynamics and kinetics for describing reactions in the soil solution. Therefore no account is given of phenomena in the gaseous and solid portions of soil unless they impinge directly on the properties of the aqueous phase, a restriction conducive to clarity in presentation and relevance to the interests of most soil chemists. Although the discussion in this book is self-contained, it does presume exposure to thermodynamics and kinetics as taught in basic courses on physical chemistry. Since most of the examples discussed relate to *soil* chemistry, a background in that discipline at the level of *The Chemistry of Soils* (Oxford University Press, 1989) will be of direct help in understanding the applications presented.

I should like to express my deep appreciation to William Casey, Wayne Robarge, and Samuel Traina for their forthright, careful review of the manuscript for this book, and to Luc Derrendinger for his commentary on Chapter 6. Their critical questions helped to exorcise numerous unclear passages and errors in the text. Finally, I thank Mary Campbell-Sposito for her assistance in preparing the index; Frank Murillo for his great skill in drawing the figures; and Danny Heap, Joan Van Horn, and Terri DeLuca for their patience in making a clear typescript from a great pile of handwritten yellow sheets. None of these persons, of course, is responsible for errors or obscurities that may remain in this book. Each only deserves my gratitude for keeping *les sottises* to a relative minimum.

NOTES

1. G. N. Lewis and M. Randall, *Thermodynamics and the Free Energy of Chemical Substances*, McGraw-Hill, New York, 1923.
2. M. L. McGlashan, The scope of chemical thermodynamics, *Chemical Thermodynamics, Spec. Periodical Rpt.* **1**:1-30 (1973).

Berkeley
January 1994

G. S.

CONTENTS

4 Surface Reactions 138

5 Ion Exchange Reactions 181

6 Colloidal Processes 222

CHEMICAL EQUILIBRIA AND KINETICS IN SOILS

όδὸς ἄνω κάτω μία καὶ ώυτή

T. S. Eliot
Burnt Norton

Je ne sais en vérité ce qu'il faut le plus admirer, de l'excès de bonté des hommes qui accueillent de si pauvres essais, ou de mon incroyable assurance à lancer de pareilles sottises dans le monde.

Marcel Bénabou
Pourquoi je n'ai écrit aucun des mes livres

1

CHEMICAL EQUILIBRIUM AND KINETICS

1.1 Chemical Reactions in Soils

Soils are multicomponent, multiphase, open systems that sustain a myriad of interconnected chemical reactions, including those involving the soil biota. The *multiphase* nature of soil derives from its being a porous material whose void spaces contain air and aqueous solution. The solid matrix (which itself is multiphase), soil air, and soil solution—each is a mixture of reactive chemical compounds—hence the *multicomponent* nature of soil. Transformations among these compounds can be driven by flows of matter and energy to and from the vicinal atmosphere, biosphere, and hydrosphere. These external flows, as well as the chemical composition of soil, vary in both space and time over a broad range of scales.

The complexity of soil notwithstanding, the principal features of its chemical behavior can be understood on the basis of well-established principles and methods for the description of reactions in *aqueous systems*. Reactions that occur exclusively in the gaseous phase or the solid matrix of soil less often control its chemical behavior than reactions involving the aqueous phase. The basic terminology associated with the latter chemical reactions will be reviewed in the present chapter to provide an initial context for the discussion of equilibria and kinetics to follow.

A chemical reaction is termed *elementary* if it occurs in a single step, with no intermediate species appearing before the products of the reaction have formed. An elementary reaction takes place on the molecular level *exactly as written* in terms of reactants and products. A reaction that is not elementary is *composite* or *overall*.[1] An example of an elementary reaction is the hydration of dissolved carbon dioxide in a soil solution to form the neutral species $H_2CO_3^0$ ("true carbonic acid"):

$$CO_2(aq) + H_2O(\ell) \rightarrow H_2CO_3^0(aq) \qquad (1.1)$$

where *aq* refers to an aqueous solution phase and ℓ to the liquid phase. In this elementary reaction, one CO_2 molecule combines with one H_2O molecule to form directly one molecule of H_2CO_3. The *molecularity* of this reaction is 2 (i.e., it is a bimolecular reaction), since that is the total number of reactant species that come together to form the product.[1] This product, incidentally, is to be distinguished conceptually from "loosely solvated CO_2," sometimes denoted as a "species" by $CO_2{\bullet}H_2O$, which, at equilibrium, makes up about 99.7% of the "nominal carbonic acid" (usually denoted $H_2CO_3^*$) in aqueous solutions.[2]

Another elementary reaction of molecularity 2 is the combination of true carbonic acid with hydroxide ion to form bicarbonate ion and water:

$$H_2CO_3^0(aq) + OH^-(aq) \rightarrow HCO_3^-(aq) + H_2O(\ell) \qquad (1.2)$$

Evidently, the overall reaction:

$$CO_2(aq) + H_2O(\ell) \rightarrow H^+(aq) + HCO_3^-(aq) \qquad (1.3)$$

can be developed by adding the two elementary reactions in Eqs. 1.1 and 1.2 to the elementary unimolecular reaction that describes the dissociation of the water molecule:

$$H_2O(\ell) \rightarrow H^+(aq) + OH^-(aq) \qquad (1.4)$$

The concept of molecularity thus is not applied to the reaction in Eq. 1.3, since it does not display the actual molecular mechanism involving the intermediate species, $H_2CO_3^0$, OH^-, and H_2O. Therefore, it would be incorrect to interpret the reaction in Eq. 1.3 as the combination of one CO_2 *molecule* with a water *molecule* to form H^+ and HCO_3^- *ions*. The error in this line of reasoning is brought into sharper focus after noting that the elementary bimolecular reaction:

$$CO_2(aq) + OH^-(aq) \rightarrow HCO_3^-(aq) \qquad (1.5)$$

can be added to the elementary unimolecular reaction in Eq. 1.4 to produce again the composite reaction in Eq. 1.3 by a completely different pathway from that obtained by the synthesis of Eqs. 1.1, 1.2, and 1.4. Experiment shows that both pathways are operable in the pH range 8–10.[2] This example illustrates how Eq. 1.3, like all other overall reactions, cannot be interpreted *prima facie* in molecular mechanistic terms. All that can be said is that 1 mol CO_2 when reacted with 1 mol H_2O yields a mole each of protons and bicarbonate ions in solution.

The development of chemical reactions to describe the transformations of material substances and the determination of which chemical reactions are elementary (i.e., the determination of reaction mechanisms) are principal research objectives in chemical science and in soil chemistry. Elementary reactions are always interpreted at the molecular level; therefore, experimental methods that probe at molecular space and time scales, notably spectroscopy, must be applied to characterize reaction mechanisms. By contrast, overall reactions have no unique molecular interpretation and therefore can be investigated with macroscopic methods that provide information only about changes in chemical composition as influenced, for example, by temperature, pressure, or time. The great complexity of soil chemical behavior has perforce dictated that *most transformations of soil constituents be described by overall reactions*. The rapid improvement in noninvasive spectroscopic techniques during the past decades suggests, however, that ultimately the description of soil chemistry in terms of elementary reactions is a realizable goal. This possibility is enhanced by the simplifying fact that all elementary reactions can be classified as acid–base (in the Lewis sense), oxidation–reduction, or free radical reactions.[3]

The reaction of dissolved CO_2 with hydroxide ions depicted in Eq. 1.4 takes place entirely in the aqueous solution phase and so is termed *homogeneous*.[1] Another example of a homogeneous reaction is the formation of an outer-sphere complex by Mn^{2+} and Cl^- in a soil solution:[4]

$$Mn^{2+}(H_2O)_6(aq) + Cl^-(aq) \rightarrow Mn^{2+}(H_2O)_6Cl^-(aq) \qquad (1.6)$$

where $Mn^{2+}(H_2O)_6$ represents an octahedral solvation complex (inner-sphere) and $Mn^{2+}(H_2O)_6Cl^-$ is an outer-sphere manganese–chloride complex. The weakly associated chloride complex is proposed to transform to an inner-sphere chloride complex by Cl^- exchange for a water molecule in the first solvation shell around Mn^{2+}:[5]

$$Mn^{2+}(H_2O)_6Cl^-(aq) \rightarrow MnCl^+(H_2O)_5(aq) + H_2O(\ell) \qquad (1.7)$$

This pair of homogeneous reactions can be added to derive the following overall reaction:

$$Mn^{2+}(aq) + Cl^-(aq) \rightarrow MnCl^+(aq) \qquad (1.8)$$

in which the water species are now suppressed to emphasize the overall nature of the complexation process depicted.

A reaction that involves chemical species in more than one phase is termed *heterogeneous*.[1] An example is the composite reaction describing the reductive dissolution of the common soil mineral hematite (α-Fe_2O_3) in the presence of visible light by oxalic acid ($H_2C_2O_4$), a ubiquitous plant litter degradation product:

$$Fe_2O_3(s) + H_2C_2O_4(aq) + 4\,H^+(aq) \overset{h\upsilon}{\rightarrow} 2\,Fe^{2+}(aq) + 2\,CO_2(aq) + 3\,H_2O(\ell) \qquad (1.9)$$

where *s* refers to the solid phase and $h\upsilon$ denotes a quantum of visible light. The sequence of elementary reactions underlying this mineral dissolution process is a topic of current research. In one scenario[6] (Fig. 1.1), the oxalate anion forms an inner-sphere complex with a Fe^{3+} cation exposed at the surface of the mineral, is subsequently excited by a photon of visible light, transfers an electron to the complexed Fe^{3+} ion to reduce it to Fe^{2+}, and finally decomposes into CO_2 species. The surface Fe^{2+} cation then detaches from the mineral as a solvation complex and equilibrates with the aqueous solution phase at the ambient pH value. This mechanistic sequence—which would be very different, for example, in the absence of photons or in the presence of oxygen—is no more than implicit in Eq. 1.9. Without the underlying elementary reactions, Eq. 1.9 states simply that 1 mol hematite combined with 1 mol oxalic acid in the presence of free protons can produce 2 mol Fe^{2+} and CO_2, plus 3 mol water.

Macroscopic chemical techniques can be used to characterize overall reactions like those in Eqs. 1.3, 1.8, and 1.9. Given the complexity of reaction mechanisms, however, measurements of the composition of the aqueous system in which an overall reaction occurs over the course of time may not always yield data that conform to the expected stoichiometry. For example, if the reaction of carbon dioxide and water to produce protons and bicarbonate ions is initiated at high pH (very low proton concentration), the disappearance of 1 mol CO_2 need not be accompanied by the disappearance of 1 mol H_2O (because of Eq. 1.5) or by the appearance of 1 mol H^+ (because of Eq. 1.1).[2,7] The unaccounted-for presence of intermediate species (like $H_2CO_3^0$ in Eq. 1.1) can lead typically to a delay in the formation of one or more final product species relative to the others, such that the expected stoichiometry in an overall reaction is violated when the reaction progress is monitored. This transient feature of mole balance in overall reactions has important ramifications when the kinetics of soil chemical processes are investigated (Section 1.3).

1.2 The Equilibrium Constant

If the reactants and products in Eq. 1.3 are at equilibrium, the reaction can be expressed in the following equation:

$$CO_2(aq) + H_2O(\ell) = H^+(aq) + HCO_3^-(aq) \qquad (1.10)$$

where the equals sign signifies the equilibrium condition. A *thermodynamic equilibrium constant* can be defined for this reaction at a chosen temperature and pressure, usually 25°C (298.15 K) and 1 atm (101.325 kPa):

$$K \equiv \mathbf{(H^+)\ (HCO_3^-)/(CO_2)\ (H_2O)} \qquad (1.11)$$

where boldface refers to the thermodynamic *activity* of the chemical species, as described in Special Topic 1 at the end of this chapter. The parameter K has a fixed value, regardless of the composition of the soil solution. To make this

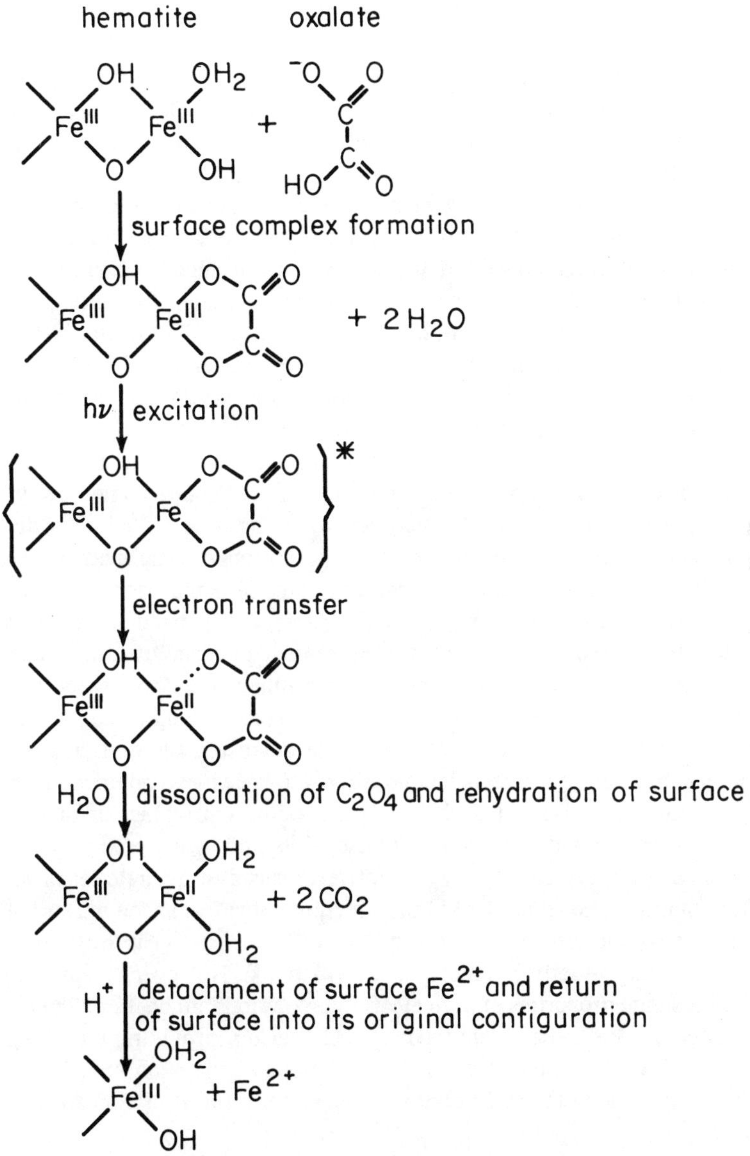

FIG. 1.1. A possible mechanism for the reductive dissolution of hematite by oxalic acid in the presence of light (after Stumm et al.[6]). See Section 3.4 for additional discussion of reductive dissolution reactions.

assertion a reality, the activity of a species is related to its molality (moles per kilogram of water) or its concentration (in moles per cubic decimeter) through an *activity coefficient*:

$$(i) \equiv \gamma_i \, [i] \tag{1.12}$$

where i is some chemical species, like H^+ or CO_2, of concentration $[i]$. The activity coefficient γ_i has the units kg mol^{-1} (or dm^3 mol^{-1}), such that the activity has *no units* and the thermodynamic equilibrium constant is *dimensionless* (see Special Topic 1).

Conventions and laboratory methods have been developed to measure γ_i, (i), and K in aqueous solutions.[8] All species activity coefficients, for example, are required to approach the value 1.0 (kg mol^{-1} or dm^3 mol^{-1}) when the species is in its *Reference State*. There are two principal definitions of the Reference State for a solute in aqueous solution, like H^+, HCO_3^-, and CO_2 in Eq. 1.10. One is the *Infinite Dilution Reference State*, wherein the activity coefficient of a solute is defined to approach unit value as the concentration approaches zero for *each* dissolved component of an aqueous solution at T = 298.15 K and P = 1 atm. The other is the *Constant Ionic Medium Reference State*, wherein the activity coefficient of a solute approaches unit value as the concentration of *only* that solute approaches zero, while the concentrations of all the other dissolved components of the aqueous solution (the "background ionic medium") remain fixed. Both definitions are valid thermodynamically, and each has advantages and disadvantages. For example, in the case of the proton, the use of the Constant Ionic Medium Reference State means that the activity coefficient of H^+ in most soil solutions will very nearly have unit value and, therefore, that a glass electrode will measure directly the proton *concentration* in these solutions. There is no need to calibrate the electrode against a set of standard buffer solutions, since one may, in principle, simply make known additions of protons to a reference solution and read the corresponding emf values of the electrode in order to calibrate it. On the other hand, this kind of calibration would have to be done for every soil solution of interest instead of a single set of standard buffer solutions (assuming that liquid junction potentials in the buffer solutions are negligibly different from those in the soil solutions). The Infinite Dilution Reference State usually is employed in this book. However, many published thermodynamic properties of pure electrolyte solutions are based on the Constant Ionic Medium Reference State (usually with $NaClO_4$ providing the background ionic medium), and this choice of Reference State is popular among those who study seawater and other saline natural waters whose composition does not vary greatly.

Even with the definition of the Reference State, chemical thermodynamics alone cannot provide a unique methodology for the measurement of single-ion activity coefficients. An infinitude of possibilities exists, each of that calls upon its own *extra*thermodynamic set of conventions according to criteria of experimental convenience and intended application. However, chemical thermodynamics does provide general constraints that limit any set of arbitrary conventions defining single-ion activities.[9]

Consider an aqueous solution containing, among others, the electrolyte M_aL_b(aq), where M refers to a metal, L refers to a ligand, and a and b are stoichiometric coefficients. The activity of the electrolyte M_aL_b is measurable by well-established methods.[8,10] Experimental data pertaining to electrolyte activities usually are catalogued in terms of the *mean ionic activity coefficient* $\gamma_\pm$: [10]

$$(M_aL_b) = \gamma_\pm^{(a+b)} m_{TM}^a m_{TL}^b \tag{1.13}$$

where m_{TM} and m_{TL} are *total* molalities of the metal and ligand, respectively. If only a single electrolyte were present in the aqueous solution to which $\gamma_\pm$ refers, then the product of molalities on the right side of Eq. 1.13 would reduce to a power of the mean ionic molality:[8]

$$m_\pm = (a^a b^b)^{1/(a+b)} m_T \tag{1.14}$$

where m_T is the molality of the electrolyte. The molalities m_{TM}, m_{TL}, and m_T are wholly macroscopic quantities that can be measured by standard spectroscopic, complexometric, or gravimetric methods.[8] Thus $\gamma_\pm$ can be calculated unambiguously with Eq. 1.13 after the activity of the electrolyte M_aL_b(aq) has been determined. It is evident that the mean ionic activity coefficient has a strict chemical thermodynamic significance.

By analogy with Eq. 1.13, one can define *single-ion activity coefficients*:[9]

$$(M_aL_b) \equiv \gamma_M^a \gamma_L^b \, m_M^a m_L^b \tag{1.15}$$

where γ is a single-ion activity coefficient, m_M is the molality of the *species* M^{m+}(aq), and m_L is the molality of the *species* L^{l-}(aq). For γ_M and γ_L to have chemical significance, the species molalities, m_M and m_L, must have a well-defined operational meaning (see Section 2.4). Thus *the single-ion activity coefficient has no meaning apart from the set of operational procedures used to define ionic species and to determine their concentrations in an aqueous solution.* Although the left sides of Eqs. 1.13 and 1.15 always must be the same, it is *not* possible in general to equate total molalities with species molalities, nor to equate $\gamma_\pm$ with $[\gamma_M^a \gamma_L^b]^{1/(a+b)}$.

The mean ionic and single-ion activity coefficients are conceptually different parameters, but both must conform to the Debye-Hückel infinite-dilution limit. This theoretical constraint on activity coefficients takes on a particular mathematical form, depending upon the way in which an electrolyte solution is characterized. In a strictly thermodynamic picture of aqueous solutions, the Debye-Hückel limit can be expressed as follows:[9]

$$\lim_{I_s \downarrow 0} \ln \gamma_\pm = -pqA_{DH}\sqrt{I_s} \tag{1.16}$$

where ln is logarithm to the base e, p and q are the valences of M and L in Eq. 1.13, A_{DH} is the Debye-Hückel limiting law parameter (A_{DH} = 1.1762 $kg^{1/2}$ $mol^{-1/2}$, or 1.1780 $dm^{3/2}$ $mol^{-1/2}$ at 298 K). The parameter I_s is the *stoichiometric ionic strength*:

$$I_s = \tfrac{1}{2} \Sigma_i Z_i^2 m_{Ti} \tag{1.17}$$

where Z_i is the valence of the metal or ligand whose total molality is m_{Ti} and the sum is over all metals (including hydrogen) and ligands (including hydroxide) in solution. The Debye-Hückel limit for the single-ion activity coefficients of an electrolyte is similar to Eq. 1.16:

$$\lim_{I_{ef}\downarrow 0} \frac{1}{a+b} \ln\left[\gamma_M^a \gamma_L^b\right] = -pqA_{DH}\sqrt{I_{ef}} \tag{1.18}$$

where I_{ef} is the *effective ionic strength*:

$$I_{ef} = \tfrac{1}{2} \Sigma_i Z_i^2 m_i \tag{1.19}$$

In Eq. 1.19, the sum is over all charged *species* in the solution. In the limit of infinite dilution, soluble complexes should make a negligible contribution to I_{ef}. If this is true, then Eqs. 1.16 and 1.18 can be combined into the following single equation:

$$\lim_{I_{ef}\downarrow 0} \frac{1}{a+b} \ln\left[\gamma_M^a \gamma_L^b\right] = \lim_{I_s\downarrow 0} \ln \gamma_\pm \tag{1.20}$$

This equation represents a general theoretical constraint on single-ion activity coefficients.

A general empirical constraint on single-ion activity coefficients also can be imposed by means of *Young's rules.*[10] For dilute solutions, Young's rules are equivalent to the statement that pairwise interactions between ions of opposite charges make the dominant contribution to $\gamma_\pm$, γ_M, and γ_L. With respect to $\gamma_\pm$, this empirically based conclusion is often specialized to the *Principle of Specific Interaction.*[10] Equations 1.16 and 1.18 are expressions of Young's rules in the Debye-Hückel limit, in the sense that the ionic strength parameter accounts for the effect of pairwise interactions between ions of opposite charge. At finite ionic strength, Young's rules suggest that any mathematical expression for $\ln \gamma_\pm$ (or $\ln \gamma_M$ and $\ln \gamma_L$) should include both linear and bilinear terms in the molalities of all metals and ligands (or all charged species) in an aqueous solution.[10]

The *Davies equation* is a semiempirical expression for calculating single-ion activity coefficients in soil solutions having effective ionic strengths up to about 0.5 mol kg^{-1}. Other equations for γ_M or γ_L exist, but the Davies equation has the distinct advantages of reliability in mixed electrolyte solutions and of exhibiting only one adjustable parameter whose value is independent of the chemical nature of a charged species. The Davies equation for the activity coefficient of a charged species J is expressed as follows:[9]

$$\ln \gamma_J = -A_{DH} Z_J^2 \left[\frac{\sqrt{I_{ef}}}{1 + \sqrt{I_{ef}}} - 0.3 I_{ef} \right] \tag{1.21}$$

where Z_J is the valence of species J. The adjustable parameter in the Davies equation is the coefficient of I_{ef}, which has the value $0.3A_{DH}Z_J^2$.

For *uncharged* monovalent metal–ligand complexes, proton–ligand complexes (or dissolved gases), and bivalent metal–ligand complexes, some model semiempirical equations for γ_i are the following:[11]

$$\log \gamma_{ML} = \frac{-0.192 I_{ef}}{0.0164 + I_{ef}} \qquad (M = Na^+, K^+, \text{etc.}) \tag{1.22}$$

$$\log \gamma_{HL} = 0.1 I_{ef} \tag{1.23}$$

$$\log \gamma_{ML} = -0.3 I_{ef} \quad (M = Ca^{2+}, Mg^{2+}, \text{etc.}) \tag{1.24}$$

for $I_{ef} < 0.1$ mol dm^{-3}, where log is logarithm to the base 10. These expressions conform to a theoretical requirement for *neutral* species, that log γ become proportional to I_{ef} in the infinite-dilution limit.[11]

The expressions for single-species activity coefficients in Eqs. 1.21–1.24 suffice to calculate activities of dissolved solutes like H^+ or CO_2 in Eq. 1.11. For the solvent, H_2O, it is still necessary to define a Reference State, which is that of the pure liquid at 298.15 K under 1 atm pressure.[12] The activity of the solvent is conventionally set equal to the product of a *rational activity coefficient* f and the mole fraction of the solvent x:[12]

$$\mathbf{(H_2O)} \equiv f_{H_2O}\, x_{H_2O} \tag{1.25}$$

where x_{H_2O} is the ratio of the moles of water to the total moles of water and solutes in an aqueous solution. For most soil solutions, $x_{H_2O} \approx 1.0$ and, therefore, $f \approx 1.0$, making $\mathbf{(H_2O)}$ correspondingly close to the value 1.0.

The combination of Eqs. 1.11 and 1.12 under the condition $(H_2O) \approx 1$ leads to the following expression:

$$\begin{aligned} K &= \mathbf{(H^+)(HCO_3^-)/(CO_2)(H_2O)} \approx \mathbf{(H^+)(HCO_3^-)/(CO_2)} \\ &= \gamma_H [H^+]_e \gamma_{HCO_3} [HCO_3^-]_e / \gamma_{CO_2} [CO_2]_e \\ &\equiv (\gamma_H \gamma_{HCO_3} / \gamma_{CO_2}) K_c \end{aligned} \tag{1.26}$$

where[1]

$$K_c \equiv [H^+]_e [HCO_3^-]_e / [CO_2]_e \tag{1.27}$$

is a *conditional equilibrium constant* for the reaction in Eq. 1.10 and the γ are prescribed by Eqs. 1.21–1.24. The conditional equilibrium constant is defined in terms of equilibrium species *concentrations*, $[\ ]_e$, which makes it less abstract than K in Eq. 1.11, but also renders it composition dependent. Moreover, K_c has units (in this case, either molality or mol dm^{-3}), whereas K has no units.

The conceptual meaning of the activity of a chemical species stems from the formal similarity between K and K_c. The conditional equilibrium constant is a more direct parameter with which to characterize equilibria, but it depends on composition, in that it contains species concentrations only, and therefore it does not correct for the interactions among species that occur as their concentrations change. In the limit of infinite dilution, these interactions must die out, and the extrapolated value of K_c must represent chemical equilibrium in an ideal solution wherein species interactions (other than those involved to form a complex like HCO_3^-) are unimportant. The concentrations in K_c become equal numerically to activities in the limit of either no interactions among species (Infinite Dilution Reference State) or an invariant set of interactions among species (Constant Ionic Medium Reference State). Thus *the activity factors in K play the role of hypothetical concentrations of species in an ideal solution*. But the real solution is not ideal as species concentrations increase because the species are brought closer together to interact more strongly. When this occurs, K_c must begin to deviate from K. The activity coefficient is introduced to "correct" the concentration factors in K_c for this nonideal species behavior and thereby restore the value of K via Eq. 1.26. This correction is expected to be larger for charged species than for neutral complexes (dipoles), and larger as the species valence increases. These trends are reflected in the model expressions in Eqs. 1.21–1.24.

1.3 Reaction Rate Laws

For the chemical reaction in Eq. 1.3, the *extent of reaction* ξ is defined by the following differential expression:[1]

$$d\xi \equiv -dn_{CO_2} = -dn_{H_2O} = dn_H = dn_{HCO_3} \qquad (1.28)$$

where n is the number of moles of a substance and the assumption is made that *the reaction stoichiometry is known and is constant* over the time period during which the reaction is investigated. More generally, if A represents a reactant with stoichiometric coefficient $-\upsilon_A$ and B represents a product with stoichiometric coefficient υ_B, then

$$d\xi \equiv dn_A/\upsilon_A = dn_B/\upsilon_B \qquad (1.29)$$

for any substances A and B, where $\upsilon_A < 0$ and $\upsilon_B > 0$ by IUPAC convention.[1]

In Eq. 1.3, $\upsilon_A = -1$ for any A and $\upsilon_B = +1$ for any B. Since Eq. 1.3 is an overall reaction, the assumption of constant stoichiometry underlying the definition of ξ is not trivial, as discussed in Section 1.1. For example, at high pH, Eq. 1.28 would not always be applicable because of the influence of the reactions in Eqs. 1.1 and 1.5. On the other hand, at equilibrium, when the hydration reaction is described by Eq. 1.10, the application of Eq. 1.28 is possible. This fact serves to emphasize the difference between *equilibrium* chemical species that figure in *thermodynamic* parameters (e.g., Eq. 1.11) and *kinetic* species that figure in the *mechanism* of a reaction. The set of kinetic species is in general larger than the set of equilibrium species for any overall chemical reaction.

The *rate of conversion* is the time derivative of the extent of reaction.[1] Thus the rate is

$$\frac{d\xi}{dt} = \upsilon_A^{-1}\frac{dn_A}{dt} = \upsilon_B^{-1}\frac{dn_B}{dt} \tag{1.30}$$

for any substances A and B in a reaction. If the volume V of the phase in which a reaction occurs is constant, then[1]

$$V^{-1}\frac{d\xi}{dt} = \upsilon_A^{-1}\frac{dc_A}{dt} = \upsilon_B^{-1}\frac{dc_B}{dt} \tag{1.31}$$

is the *rate of reaction based on concentration,* where $c = n/V$ is concentration.[1] Sometimes the left side of Eq. 1.31 is also termed the rate of concentration increase. If the volume V is not constant or is not the volume of a particular phase, the terms on the right side of Eq. 1.31 are expressed as $\upsilon_A^{-1}V^{-1}\,dn_A/dt$ and $\upsilon_B^{-1}\,V^{-1}\,dn_B/dt$. This kind of generalization is needed for reactions occurring in open systems or in multiphase systems (heterogeneous reactions). Note that the rate of reaction has the same numerical value for all species involved, as long as Eq. 1.29 can be applied. Thus the rate of the reaction in Eq. 1.3 can be measured by monitoring the moles of CO_2, water, protons, or bicarbonate over time, as long as the stoichiometry of the reaction does not change.

The rate of a chemical reaction in aqueous solution typically is assumed to depend in some way on the composition of the solution. As an example, consider the following overall reaction to form a neutral sulfate complex with a bivalent metal cation as the central group:

$$M^{2+}(aq) + SO_4^{2-}(aq) \rightarrow MSO_4^0(aq) \tag{1.32}$$

where the metal M can be Ca, Mg, Mn, Cu, etc. Detailed spectroscopic investigation shows that $MSO_4^0(aq)$ can be either an inner- or outer-sphere complex, with the latter species dominant. The rate at which MSO_4^0 forms is quite high, as is usual for metal–ligand complexes.[5] In mathematical terms, this rate of formation can be expressed by the time derivative of $[MSO_4^0]$, where, as in Eq. 1.12, the square brackets represent a concentration in moles per liter (moles per cubic decimeter). The rate of increase of soluble complex concentration can be measured by a variety of spectroscopic and electrochemical techniques.[5,8]

It is common to *assume* that the observed rate can be represented mathematically by the difference of two terms:[13]

$$\frac{d[MSO_4^0]}{dt} = R_f - R_b \tag{1.33}$$

where R_f and R_b each are functions of the composition of the solution in which the reaction in Eq. 1.32 takes place, as well as of the temperature and pressure. Because the reaction in Eq. 1.32 is not elementary, Eq. 1.33 need not have any direct relationship to the molecular mechanism by which MSO_4^0 forms. For example, there could be intermediate species that do not appear in the reaction in Eq. 1.32 but that help to determine the observed rate and prevent it from being a simple difference expression. Whenever Eq. 1.33 is appropriate, however, R_f and R_b usually are interpreted as the respective rates of formation ("forward reaction") and dissociation ("backward reaction") of MSO_4^0. It is then common to *assume* that R_f depends on powers of the concentrations of the reactants and that R_b depends on powers of the concentrations of the products:[13]

$$\frac{d[MSO_4^0]}{dt} = k_f[M^{2+}]^{\alpha}[SO_4^{2-}]^{\beta} - k_b[MSO_4^0]^{\delta} \tag{1.34}$$

where k_f, k_b, α, β, and δ are empirical parameters. The exponents α, β, and δ—which need not be integers—are the *partial orders* of the reaction with respect to the associated species [e.g., αth order with respect to M^{2+}(aq)]. The parameters k_f and k_b are the *rate coefficients* for the formation ("forward") and dissociation ("backward") reactions, respectively. Each of the five parameters in Eq. 1.34 may be functions of composition, temperature, and pressure.[1] Note that the SI units of the rate coefficients will depend on the partial orders of the reaction with respect to reactants or products and that *there is no necessary relationship between order and molecularity*. Equations 1.33 and 1.34 thus are empirical models of the overall reaction rate whose relevance to molecular mechanism must be demonstrated, not assumed. Any such model of an overall reaction is required only to yield a positive rate when the direction of the reaction is consistent with a decrease in Gibbs potential, and a zero rate when chemical equilibrium is established.[13]

Equation 1.34 is an example of a *reaction rate law*. Its mathematical form and five associated empirical parameters are objects for experimental study. To facilitate this study, Eq. 1.34 might be reformulated as the specific rate law:

$$\frac{d[MSO_4^0]}{dt} = k_f[M^{2+}][SO_4^{2-}] = k_f[M^{2+}]^2 \tag{1.35}$$

under the conditions that (a) the rate of dissociation of the complex is negligible; (b) the reaction orders with respect to M^{2+} and SO_4^{2-} are the same as the stoichiometric coefficients of these two species in Eq. 1.32; and (c) $[M^{2+}] = [SO_4^{2-}]$. The experimental value of k_f now will have the units $dm^3\ mol^{-1}\ s^{-1}$ and the *overall* order of the reaction ($\equiv \alpha + \beta$ in Eq. 1.34) will be 2 [irrespective of assumption (c)]. Equation 1.35 can be simplified further and solved explicitly for $[M^{2+}]$ as a function of time[14] after rewriting the left side as $-d[M^{2+}]/dt$ (by Eq. 1.32) and prescribing an initial condition on the concentration of M^{2+}. The mathematical expression that results from solution then can be fitted to experimental rate data in order to test Eq. 1.35 and determine the value of the formation rate coefficient k_f.

Alternatively, if the reaction in Eq. 1.32 is at equilibrium, thereby eliminating conditions (a) and (c), then the condition $R_f = R_b$ can be imposed along with assumption (b) (i.e., $\alpha = \beta = \delta = 1$) and Eq. 1.34 leads to the following expression:

$$\frac{k_f}{k_b} = \frac{[MSO_4^0]_e}{[M^{2+}]_e[SO_4^{2-}]_e} \equiv K_{sc} \tag{1.36}$$

as applied to the reaction in Eq. 1.32, where $[\ ]_e$ is the concentration of a species *at equilibrium*. The parameter K_{sc} defined by the right side of Eq. 1.36 has the units of inverse concentration and is the *conditional stability constant* for the formation of the complex MSO_4^0. It is "conditional" because it is equal numerically to k_f/k_b, a function of composition, temperature, and pressure. Equation 1.36 shows that K_{sc} can be calculated either with kinetics data (k_f and k_b) or with equilibrium data (the $[\ ]_e$). An alternative possibility is that one of the rate coefficients can be calculated by measuring the other rate coefficient along with the equilibrium concentrations.

The facile line of reasoning that leads to Eqs. 1.33–1.36 is so abundant in the literature of soil chemical kinetics that the rather arbitrary nature of the underlying assumptions often is forgotten.[13] For example, Eq. 1.36 is not a unique consequence of Eq. 1.33. If the rate law

$$\frac{d[MSO_4^0]}{dt} = g([M^{2+}],[SO_4^{2-}],[MSO_4^0]) \left\{ K_{sc} - \frac{[MSO_4^0]}{[M^{2+}][SO_4^{2-}]} \right\}^p \tag{1.37}$$

were to replace Eq. 1.33, where $p > 1$ and $g(\cdot)$ is any positive-valued function of species concentrations, then the direction of the arrow in Eq. 1.32 still would be respected and Eq. 1.36 still could be derived, but the rate of increase of

$[MSO_4^0]$ would not be equal to a simple difference between rates of formation and dissociation of the complex. This example shows that the thermodynamic requirements of a positive overall reaction rate when conditions are favorable and a zero rate at equilibrium are not sufficient to invoke Eq. 1.33 as the unique rate law. On the other hand, even if Eq. 1.33 is assumed, Eq. 1.36 is not a necessary consequence. The rate law in Eq. 1.34 leads, at equilibrium, to the expression

$$\frac{k_f}{k_b} = \frac{[MSO_4^0]_e^\delta}{[M^{2+}]_e^\alpha [SO_4^{2-}]_e^\beta} \tag{1.38}$$

This equation reduces to Eq. 1.36 only if (1) the ratio of rate coefficients on the left side depends on a power of K_{sc}; (2) the partial reaction orders α, β, δ each are the same multiple of that power; and (3) the power is equal to 1. If the third condition is not met, k_f/k_b will be equal to some power of K_{sc}, but not to K_{sc} itself. More generally,[13] for the overall formation reaction

$$aA + bB \rightarrow cC \tag{1.39}$$

the necessary conditions are

$$k_f/k_b = f(K_{sc}^p) \tag{1.40a}$$

$$\frac{\delta}{c} = \frac{\alpha}{-a} = \frac{\beta}{-b} = p \tag{1.40b}$$

and $p = 1$, if the rate law in Eq. 1.34 is used. Evidently, the same result would be obtained even if the right side of Eq. 1.34 were multiplied by some positive function like $g(\cdot)$ in Eq. 1.37. On the other hand, Eq. 1.36 will *never* be a correct expression for the conditional stability constant if the rate coefficients are measured under conditions far from equilibrium and intermediate species figure importantly in the reaction mechanism near equilibrium.[13] In that case, the rate coefficients associated with the formation and loss of the intermediate species must enter into the rate law and in part determine the value of K_{sc}.

1.4 Temperature Effects

The effect of temperature on a chemical reaction at equilibrium can be described quantitatively by considering the change in a thermodynamic equilibrium constant with temperature. As demonstrated in Special Topic 1 at the end of this chapter, the thermodynamic equilibrium constant is related formally to the Gibbs energy change for a reaction, with all reactants and products in their Standard States:

$$\Delta_r G^0 = -RT \ln K \tag{s1.14}$$

where $\Delta_r G^0$ is the *standard Gibbs energy change*, R is the molar gas constant, and T is absolute temperature. The conceptual meaning and numerical calculation of $\Delta_r G^0$ are discussed in Special Topic 1. Suffice it to say here that $\Delta_r G^0$ can be expressed formally as the algebraic sum of two other Standard-State thermodynamic functions:[12,15]

$$\Delta_r G^0 = \Delta_r H^0 - T\Delta_r S^0 \tag{1.41}$$

where $\Delta_r H^0$ is the *standard enthalpy change* and $\Delta_r S^0$ is the *standard entropy change* for the reaction to which K refers.

Strictly, both T and K in Eq. 1.41 should be written T^0 and K^0 to denote the fact that they refer to the Standard States chosen for the reactants and products in a chemical reaction. As discussed in Special Topic 1, Standard states include a prescription of both temperature and applied pressure [usually $T^0 = 298.15$ K and $P^0 = 0.1$ MPa (1 bar) or 101.325 kPa (1 atm)], and it is under this condition that the chemical reaction described by K is investigated at equilibrium. The issue of temperature effects on K, then, is actually the problem of finding how K changes when the Standard-State temperature is changed at fixed Standard-State pressure. Evidently, according to Eqs. 1.41 and 1.42,

$$\begin{aligned}\left(\frac{\partial \ln K}{\partial T}\right)_{P^0} &= -\frac{1}{RT}\left(\frac{\partial \Delta_r G^0}{\partial T}\right)_{P^0} - (\ln K/T) \\ &= (\Delta_r S^0/RT) + (\Delta_r G^0/RT^2) \\ &= \Delta_r H^0/RT^2\end{aligned} \tag{1.42}$$

The value of $\Delta_r H^0$ for a given chemical reaction can be calculated with tabulated thermodynamic data (see Special Topic 1). As an example of the use of Eq. 1.42, the temperature coefficient for ln K relating to the CO_2 hydration reaction in Eq. 1.10 can be calculated. For this reaction, with protons and bicarbonate ions taken as products, $\Delta_r H^0 = 7.64$ kJ mol^{-1} and

$$\left(\frac{\partial \ln K}{\partial T}\right)_{P^0} = \frac{7.64 \text{ kJ mol}^{-1} \times 10^3 \text{J (kJ)}^{-1}}{8.3144 \text{ J mol}^{-1} \times (298.15 \text{ K})^2} = 0.010 \text{ K}^{-1}$$

which implies a change in ln K of 0.01 per degree of absolute temperature change near $T = 298$ K.

Temperature effects on the rates of chemical reactions cannot be calculated in terms of the temperature dependence of reaction rate coefficients as easily as demonstrated in Eq. 1.42 because of the model-specific nature of overall reaction rate laws (Section 1.3). One empirical approach that is in widespread use applies to rate laws that are expressed as the difference between two power-law terms in species concentrations. As a concrete example, such a rate law can be written for the composite CO_2 hydration reaction in Eq. 1.3:

$$-\frac{d[CO_2]}{dt} = k_f^*[CO_2] - k_b[H^+][HCO_3^-] \tag{1.43}$$

where the concentration of the reactant H_2O has been taken effectively constant and absorbed into the forward-reaction rate coefficient (then denoted k_f^*). At equilibrium the left side of Eq. 1.43 vanishes and the conditional equilibrium constant in Eq. 1.27 applies, with $K_c = k_f^*/k_b$, similar to Eq. 1.36. This model of the rate of decrease of dissolved CO_2 concentration and the mathematical form of Eq. 1.42 for the temperature coefficient of the corresponding thermodynamic equilibrium constant (Eq. 1.26) suggest that the temperature dependence of rate coefficients might be expressed as follows:

$$\left(\frac{\partial \ln k}{\partial T}\right)_{P^0} = \frac{E(T)}{RT^2} \tag{1.44}$$

where E(T) is some empirical function of temperature. In the case of the CO_2 hydration reaction in Eq. 1.3, the application of Eqs. 1.26, 1.27, and 1.42–1.44 produces the following relationship:

$$\left(\frac{\partial \ln k_f}{\partial T}\right)_{P^0} - \left(\frac{\partial \ln k_b}{\partial T}\right)_{P^0} = \frac{E_f(T) - E_b(T)}{RT^2}$$

$$= \left(\frac{\partial \ln K_c}{\partial T}\right)_{P^0} = \left(\frac{\partial \ln [\gamma_{CO_2}/\gamma_H \gamma_{HCO_3}]}{\partial T}\right)_{P^0} + \frac{\Delta_r H^0}{RT^2} \tag{1.45}$$

which indicates that the temperature dependence of E(T) comes from that of the aqueous species activity coefficients. This illustrative example can be generalized to define the *Arrhenius equation* for the temperature dependence of rate constants:

$$\left(\frac{\partial \ln k}{\partial T}\right)_{P^0} \equiv \frac{E_a}{RT^2} \tag{1.46}$$

where E_a is termed the *Arrhenius activation energy* by analogy with the interpretation of $\Delta_r H^0$ when applied to elementary reactions.[14] If Eq. 1.46 is invoked to describe the temperature dependence of a rate coefficient over a small enough temperature range, the parameter E_a will be approximately constant. The integrated form of the Arrhenius equation is

$$\ln k = \ln A - E_a/RT \tag{1.47}$$

where ln A is a constant of integration. Common applications of the Arrhenius equation yield a graph of ln k against 1/T, which should be a straight line with slope $-E_a/R$, within experimental variability (Fig. 1.2). The value of Eq. 1.46 as a model of the effect of temperature on reaction rates must be assessed in this way for each application.

1.5 Coupled Rate Laws

Taken together, Eqs. 1.1 and 1.2 constitute a sequential chemical reaction to form bicarbonate from dissolved carbon dioxide:

$$\left.\begin{array}{r} CO_2(aq) + H_2O(\ell) \rightarrow H_2CO_3^0(aq) \\ H_2CO_3^0(aq) + OH^-(aq) \rightarrow HCO_3^-(aq) + H_2O(\ell) \end{array}\right\} \tag{1.48}$$

This alternate way of writing the *overall* reaction

$$CO_2(aq) + OH^-(aq) \rightarrow HCO_3^-(aq) \tag{1.49}$$

exposes its mechanism to distinguish it from the *elementary* reaction in Eq. 1.5. Similarly, Eqs. 1.6 and 1.7 combine to yield the sequential reaction

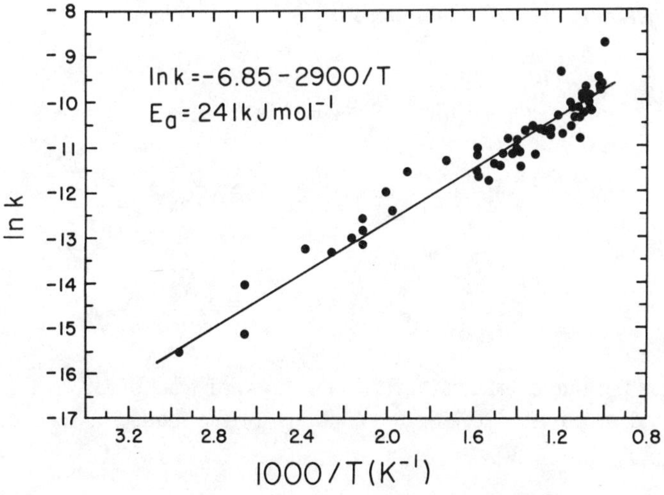

FIG. 1.2. An Arrhenius plot of a zeroth-order reaction rate coefficient (normalized to unit surface area and the unit cell) for the dissolution of a variety of silicate minerals [data from B. J. Wood and J. V. Walther, Rates of hydrothermal reaction, *Science* **222**:413 (1983)]. See Section 3.1 for additional discussion of rate coefficients for dissolution reactions.

$$Mn^{2+}(H_2O)_6(aq) + Cl^-(aq) \rightarrow Mn^{2+}(H_2O)_6Cl^-(aq)$$
$$\rightarrow MnCl^+(H_2O)_5(aq) + H_2O(\ell) \qquad (1.50)$$

for the formation of an inner-sphere complex between manganous ion and chloride ion. Sequential chemical reactions like these are the common result of investigating the kinetic species involved in the mechanism of an overall reaction. They often are described by a set of rate laws that are *coupled* because of the sharing of one or more species concentrations among the rate equations. This coupling increases the complexity of the mathematical analysis of the rate equations.[16]

The sequential reactions in Eqs. 1.48 and 1.50 are special cases of the two abstract reaction schemes

$$\left.\begin{array}{l} A + B \rightleftarrows C \\ C + D \rightleftarrows E + B \end{array}\right\} \qquad (1.51)$$

$$A + B \rightleftarrows C \rightleftarrows D + E \qquad (1.52)$$

where A, B, and so on, are chemical species and the symbol $\rightleftarrows$ denotes the possible occurrence of both forward and backward reactions. A set of rate laws for these two schemes can be formulated after making the *assumptions* that (a) the rate law is a difference expression, as in Eq. 1.33; (b) each term in the rate law depends on powers of concentrations of reactants and products, as in Eq. 1.34; and (c) reaction rate order and reaction stoichiometry are identical.

For the sequential reaction in Eq. 1.51, the set of rate equations generated through these simplifying assumptions is

$$\frac{dc_A}{dt} = -k_f c_A c_B + k_b c_c \qquad (1.53a)$$

$$\frac{dc_B}{dt} = -k_f c_A c_B + k_b c_C + k'_f c_C c_D - k'_b c_E c_B = \frac{-dc_C}{dt} \qquad (1.53b)$$

$$\frac{dc_D}{dt} = -k'_f c_C c_D + k'_b c_E c_B = \frac{-dc_E}{dt} \qquad (1.53c)$$

where k_f is the rate coefficient for the first forward reaction, k'_f is that for the second, and so on. The set of rate equations for the sequential reaction in Eq. 1.52 is

$$\frac{dc_A}{dt} = \frac{dc_B}{dt} = -k_f c_A c_B + k_b c_C \qquad (1.54a)$$

$$\frac{dc_C}{dt} = k_f c_A c_B - (k_b + k'_f)c_C + k'_b c_D c_E \tag{1.54b}$$

$$\frac{dc_D}{dt} = \frac{dc_E}{dt} = k'_f c_C - k'_b c_D c_E \tag{1.54c}$$

Because of the stoichiometric constraints implied by Eqs. 1.51 and 1.52, not all the rate equations in Eqs. 1.53 and 1.54 are independent. In Eq. 1.53, the rate at which the concentration of species C increases must be the same as the combined rates of decrease in the concentrations of species A and E, such that Eq. 1.53b can be derived by adding Eqs. 1.53a and 1.53c and changing the sign of the sum. Similarly, in Eq. 1.54, the rate at which the concentration of species C increases must equal the sum of the rates of decrease of c_A and c_D, such that Eq. 1.54b can be derived by adding Eq. 1.51a to Eq. 1.51c and reversing all signs. Thus any two of Eq. 1.53 or 1.54 are sufficient to describe the kinetics of the reaction scheme in Eq. 1.51 or 1.52. If the expressions for species A and D are selected, the equations

$$\frac{dc_A}{dt} = -k_f c_A c_B + k_b c_C \tag{1.53a}$$

$$\frac{dc_D}{dt} = -k'_f c_C c_D + k'_b c_E c_B \tag{1.53c}$$

or

$$\frac{dc_D}{dt} = k'_f c_C - k'_b c_D c_E \tag{1.54c}$$

constitute the coupled-rate laws sufficient to describe the sequential reactions in Eq. 1.51 or 1.52. The coupling, of course, arises from the sharing of the concentration of one or more species between Eqs. 1.53a and either 1.53c or 1.54c.

Mathematical solutions of coupled rate equations are available for a variety of special cases,[16] but approximate solutions informed by experimental data concerning the relative rates of contributing reactions are more the rule. For the reactions in Eq. 1.48, as an example, it is known[2,7] that the second reaction comes to equilibrium very much faster than the first and that, in the first reaction, the forward rate is much smaller than the backward rate. Thus the rate of formation of bicarbonate from the hydration of CO_2 is limited by the rate of formation of true carbonic acid (at pH < 8). With respect to Eqs. 1.53a and 1.53c, this means that, on the time scale of formation of species C ($H_2CO_3^0$), the rate of increase of the concentration of species D (OH^-) is nil. Moreover, the concentration of species B (H_2O) is effectively constant in aqueous solution and can be absorbed into the rate coefficients k_f and k'_b. Thus the rate laws reduce to the equation

$$\frac{dc_A}{dt} = - k_f^* c_A + k_b c_C^{eq} \tag{1.55}$$

where $k_f^* \equiv k_f c_B$ and $c_C^{eq} = k'_b c_E^{eq} c_B^{eq}/k'_f c_D^{eq} \equiv k'^*_b c_E^{eq}/k'_f c_D^{eq}$. This rate expression for species A (CO_2) is now formally decoupled from that for species D (OH^-). Its mathematical solution is[16]

$$c_A(t) = c_A^{eq} + [c_A(0) - c_A^{eq}] \exp(-k_f^* t) \tag{1.56}$$

where $c_A^{eq} = k_b c_C^{eq}/k_f^*$ is the concentration of CO_2 in equilibrium with $H_2CO_3^0$. Equation 1.56 exhibits the exponential decay with time that is typical of first-order reactions. From experiment[2] it is found that $k_f^* \approx 0.051\ s^{-1}$ and $k_b \approx 20\ s^{-1}$. It follows that $[CO_2]_e / [H_2CO_3^0]_e = k_b/k_f^* \approx 389$; that is, that the molar ratio of true carbonic acid to dissolved CO_2 is only about 0.0026 at equilibrium, as noted in Section 1.1.

For the reactions in Eq. 1.50, it is known[5] that the first reaction comes to equilibrium much more quickly than the second and that in the second reaction the forward rate is much larger than the backward rate. As in the CO_2 hydration reaction, the concentration of water is effectively constant (species E in Eq. 1.52). Thus the rate of inner-sphere complex formation from the outer-sphere complex intermediate species limits the overall rate of the reaction in Eq. 1.8. The impact of these experimental facts on the coupled rate laws in Eq. 1.53a and 1.54c is to reduce them to a single equation:

$$\frac{dc_D}{dt} = k'_f c_C^{(eq)} = k'_f K_{osc} c_A^{(eq)} c_B^{(eq)} \tag{1.57}$$

where $K_{osc} \equiv k_f/k_b$ is the conditional equilibrium constant for the formation of the outer-sphere complex species, $Mn^{2+}(H_2O)_6Cl^-$. Equation 1.57 predicts a rate of increase in the concentration of species D [$MnCl^+(H_2O)_5$] that is proportional to the concentrations of species A (Mn^{2+}) and species B (Cl^-), which are in equilibrium only with the outer-sphere intermediate species [hence the *parenthetical* symbol, (eq)]. From experiment,[5] $k'_f K_{osc} \approx 1.5 \times 10^7\ dm^3\ mol^{-1}\ s^{-1}$.

Special Topic 1: Standard States

Thermodynamic properties, such as the equilibrium constant for a chemical reaction, do not have absolute values.[17] Their measurement and use in the characterization of chemical equilibria depend on a set of conventions that prescribes the conditions under which they are defined to have the value zero. All data concerning thermodynamic properties then are referenced to these conditions. For physical properties, such as absolute temperature (T) and applied pressure (P), thermodynamic methods have been developed to establish the conventions for measuring "absolute values."[18] For chemical properties, however, the situation is more complex because of the molecular nature of matter, to which chemical thermodynamics must accommodate although it cannot address itself explicitly in its fundaments.[17]

The most important chemical thermodynamic property is the *chemical potential* of a substance, denoted μ.[18] The chemical potential is the intensive property that is the criterion for equilibrium with respect to the transfer or transformation of matter. Each component in a soil has a chemical potential that determines the relative propensity of the component to be transferred from one phase to another, or to be transformed into an entirely different chemical compound in the soil. Just as thermal energy is transferred from regions of high temperature to regions of low temperature, so matter is transferred from phases or substances of high chemical potential to phases or substances of low chemical potential. Chemical potential is measured in units of joules per mole (J mol^{-1}) or joules per kilogram (J kg^{-1}).

Virtually all chemical reactions in soils are studied as isothermal, isobaric processes. It is for this reason that the measurement of the chemical potentials of soil components involves the prior designation of a set of *Standard States* that are characterized by selected values of T and P and specific conditions on the phases of matter. Unlike the situation for T and P, however, there is no strictly thermodynamic method for determining absolute values of the chemical potential of a substance. The reason for this is that μ represents an intrinsic chemical property that, by its very conception, cannot be identified with a universal scale, such as the Kelvin scale for T, which exists regardless of the chemical nature of a substance having the property. Moreover, μ cannot usefully be accorded a reference value of zero in the complete absence of a substance, as is the applied pressure, because there is no thermodynamic method for measuring μ by virtue of the creation of matter.

Therefore, it is necessary to adopt a conventional definition of a state of a substance in which the chemical potential of that substance vanishes. This definition will require a statement about T and P (solely because these properties are convenient to maintain under experimental control), as well as a specification of the phase in which a substance occurs. The conventions agreed on in thermodynamics are expressed as follows:[17,19]

The chemical potential of any chemical element in its most stable phase under Standard State conditions is, by convention, equal to the value 0.

The chemical potential of the proton and of the electron in aqueous solution in the Standard State is, by convention, equal to the value 0.

The definition of the chemical potential of an element in the Standard State applies to every entry in the Periodic Table. For a chosen chemical element, all that one must do is establish what phase is the most stable one under "Standard-State conditions." Note that the convention concerning μ does not permit a comparison of chemical potentials among the elements, nor is this kind of comparison necessary in chemical thermodynamics. On the other hand, the chemical potentials of a given element in states other than the Standard State can be compared, as can the chemical potentials of all compounds formed from elements.

A separate convention for the chemical potentials of protons and electrons in solution is required because a change in Gibbs energy with respect to the mole number of an ionic solute carried out while all other composition variables remain fixed, although sufficient to define the chemical potential of any ionic solute, is impossible to measure.[20] This particular type of change is impossible because of the requirement of electroneutrality, which stipulates that, for example, a shift in the number of moles of a cation in a solution must always be accompanied by a balancing shift in the number of moles of the anions. Therefore, one cannot determine the chemical potential of an ionic solute experimentally, even given the convention already provided for chemical potentials of the elements, without specifying arbitrarily the Standard-State chemical potential for one ionic solute as a reference. This specification is made for the proton, in the case of acid–base reactions, and for the electron, in the case of oxidation–reduction reactions (see Section 2.2).

Electrolytes pose a special problem in chemical thermodynamics because of their tendency to dissociate in water into ionic species. It proves to be less cumbersome at times to describe an electrolyte solution in thermodynamic-like terms if dissociation into ions is explicitly taken into account. The properties of ionic species in an aqueous solution cannot be thermodynamic properties because ionic species are strictly molecular concepts. Therefore the introduction of ionic components into the description of a solution is an *extra*thermodynamic innovation that must be treated with care to avoid errors and inconsistencies in formal manipulations.[20] By convention, the Standard State of an ionic solute is that of the solute at unit molality in a solution (at a designated temperature and pressure) in which *no interionic forces are operative*. This convention implies that an electrolyte solution in its Standard State is an ideal solution,[21] as mentioned in Section 1.2.

Standard-State conventions for chemical elements and dissolved solutes are summarized in Table s1.1. Note that the Standard states for gases and for solutes are hypothetical, ideal states and not actual states. For gases, this choice of Standard State is useful because the ideal gas represents a good limiting approximation to the real behavior of gases and possesses equations of state that are mathematically tractable in applications. For solutes, the choice of a hypothetical Standard State is of value because the alternative choice, consisting simply of the pure solute at unit mole fraction, is not very relevant to a solution component whose concentration must always remain small. Moreover, by making the Standard State have the property of no interactions among the solute molecules or ions, it is possible to define a useful thermodynamic parameter that accounts for differences among solutes in their solution behavior, namely, the *activity coefficient*, discussed in Section 1.2.

With the establishment of conventions for the Standard State and for the reference zero value of the chemical potential, it is possible to develop fully the thermodynamic description of chemical reactions. This development relies on the concept of thermodynamic *activity*, introduced in Section 1.2, and on the condition for chemical equilibrium in a reaction:[1,15]

For an isothermal, isobaric chemical reaction at equilibrium, denoted symbolically by the equation:

$$\Sigma_j \upsilon_j A_j = 0 \tag{s1.1}$$

where A_j is a chemical species and υ_j is a stoichiometric coefficient subject to the convention:

$$\upsilon_j > 0 \textit{ for a product} \qquad \upsilon_j < 0 \textit{ for a reactant} \tag{s1.2}$$

the thermodynamic equilibrium condition is

$$\Sigma_j \upsilon_j \mu(A_j) = 0 \quad (T,\ P \text{ fixed}) \tag{s1.3}$$

where $\mu(A_j)$ is the chemical potential of species A_j.

Table s1.1 Summary of Standard-State Conventions for Chemical Elements, Pure Compounds and Dissolved Solutes at $T^0 = 298.15$ K[19]

Chemical element or substance	Standard-State conditions[a]
H, He, N, O, F, Ne, Cl, Ar, Kr, Xe, Rn	Ideal gas
Br, Hg	Liquid
All other elements[b]	Crystalline solid
Gas (*g*)	Ideal gas
Liquid (*l*)	Liquid
Solid (*s*)	Crystalline or amorphous solid
Solvent (all solutions)	Pure substance (x = 1)
Non-electrolyte or undissociated solute	Hypothetical, ideal solution, $m = 1$ mol kg^{-1} or x = 1
Electrolyte	Hypothetical, ideal solution, $m_{\pm} = 1$ mol kg^{-1}

[a]$P^0 = 0.1$ MPa or 101.325 kPa; x = mole fraction, m = molality, $m_{\pm}$ = mean ionic molality
[b]*White* P and Sn are the Standard-State solid phases for these two chemical elements.

The connection between chemical potential and activity is made by way of the concept of *fugacity*.[12] The fugacity f of a gas can be defined by

$$\ln f \equiv \lim_{\delta \downarrow 0} \left[\ln \delta + \int_{\delta}^{P} \left(\frac{V}{nRT} \right) dP' \right] \tag{s1.4}$$

where $\delta \downarrow 0$ means "δ goes to 0 through positive values," and R = 8.3144 J mol^{-1} K^{-1} is the molar gas constant. According to Eq. s1.4, the value of f at the designated temperature T and a chosen value of P will depend on the particular equation of state, V(T,P,n), that describes the gas. If the ideal gas expression

$$V^{id}(T,P,n) = nRT/P$$

is employed in Eq. s1.4, *the fugacity is found to be equal to the pressure*

$$\ln f^{id} = \lim_{\delta \downarrow 0} \left[\ln \delta + \int_{\delta}^{P} (1/P')\,dP' \right] = \ln P$$

On the other hand, if V(T,P,n) was given by the expression

$$V(T,P,n) = (nRT/P) + n\,B_2(T)$$

where $B_2(T)$ is a parameter termed the second virial coefficient, the fugacity would be given by the equation

$$\ln f = \lim_{\delta \downarrow 0} \left\{ \ln \delta + \int_{\delta}^{P} d\ln P' + \left[\frac{B_2(T)}{RT} \right] \int_{\delta}^{P} dP' \right\}$$

$$= \ln P + B_2(T)P/RT$$

In summary, an experimentally determined equation of state for V leads to a well-defined fugacity for every T and P. If a gas shows ideal behavior to a good approximation (as many gases do for $P < 10$ atm), the fugacity of the gas is equal to its pressure. This fact, in turn, means that the Standard-State fugacity of a gas is equal to 1 atm, since the Standard-State pressure is 1 atm (Table s1.1).

The fugacity of a pure liquid or solid can be defined by applying Eq. s1.4 to the vapor in equilibrium with the substance in either condensed phase. Usually, the volume of the vapor will follow the ideal gas equation of state very closely, and the fugacity of the vapor may be set equal to the equilibrium vapor pressure. The thermodynamic basis of associating the fugacity of a condensed substance with that of its equilibrium vapor may be seen by combining Eq. s1.4 with the Gibbs-Duhem equation[12] applied to the vapor under the condition of fixed T:

$$\ln f_{vap} = \lim_{\delta \downarrow 0} \left[\ln \delta + \left(\frac{1}{RT}\right) \int_{\mu(\delta)}^{\mu(P)} d\mu \right]$$

where the form of the integral term comes from replacing V dP by n dμ. It follows from this expression that

$$\ln f_{vap} = \mu(P)/RT + C(T) \tag{s1.5}$$

where

$$C(T) \equiv \lim_{\delta \downarrow 0} [\ln \delta - \mu(\delta)/RT] \tag{s1.6}$$

Equation s1.5 demonstrates that f_{vap} is related directly to the chemical potential of the equilibrium vapor. But this latter quantity, in turn, equals the chemical potential of the substance in the condensed phase. Therefore the fugacity of a condensed phase may be defined by the expression

$$\ln f \equiv \mu/RT + C(T) \tag{s1.7}$$

where C(T) is defined in Eq. s1.6 and μ is the chemical potential of the substance in the condensed phase. Alternatively, Eq. s1.4 may be used to define the fugacity of a substance in any phase since, as the pressure $\delta \downarrow 0$, a condensed phase will vaporize to become a gas and C(T) in Eq. s1.6 will have the same numerical value regardless of what phase actually exists when the applied pressure equals P. It follows that Eqs. s1.4 and s1.7 are completely equivalent. Moreover, *the fugacities of a substance coexisting in two phases that are in equilibrium are the same.*

If Eq. s1.7 is applied to both the equilibrium state of a substance and its Standard State, and the two resulting equations are subtracted, one obtains the expression

$$\ln f - \ln f^0 = \mu/RT - \mu^0/RT$$

or

$$\mu = \mu^0 + RT \ln f/f^0 \tag{s1.8}$$

Equation s1.8 applies to any substance, in any phase, in any kind of mixture at equilibrium. It expresses the idea that the chemical potential of a substance always may be written as equal to the Standard-State chemical potential plus a logarithmic term in the ratio of equilibrium-to-Standard-State fugacities. This last ratio, represented by bold parentheses, is defined to be the relative activity or, more commonly, the *thermodynamic activity* of the substance:

$$\mathbf{(\ )} \equiv f/f^0 \tag{s1.9}$$

The activity, therefore, is a *dimensionless* quantity that serves as a measure of the deviation of the chemical potential from its value in the Standard State. By definition, the activity of any substance in its Standard State is equal to 1. Thus, for example, the activity of pure $CaCO_3(s)$ at 298.15 K and under a pressure of 1 atm is 1.0, as is the activity of pure liquid water in a beaker under the same conditions.

Although activity and fugacity are closely related, they have quite different characteristics in regard to phase equilibria. Consider, for example, the equilibrium between liquid water and water vapor in the interstices of an unsaturated soil. At a given temperature and pressure, the principles of thermodynamic equilibrium demand that the chemical potentials and fugacities of water in the two phases be equal. However, the activities of water in the two phases will *not* be the same because the Standard State for the two phases is not the same. Indeed, f^0 = 1 atm for the water vapor, so its activity is numerically equal to its own vapor pressure (assuming ideal gas behavior). In the case of the liquid soil water, f^0 is not equal to 1 atm but instead is equal (approximately) to the much smaller equilibrium vapor pressure over pure liquid water at T = 298.15 K and P = 1 atm. Therefore the activity of the liquid water is (approximately) equal to its relative humidity divided by 100. The general conclusion to be drawn here is that *the activity of a substance coexisting in two phases at equilibrium cannot be the same in both phases unless the Standard State for both is the same.* On the other hand, the chemical potential and fugacity are always the same in the two phases at equilibrium.

The combination of Eqs. s1.8 and s1.9 produces the equation

$$\mu = \mu^0 + \mathrm{RT}\ \ln\mathbf{(\ )} \tag{s1.10}$$

for the chemical potential of any substance at equilibrium. This expression may be inserted into Eq. s1.3 to derive the identity

$$\Delta_r G^0 + \mathrm{RT}\ \Sigma_j \mu_j \ln\mathbf{(\ )}_j = 0 \tag{s1.11}$$

where

$$\Delta_r G^0 \equiv \Sigma_j \nu_j \mu_j^0 \tag{s1.12}$$

is called the *standard Gibbs energy change* of the reaction. The second term on the left side of Eq. s1.11 may be collected in part into the parameter

$$K \equiv \Pi_j(\ )_j^{\nu j} \tag{s1.13}$$

which is called the *thermodynamic equilibrium constant* for the reaction. Thus Eq. s1.11 can be written in the form

$$\Delta_r G^0 = -RT \ln K \tag{s1.14}$$

upon noting that, according to the mathematics of logarithms,

$$\Sigma_j v_j \ln(\ )_j = \ln [\Pi j(\ )_j^{\nu j}] \tag{s1.15}$$

It should be stressed that the chemical significance of Eq. s1.14 is quite the same as that of Eq. s1.3. The definitions in Eqs. s1.12 and s1.13 are made solely for convenience in applications. Also, in a strict sense, K should be written K^0 to denote the Standard-State T^0 and P^0, as mentioned in Section 1.4.

The principal utility of $\Delta_r G^0$ is that it may be employed to calculate K and to determine the thermodynamic stability of products relative to reactants *when all these species are in their Standard States* (cf. Eqs. s1.3 and s1.11). The criteria for stability are[21]

> *Products are more stable than reactants if $\Delta_r G^0 < 0$. Reactants are more stable than products if $\Delta_r G^0 > 0$.*

The more important, quantitative use of $\Delta_r G^0$ is the calculation of thermodynamic equilibrium constants based on Eq. s1.14. The equilibrium constant is a *dimensionless* quantity equal to the weighted ratio of the activities of the products ($\upsilon_j > 0$) to those of the reactants ($\upsilon_j < 0$) in a chemical reaction. Often this important thermodynamic parameter may be determined by direct or indirect measurements of activities themselves. In general, however, if tabulated values of the Standard-State chemical potentials for the reactants and products are available, the equilibrium constant for any reaction may be calculated at once (at T^0 and P^0) from Eqs. s1.12 and s1.14. The latter of these, at $T^0 = 298.15$ K, may be written in the practical form:

$$\Delta_r G^0 = -5.708 \log K \tag{s1.16}$$

where $\Delta_r G^0$ is measured in kilojoules per mole.

The Standard-State chemical potentials of substances in the gas, liquid, and solid phases, as well as of solutes in aqueous solution, can be determined by a variety of experimental methods, among them spectroscopic, colorimetric, solubility, colligative-property, and electrochemical techniques.[8,17] The accepted values of these fundamental thermodynamic properties are and should be undergoing constant revision under the critical eyes of specialists. It is not the purpose of this book to discuss the practice of determining values of μ^0 for all compounds of interest in soils. This is best left to specialized works on experimental thermodynamics. Suffice it to say that Standard-State chemical potentials must be selected from the literature on the basis of precision of measurement, internal experimental consistency, and consistency with other, currently accepted μ^0 values.[19]

Because of the continual revision in μ^0 values, no attempt will be made to present a list of critically compiled data, even for the compounds of principal interest in soils. In this and subsequent chapters, Standard-State chemical potentials for gases, liquids, solids, and solutes usually will be taken from data in the following critical compilations.

1. D. D. Wagman et al., The NBS tables of chemical thermodynamic properties, *J. Phys. Chem. Ref. Data* **11**: Suppl. No. 2 (1982).
2. R. A. Robie, B. S. Hemingway, and J. R. Fisher, Thermodynamic properties of minerals and related substances at 298.15 K and 1 bar (10^5 Pa) pressure and at higher temperatures, *Geol. Survey Bull. 1452*, U.S. Government Printing Office, Washington, D.C., 1978.
3. A. E. Martell and R. M. Smith, *Critical Stability Constants*, 6 vols., Plenum Press, New York, 1974–1989.
4. L. D. Pettit and H. K. J. Powell, *IUPAC Stability Constants Database,* Academic Software, Timble, Otley, England, 1993.

The compilations by Wagman et al. and Robie et al. are quite extensive, including many solids as well as ionic solutes in aqueous solution. Since a compound may be written as the product of a chemical reaction that involves only chemical elements as reactants, and since μ^0 for an element is equal to zero, μ^0 for a compound can be considered to be a special example of $\Delta_r G^0$ for a reaction that forms the compound from its constituent chemical elements. Thus μ^0 values also are termed *standard Gibbs energies of formation* and given the symbol $\Delta_f G^0$. In addition to μ^0 (or $\Delta_f G^0$) values, Wagman et al. and Robie et al. list $\overline{H}^0$ and $\overline{S}^0$ for many substances. These Standard-State thermodynamic properties are related to $\Delta_r H^0$ and $\Delta_r S^0$ in Eq. 1.42:[15]

$$\Delta_r H^0 \equiv \Sigma_j \nu_j \overline{H}_j^0 \qquad \Delta_r S^0 \equiv \Sigma_j \nu_j \overline{S}_j^0 \tag{s1.17}$$

under the conventions in Eq. s1.2 for the stoichiometric coefficients in a chemical reaction.

As an example of the use of Eq. s1.16, the equilibrium constant for the CO_2 hydration reaction in Eq. 1.16 will be calculated at 298 K. The Standard-State chemical potentials that contribute to $\Delta_r G^0$ for this reaction are, together with their reported uncertainties,

$$\begin{aligned}
\mu^0(CO_2(aq)) &= -386.0 \pm 0.1 \text{ kJ mol}^{-1} \\
\mu^0(H_2O(\ell)) &= -237.1 \pm 0.1 \text{ kJ mol}^{-1} \\
\mu^0(H^+(aq)) &\equiv 0 \\
\mu^0(HCO_3^-(aq)) &= -586.9 \pm 0.1 \text{ kJ mol}^{-1}
\end{aligned}$$

These values are taken from the compilation by Robie et al. except for $\mu^0(CO_2(aq))$, which comes from Wagman et al. The other μ^0 values do not differ in the two compilations within experimental precision. With these data and Eq. s1.12, the result for Δ_rG^0 is

$$\Delta_rG^0 = \mu^0(H^+) + \mu^0(HCO_3^-) - \mu^0(CO_2) - \mu^0(H_20) = 36.2 \pm 0.2 \text{ kJ mol}^{-1}$$

where the uncertainty in Δ_rG^0 has been computed with the help of the equation

$$\sigma(\Delta_rG^0) = [\Sigma_j(\upsilon_j\sigma_j)^2]^{1/2} \qquad \text{(s1.18)}$$

υ_j is the stoichiometric coefficient, and σ_j is the uncertainty in the μ^0 value for the jth reactant or product in the reaction described by Δ_rG^0. The corresponding equilibrium constant is, by Eq. s1.16,

$$\log K = -6.34 \pm 0.04$$

This result may be compared with log K = -6.35 ± 0.01, as derived from direct measurement of activities (Eq. 1.17) and compiled by Martell and Smith.

In connection with this example, it must be emphasized that although a table of experimentally determined μ^0 values for a variety of compounds and ionic solutes is sufficient to calculate the equilibrium constant for any conceivable chemical reaction among some species in a set chosen from the table, this most useful possibility is tempered by the requirement that the μ^0 data be of *high precision*, since the percentage error in the computed K, according to elementary calculus, will be approximately equal to the experimental error in $|\Delta_rG^0|$. For example, a reasonable error of ± 1 kJ mol^{-1} in Δ_rG^0, typical of reactions involving solids, is about 40% of RT at 298.15 K, and this is the estimated percentage error in the corresponding K value. On the other hand, if log K values were desired instead of K values, the percentage error would be equal to the percentage error in $|\Delta_rG^0|$. For example, if there is an error of ± 1 kJ mol^{-1} in a Δ_rG^0 value of 100 kJ mol^{-1}, this amounts to a 1% error in the corresponding log K value.

NOTES

1. The IUPAC nomenclature used in this chapter is described by K. J. Laidler, *Chemical Kinetics*, Harper & Row, New York, 1987, Chap. 1. See also I. Mills, T. Cvitaš , K. Homann, N. Kallay, and K. Kuchitsu, *Quantities, Units, and Symbols in Physical Chemistry*, Blackwell, Oxford, 1988. For an introduction to the concepts of chemical species, including outer-sphere and inner-sphere complexes, see, for example, G. Sposito, *The Chemistry of Soils*, Oxford University Press, New York, 1989, Chap. 4.

2. See, for example, R. G. Compton and P. R. Unwin, The dissolution of calcite in aqueous solution at pH < 4: Kinetics and mechanism, *Phil. Trans. R. Soc. Lond.*

A330:1 (1990).

3. See, for example, W. B. Jensen, The Lewis acid–base definitions: A status report, *Chem. Rev.* **78**:1 (1978).

4. The chloride ion is depicted as an unsolvated species in Eq. 1.6 because of experimental evidence that the residence time of a water molecule coordinated to Cl^- is about 1 ps, the same as the residence time of the molecule in the structure of bulk water. See, for example, J. E. Enderby and G. W. Neilson, The structure of electrolyte solutions, *Rep. Prog. Phys.* **44**:38 (1981).

5. J. Burgess, *Metal Ions in Solution*, Wiley, New York, 1978, Chap. 12.

6. W. Stumm, B. Sulzberger, and J. Sinniger, The coordination chemistry of the oxide–electrolyte interface; The dependence of surface reactivity (dissolution, redox reactions) on surface structure, *Croatica Chem. Acta* **63**:277 (1990).

7. P. Jones, M. Haggett, J. L. Longridge, The hydration of carbon dioxide, *J. Chem. Educ.* **41**:610 (1964).

8. R. A. Robinson and R. H. Stokes, *Electrolyte Solutions*, 2nd rev. ed., Butterworths, London, 1970.

9. For a discussion of single-ion activity coefficients and their calculation, see G. Sposito, The future of an illusion: Ion activities in soil solutions, *Soil Sci. Soc. Am. J.* **48**:531 (1984).

10. M. Whitfield, Activity coefficients in natural waters, pp. 153–299 in *Activity Coefficients in Electrolyte Solutions*, Vol. II, ed. by R. M. Pytkowicz, CRC Press, Boca Raton, FL, 1979.

11. G. Sposito and S. J. Traina, An ion-association model for highly saline, sodium chloride-dominated waters, *J. Environ. Qual.* **16**:80 (1987).

12. See, for example, Chaps. 1 and 2 in G. Sposito, *The Thermodynamics of Soil Solutions*, Clarendon Press, Oxford, 1981.

13. These issues are discussed in an especially cogent way in Chap. 15 of K. Denbigh, *The Principles of Chemical Equilibrium*, Cambridge University Press, Cambridge, 1981. See also Chap. 8 in K. J. Laidler, *op. cit.*[1]

14. See, for example, Chap. 2 in K. J. Laidler, op. cit.[1]

15. Standard-State thermodynamic functions and reaction equilibria are discussed, for example, in Chap. 10 of K. Denbigh, op. cit.[13]

16. Mathematical solutions of rate equations are compiled in Chap. 1 of C. H. Bamford and C. F. H. Tipper, *Comprehensive Chemical Kinetics*, Vol. 2, *The Theory of Kinetics*, Elsevier, Amsterdam, 1969.

17. The fundamental concepts of experimental chemical thermodynamics and Standard States are discussed superbly by M. L. McGlashan, The scope of chemical thermodynamics, *Chemical Thermodynamics, Spec. Periodical Rpt.* **1**:1 (1973).

18. See, for example, Chap. 3 in E. A. Guggenheim, *Thermodynamics*, North-Holland, Amsterdam, 1967.

19. An excellent discussion of Standard-State conventions is given by D. D. Wagman et al., The NBS tables of chemical thermodynamic properties, *J. Phys. Chem. Ref. Data* **11**:Supplement No. 2 (1982).

20. See, for example, Chap. 8 in E. A. Guggenheim, op. cit.[18]

21. Ideal solutions are discussed in Chap. 8 of K. Denbigh, op. cit.,[13] and in Chap. 7 of E. A. Guggenheim, op. cit.[18]

FOR FURTHER READING

Denbigh, K., *The Principles of Chemical Equilibrium*, Cambridge University Press, London, 1981. A fundamental reference book on chemical thermodynamics that should be in the library of all soil chemists. Chapters 1, 2, 10, and 15 are especially useful.

Laidler, K. J., Chemical kinetics and the origins of physical chemistry, *Arch. Hist. Exact Sci.* **32**:43 (1985). A historical account of the seminal contributions of J. H. van't Hoff, S. Arrhenius, and W. Ostwald to chemical kinetics—well worth the time for anyone interested in how the basic ideas behind reaction rate laws were conceived.

Laidler, K. J., *Chemical Kinetics,* Harper, New York, 1987. The first two chapters of this standard textbook provide a fine discussion of kinetics concepts and experimental methodology.

Lasaga, A. C., Rate laws of chemical reactions, *Rev. Mineral.* **8**:1 (1981). A lively review account of geochemical kinetics by a gifted modern practitioner; parallel reading for the present chapter.

McGlashan, M. L., The scope of chemical thermodynamics, *Chem. Thermodynami., Spec. Periodical Rept.* **1**:1 (1973). A pert and precise summary of what chemical thermodynamics can (and *cannot*) do and how it does it: "must reading" to accompany the present chapter.

Wilkins, R. G., *Kinetics and Mechanism of Reactions of Transition Metal Complexes,* VCH, New York, 1991. A superb introduction to chemical kinetics and elementary complexation reactions, presented from the experimentalist's point of view.

PROBLEMS

1. Define and distinguish carefully among the species, $CO_2(aq)$, $CO_2 \bullet H_2O$, $H_2CO_3^0$, and $H_2CO_3^*$. Explain why $\Delta_r G^0 \equiv 0$ for the "reaction"

$$CO_2(aq) + H_2O(\ell) = CO_2 \bullet H_2O(aq)$$

(*Hint*: No chemical bonds are created when "loosely solvated CO_2" is formed. Since $H_2CO_3^*$ represents the combined totality of hydrated CO_2 in aqueous solution, it is not a true chemical species despite its ubiquitous appearance in reactions involving carbonic acid as described in textbooks.)

2. Show that log K = −3.76 at 298 K for the dissociation reaction:

$$H_2CO_3^0(aq) = H^+(aq) + HCO_3^-(aq)$$

Check your result, given $\mu^0(H_2CO_3^0(aq)) = -608.4$ kJ mol^{-1}. (*Hint*: Use the fact that log K = −6.35 at 298 K for the reaction in Eq. 1.10, along with the ratio, $[CO_2(aq)]_e/[H_2CO_3^0(aq)]_e \approx 389$, calculated in Section 1.5. Follow the method described at the end of Special Topic 1 to find $\Delta_r G^0 = 21.5$ kJ mol^{-1} for the dissociation of $H_2CO_3^0(aq)$.)

3. Formulate a rate law for changes in CO_2 concentration caused by the hydration reaction:

$$CO_2(aq) + H_2O(\ell) \underset{k_b}{\overset{k_f}{\rightleftharpoons}} H_2CO_3^0(aq)$$

using the approach exemplified in Eqs. 1.33 and 1.34 under the assumption that reaction order and stoichiometry are identical. Given that the first-order rate constants for the reaction are $k_f^* = 0.051$ s^{-1} and $k_b = 20$ s^{-1} at 298 K, calculate a conditional equilibrium constant. (*Hint*: Absorb the concentration of $H_2O(\ell)$ into k_f to define k_f^* and follow the model in Eq. 1.53a. At equilibrium, $K_c = [H_2CO_3^0(aq)]_e/[CO_2(aq)]_e = 2.6 \times 10^{-3}$.)

4. Calculate the conditional equilibrium constant for the reaction in Eq. 1.10 at an effective ionic strength of 50 mol m^{-3} (T = 298 K). (*Answer*:

$$\log K_c = -6.35 + 0.1 I_{ef} + 1.02\left[\frac{\sqrt{I_{ef}}}{1 + \sqrt{I_{ef}}} - 0.3 I_{ef}\right] = -6.17,$$

with K_c in mol dm^{-3}.)

5. Derive the equation

$$\log K_{sc} = \log K_s + 1.53\, I_e - 4.10\left\{\frac{\sqrt{I_{ef}}}{1 + \sqrt{I_{ef}}}\right\}$$

for the conditional equilibrium constant defined in Eq. 1.36, where K_s is the corresponding thermodynamic equilibrium constant and I_{ef} is effective ionic strength. (*Hint*: Use Eqs. 1.21–1.24 in the derivation.)

6. Given Arrhenius activation energies of 62.8 and 66.9 kJ mol^{-1} for k_f^* and k_b, respectively, in Problem 3, estimate $(\partial \ln K_c/\partial T)_{P^0}$. (*Answer*: -0.006 K^{-1}.)

7. The mechanism of the complexation reaction

$$Al^{3+}(aq) + F^-(aq) \xrightarrow{k_f'} AlF^{2+}(aq)$$

can be described approximately by the pathways in Eqs. 1.6 and 1.7 along with the rate law in Eq. 1.57 (with D = AlF^{2+}, A = Al^{3+}, and B = F^-). At 293 K, $k_f' = 3.4\ s^{-1}$, and at 298 K, $k_f' = 6.5\ s^{-1}$. Estimate the Arrhenius activation energy for k_f'. (*Answer*: $E_a = 94$ kJ mol^{-1}.)

8. Formulate a rate law for changes in CO_2 concentration caused by the reaction

$$CO_2(aq) + OH^-(aq) \underset{k_{HCO_3}}{\overset{k_{OH}}{\rightleftharpoons}} HCO_3^-(aq)$$

Use Eqs. 1.33 and 1.34 as the basis for developing an equation for $d[CO_2]/dt$. Given $k_{OH} = 8500\ dm^3\ mol^{-1}\ s^{-1}$ and $k_{HCO_3} = 2 \times 10^{-4}\ s^{-1}$, estimate the pH value at which the rate of decrease in CO_2 concentration caused by the reaction to form HCO_3^- would equal that caused by the reaction in Problem 3. (*Answer*: pH $\approx 14 + \log (k_f^*/k_{OH}) = 8.55$.)

9. The reactions in Problems 3 and 8 are special cases of the abstract first/second-order reaction

$$A + B \underset{k_b}{\overset{k_f}{\rightleftharpoons}} C$$

where A, B, and C are chemical species and k_f, k_b are rate constants. Show that

$$\frac{dc_A}{dt} = \frac{k_f}{K_c}(c_C - K_c c_A c_B)$$

under the three assumptions stated in connection with Eqs. 1.51 and 1.52, where $K_c = c_C^{eq}/c_A^{eq}\ c_B^{eq}$. What is the mathematical form of $c_A(t)$ when c_B and c_C are constant at their equilibrium values? (*Answer*: $c_A(t) = c_A^{eq} + [c_A(0) - c_A^{eq}] \exp(-k_f c_B^{eq}\ t)$.)

10. Formulate a rate law for changes in CO_2 concentration caused by the reactions in Problems 3 and 8 occurring simultaneously. Show that this rate law simplifies to the expression

$$\frac{d[CO_2]}{dt} = -0.060[CO_2] + 1.36 \times 10^{-3}[HCO_3^-]_e$$

at pH 8 whenever $H_2CO_3^0$ and HCO_3^- equilibrate rapidly on the time scale of the CO_2 kinetics. What is the mathematical form of $[CO_2]$ as a function of time? (*Hint*: The rate law for a species involved in "parallel" reactions is the sum of the rate laws for the individual reactions. According to the data in Problem 2, $[H_2CO_3^0]_e/[HCO_3^-]_e \approx 5.8 \times 10^{-5}$. Note the formal similarity between the approximate rate law and Eq. 1.55.)

2

CHEMICAL SPECIATION IN AQUEOUS SOLUTIONS

2.1 Complexation Reactions

A *complex* is a unit in which an ion, atom, or molecule binds to other ions, atoms, or molecules. The binding species is termed a *central group* and a bound species is termed a *ligand*.[1] The aqueous ions, $MnCl^+$, HCO_3^-, and $Al(H_2O)_6^{3+}$, are complexes, with Mn^{2+}, CO_3^{2-}, and Al^{3+}, respectively, usually chosen as central groups and Cl^-, H^+, and H_2O, respectively, usually chosen as ligands. If the ligands in a complex are water molecules, the unit is called a *solvation complex*. The "free" cations and anions, represented in the chemical reactions discussed in Chapter 1 by, for example, $Mn^{2+}(aq)$ or $OH^-(aq)$, typically are solvation complexes, since ion charge serves to attract and bind water dipoles in aqueous solution. The term *ligand* is applied conventionally to anions or neutral electron-donor molecules bound to a metal cation, but it has been applied as well to such cations as protons coordinated to such anions as carbonate. When two or more distinct chemical bonds are formed between a ligand and the central group, the ligand (and the complex) is termed *multidentate*, and when two or more central groups are involved, the complex is *multinuclear*. Multidentate complexes that form heterocyclic molecular structures are termed *chelates*.

If the central group and ligands in a complex are in direct contact, the complex is called *inner-sphere*. If one or more water molecules is interposed between the central group and a ligand, the complex is *outer-sphere*. Inner-sphere complexes usually are much more stable than outer-sphere complexes because the latter cannot easily involve the sharing of electrons between the central group and the ligand, whereas the former can. Thus the "driving force" for inner-sphere complexes is the heat evolved through strong bond formation between a central group and ligand. For outer-sphere complexes, the heat of bond formation is not so large, and the "driving force" involves favorable *stereochemical* aspects of coordination about the central group.

Table 2.1 lists the principal complexes found in well-aerated soil solutions. The ordering of free-ion and complex species in each row from left to right is

Table 2.1 Representative Chemical Species in Soil Solutions

	Principal Species	
Cation	Acid Soils	Alkaline Soils
Na^+	Na^+	Na^+, $NaHCO_3^0$, $NaSO_4^-$
Mg^{2+}	Mg^{2+}, $MgSO_4^0$, org[a]	Mg^{2+}, $MgSO_4^0$, $MgCO_3^0$
Al^{3+}	org, AlF^{2+}, $AlOH^{2+}$	$Al(OH)_4^-$, org
Si^{4+}	$Si(OH)_4^0$	$Si(OH)_4^0$
K^+	K^+	K^+, KSO_4^-
Ca^{2+}	Ca^{2+}, $CaSO_4^0$, org	Ca^{2+}, $CaSO_4^0$, $CaHCO_3^+$
Cr^{3+}	$CrOH^{2+}$	$Cr(OH)_4^-$
Cr^{6+}	CrO_4^{2-}	CrO_4^{2-}
Mn^{2+}	Mn^{2+}, $MnSO_4^0$, org	Mn^{2+}, $MnSO_4^0$, $MnCO_3^0$, $MnHCO_3^+$, $MnB(OH)_4^+$
Fe^{2+}	Fe^{2+}, $FeSO_4^0$, $FeH_2PO_4^+$	$FeCO_3^0$, Fe^{2+}, $FeHCO_3^+$, $FeSO_4^0$
Fe^{3+}	$FeOH^{2+}$, $Fe(OH)_3^0$, org	$Fe(OH)_3^0$, org
Ni^{2+}	Ni^{2+}, $NiSO_4^0$, $NiHCO_3^+$, org	$NiCO_3^0$, $NiHCO_3^+$, Ni^{2+}, $NiB(OH)_2^+$
Cu^{2+}	org, Cu^{2+}	$CuCO_3^0$, org, $CuB(OH)_4^+$, $Cu[B(OH)_4]_2^0$
Zn^{2+}	Zn^{2+}, $ZnSO_4^0$, org	$ZnHCO_3^+$, $ZnCO_3^0$, org, Zn^{2+}, $ZnSO_4^0$, $ZnB(OH)_4^+$
Mo^{6+}	$H_2MoO_4^0$, $HMoO_4^-$	$HMoO_4^-$, MoO_4^{2-}
Cd^{2+}	Cd^{2+}, $CdSO_4^0$, $CdCl^+$	Cd^{2+}, $CdCl^+$, $CdSO_4^0$, $CdHCO_3^+$
Pb^{2+}	Pb^{2+}, org, $PbSO_4^0$, $PbHCO_3^+$	$PbCO_3^0$, $PbHCO_3^+$, org, $Pb(CO_3)_2^{2-}$, $PbOH^+$

[a]org = metal-organic complex

roughly according to decreasing concentration typical for either acid or alkaline soils. A soil solution can easily contain 100–200 different soluble complexes. The main effect of pH on these complexes, evident in Table 2.1, is to favor free metal cations and protonated anions at low pH and carbonate or hydroxyl complexes at high pH. Thus speciation is affected by competition for protons between solvation shell and bulk liquid water molecules.

Complexes may be pictured as addition products (*adducts*) of Lewis acids and bases.[2] A *Lewis acid* is any chemical species that employs an empty electronic orbital in initiating a complexation reaction, whereas a *Lewis base* is any chemical species that employs a doubly occupied electronic orbital in initiating a complexation reaction. Lewis acids and bases can be neutral molecules, ions, and neutral or charged macromolecules. The electronic orbitals referred to can be molecular orbitals, bands, or atomic orbitals, depending on the structure of the Lewis acid or base. The passage, "in initiating a complexation reaction," emphasizes the point that complex formation is brought on by, but may not involve only, the empty and doubly occupied orbitals that differentiate Lewis acids from Lewis bases. The proton and all of the metal cations of interest in soil solutions are Lewis acids. The Lewis bases include H_2O; oxyanions such as OH^-, COO^-, CO_3^{2-}, SO_4^{2-}, and PO_4^{3-}; and organic N, S, and P electron donors. Some common examples of Lewis acid–base reactions are the solvation of a metal cation, hydrolysis reactions, the protonation of an oxyanion, and the coordination of a metal cation with an organic functional group.

A *hard Lewis acid* is a molecular unit of relatively small size, high oxidation state, high electronegativity, and low polarizability. It tends also to possess outer electrons that are relatively difficult to excite to higher energies. (Electronegativity is a measure of the ability of a molecular species to attract electrons. Polarizability is a measure of the ease with which the electronic orbitals in a molecular species deform in the presence of an electric field.) A *soft Lewis acid*, on the other hand, is of relatively large size, low oxidation state, low electronegativity, and high polarizability. It tends to possess easily excitable outer electrons that often are *d*-orbital electrons. Examples of hard Lewis acids are the proton and cations of the metals Na, K, Mg, and Ca. Examples of soft Lewis acids are Cd^{2+}, Cu^{+}, Hg^{2+}, and pi-electron acceptors such as quinones. A list of hard, "borderline," and soft Lewis acids that are important in soils is given in Table 2.2. The borderline Lewis acids include most of the significant bivalent trace metal cations. Their existence indicates that no sharp dividing line separates hard acids from soft acids. Instead, there is only gradual change as the polarizability of a Lewis acid increases. Even within the set of hard acids, for example, there is a spectrum of hardness instead of a single degree of hardness applicable to all members. Thus, although Li^{+}, Cs^{+}, Mg^{2+}, and Ba^{2+} are all classified as hard Lewis acids, Cs^{+} and Ba^{2+} are much softer than Li^{+} and Mg^{2+}. This relationship among metal ions can be quantified with the *Misono softness* parameter (Y), which is defined by the equation[3]

$$Y \equiv 10 \frac{I_Z R}{\sqrt{Z}\, I_{Z+1}} \tag{2.1}$$

where R is the ionic radius of a metal cation whose valence is Z and whose ionization potential (*not* ionic potential) is I_z. For example, from data on ionic radii[4] and with ionization potentials taken from the *Handbook of Chemistry and Physics*, one calculates for Cu^{2+} (coordination number 6): $Y = 10(1958)(0.073)/\sqrt{2}(3554) = 0.284$ nm, where $I_2 = 1958$ kJ mol^{-1} and $I_3 = 3554$ kJ mol^{-1}. Values of Y smaller than about 0.25 nm correspond to metal ions that have high electronegativity and low polarizability; they tend to form electrostatic instead of covalent chemical bonds. Values of $Y > 0.32$ nm correspond to metal ions that have low electronegativity and high polarizability; they tend to form covalent chemical bonds. Values of Y between 0.25 and 0.32 nm correspond to "borderline" metal ions whose tendency to covalency depends on whether specific solvent, stereochemical, and electronic configurational factors are present. These divisions match closely to those in the well-known "class a–borderline–class b" classification used in inorganic chemistry.[2]

Besides Y, the *ionic potential*

$$IP \equiv Z/R \tag{2.2}$$

provides a useful parameter with which to classify metal cations as to the complexes they might form with nearest-neighbor water molecules in a dilute

Table 2.2 Representative Hard and Soft Lewis Acids and Bases[a]

Lewis Acids

Hard Acids
H^+, Li^+, Na^+, K^+, (Rb^+, Cs^+), Mg^{2+}, Ca^{2+}, Sr^{2+}, (Ba^{2+}), Ti^{4+}, Zr^{4+}, Cr^{3+}, Cr^{6+}, Mn^{4+}, Mn^{3+}, Mn^{2+}, Fe^{3+}, Co^{3+}, Al^{3+}, Si^{4+}, CO_2

Borderline Acids
Fe^{2+}, Co^{2+}, Ni^{2+}, Cu^{2+}, Zn^{2+}, (Pb^{2+})

Soft Acids
Cu^+, Ag^+, Au^+, Cd^{2+}, Hg^+, Hg^{2+}, CH_3Hg^+; pi-electron acceptors such as quinones

Lewis Bases

Hard Bases
NH_3, RNH_2, H_2O, OH^-, O^{2-}, ROH, CH_3COO^-, CO_3^{2-}, NO_3^{-1}, PO_4^{3-}, SO_4^{2-}, F^-

Borderline Bases
$C_6H_5NH_2$, C_5H_5N, N_2, NO_2^-, SO_3^{2-}, Br^-, (Cl^-)

Soft Bases
C_2H_4, C_6H_6, R_3P, $(RO)_3P$, R_3As, R_2S, RSH, $S_2O_3^-$, S^{2-}, I^-

[a]R = organic molecular unit; ( ) indicates a tendency to softness

aqueous solution at equilibrium with atmospheric CO_2 (pH 5.6). If IP < 30 nm^{-1}, a metal cation will form only solvation complexes with water molecules; if 95 > IP > 30 nm^{-1}, a metal cation can repel protons strongly enough from a solvating water molecule to form a hydrolytic species; if IP > 95 nm^{-1}, the repulsion is strong enough to allow the formation of oxyion species. Typical examples of metals placed in the three respective categories are Na^+ (9.8), Al^{3+} (56.6), and Mo^{6+} (142.9), with their ionic potentials in nm^{-1} given in parentheses. Figure 2.1 shows a number of metal cations classified according to their IP and Y values.

The chemical significance of Fig. 2.1 is perhaps brought into sharper focus by the examples in Figs. 2.2 and 2.3. The graph in Fig. 2.2 shows typical trends in the logarithm of the stability constant for complexation reactions like that in Eq. 1.8 between either soft or hard metal cations and increasingly soft inorganic anions.[5] The same trends are observed if a spectrum of O-containing organic functional groups to S-containing organic functional groups replaces the "halide spectrum" from F to I (cf. Table 2.2). Thus knowledge of the softness of a metal cation helps in estimating the stability of its complexes with a given ligand. Figure 2.3 shows the regular decline in the logarithm of the first-order rate constant (k_{wex}) for water exchange in the solvation complex of a metal cation as the ionic potential of the metal cation increases.[6] The time scale for this exchange reaction (M = metal, H_2O' = "incoming" H_2O),

$$M(H_2O)_6^{m+}(aq) + H_2O'(\ell) \xrightarrow{k_{wex}} M(H_2O)_5H_2O'^{m+}(aq) + H_2O(\ell) \tag{2.3}$$

ranges over 10 orders of magnitude while IP increases only by one order of magnitude.

The same degree of quantitative characterization as appears in Fig. 2.1 for Lewis acids has not been developed for Lewis bases. Speaking in qualitative terms, a *hard Lewis base* is a molecular unit of high electronegativity and low polarizability. It tends to be difficult to oxidize and does not possess low-energy, empty electronic orbitals. A *soft Lewis base*, on the other hand, shows a low electronegativity, a high polarizability, and a proclivity toward oxidation. Examples of Lewis bases that are of significance in soils are given also in Table 2.2. Note that almost every inorganic anion of concern in soil solutions is a hard Lewis base.

The trends exemplified by Figs. 2.2 and 2.3 can be codified in a general,

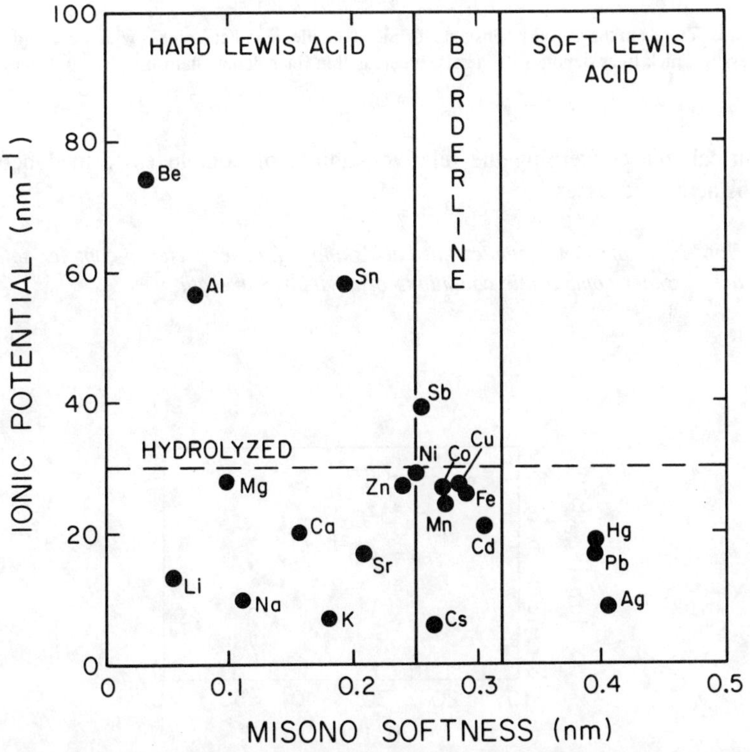

FIG. 2.1. Classification of some metals according to ionic potential and Lewis acid softness (transition metals are in oxidation state + II). (Reprinted by permission of Oxford University Press, from G. Sposito, *The Chemistry of Soils*, Oxford University Press, New York, 1989.)

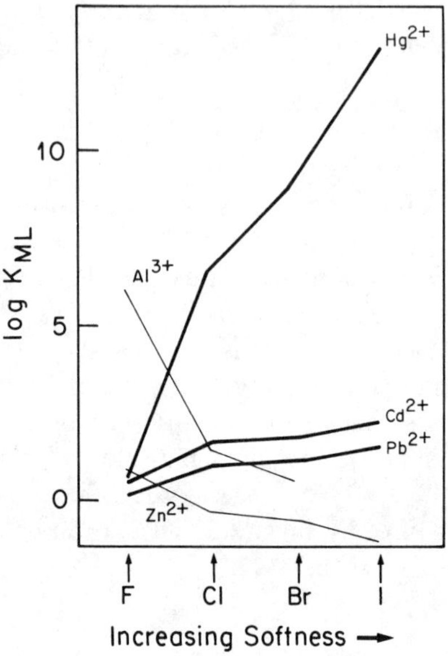

FIG. 2.2. Trends in the stability constants for metal–halide ligand complexes with increasing ligand softness.[5] Thick lines denote soft metals, whereas thin lines denote hard metals.

empirical rule concerning the relative stability of complexes formed between Lewis acids and bases:[2]

> *Hard bases prefer to complex hard acids and soft bases prefer to complex soft acids, under comparable conditions of acid–base strength.*

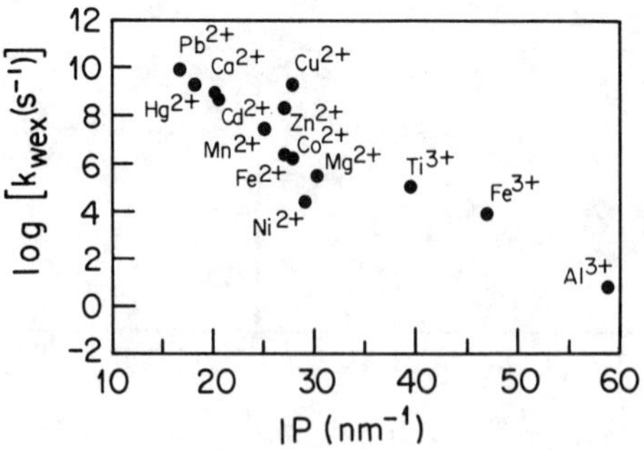

FIG. 2.3. Decreasing trend in the first-order rate coefficient k_{wex} for water exchange (Eq. 2.3) with increasing metal ionic potential. (Adapted with permission from J. G. Hering and F. M. M. Morel.[6] Copyright © 1990 by John Wiley and Sons.)

This rule is known as *the HSAB Principle*. It can be illustrated in terms of abstract chemical reactions as follows:

$$A_H + A_SB_H \rightleftarrows A_HB_H + A_S$$
$$B_S + A_SB_H \rightleftarrows A_SB_S + B_H$$
$$A_H + B_H \rightleftarrows A_HB_H$$
$$A_S + B_S \rightleftarrows A_SB_S$$
$$A_HB_S + A_SB_H \rightleftarrows A_HB_H + A_SB_S$$

where A_H is a hard acid, A_S is a soft acid, B_H is a hard base, B_S is a soft base, AB is a complex, and the longer arrow denotes the direction in which the reaction is in some sense more favorable (larger rate constant or equilibrium constant).

According to Table 2.2, the metal cations, Na^+, K^+, Mg^{2+}, and Ca^{2+}, which usually are of greatest concentration in neutral and alkaline soil solutions, and the metal cations Fe^{3+} and Al^{3+}, which are significant in acidic soil solutions, are classified as hard Lewis acids. The application of the HSAB Principle then leads to the conclusion that these cations preferentially will form complexes with ligands that contain O and F atoms. In particular, H_2O, OH^-, $RCOO^-$, and inorganic oxyanions should be much preferred to Cl^-, S^{2-}, and most organic electron donors. Because ion size and charge are the dominant properties of hard Lewis acids and bases, it is expected that they will tend to form outer-sphere complexes. Indeed, since H_2O is one of the hardest Lewis bases, the existence of an outer-sphere complex comprising a hard Lewis acid means simply that, according to the HSAB Principle, H_2O is retained in preference to the other ligand in the complex. On the other hand, if the other ligand is a sufficiently hard Lewis base, the complex formed could be inner-sphere. In that case, the solvation shells that surround the metal cation and the ligand before they react must be disrupted in order for the complex to form. Since the metal–ligand bond is dominated by ionic charge, the energy released when it forms is about the same, in absolute value, as the energy required to disrupt metal–water bonds. Thus Δ_rH^0 is small and positive for inner-sphere, hard acid–hard base complexes. However, Δ_rS^0 will be large and positive because of solvation complex disruption, with the result that $\Delta_rG^0 \approx -T\Delta_rS^0$ will be strongly negative (i.e., a large stability constant because of a large entropy change).

The bivalent transition metal cations and the cations formed by the metals in Group IB and IIB of the periodic table tend to be either borderline or soft Lewis acids. One exception is Mn^{2+}, which is a hard Lewis acid principally because it possesses no paired *d* electrons. Aside from this exception, the HSAB Principle leads to the conclusion that the bivalent and univalent trace metal cations will tend to form complexes most readily with organic N, P, and S electron donors, Cl^-, and S^{2-}. Because high polarizability is their dominant property as Lewis acids, these cations are expected to engage in more electron sharing with ligands and, therefore, to form inner-sphere complexes to a greater extent than hard Lewis acids. In any case, the solvation shells surrounding the metal cation and ligand will be easier to disrupt (because H_2O is a hard species), and $\Delta_r H^0$ will be strongly negative because of the energy release upon bond formation. The absolute value of $\Delta_r S^0$ will not be large, since the solvation shells are more weakly ordered. Thus $\Delta_r G^0 \approx \Delta_r H^0 < 0$ for inner-sphere complexation reactions between soft Lewis acids and bases.

The HSAB Principle, although empirical, does make broad statements possible concerning the extent of complexation expected for a given metal cation in a soil solution of a given composition. For example, it is clear that the speciation of trace metals will depend more sensitively on the content and type of organic matter present in a soil than will the speciation of, for example, Na, K, Ca, or Mg. Moreover, since Cl^- and S^{2-} behave as relatively soft Lewis bases, an increase in their concentrations will tend to affect the trace metals more than the alkali metals or the alkaline-earth metals. On the other hand, changes in the concentrations of the relatively hard Lewis bases, CO_3^{2-}, SO_4^{2-}, and PO_4^-, should produce important changes in the speciation of the Group IA and IIA metals. These kinds of concepts can be a helpful guide in the application of kinetics and thermodynamics to complex formation and lability in soil solutions.

In Chapter 1 the broad statement is made that the rates of metal complexation reactions are generally high. A more refined conclusion can be drawn from Table 2.3, which lists the time scales over which a number of complex formation and dissociation reactions occur that are important in soil solutions and other natural waters.[7] Perusal of these data makes clear the point that although they are usually very rapid, complexation reactions do span a time scale ranging over at least 10 orders of magnitude. Thus the kinetics of these reactions can be very important to understanding the aqueous speciation of metals and ligands in detail.

Table 2.3 Representative Time Scales[a] for Complex Formation and Dissociation Reactions at 25°

Reaction	Order[b]	Time Scale(s)
$MnSO_4^0 \rightarrow Mn^{2+} + SO_4^{2-}$	1	10^{-9}
$Fe^{3+} + H_2O \rightarrow FeOH^{2+} + H^+$	1	10^{-7}
$FeOH^{2+} + H^+ \rightarrow Fe^{3+} + H_2O$	2[c]	10^{-6}
$Mn^{2+} + SO_4^{2-} \rightarrow MnSO_4^0$	2[c]	10^{-4}
$NiC_2O_4^0 \rightarrow Ni^{2+} + C_2O_4^{2-}$	1	10^{-1}
$Ni^{2+} + C_2O_4^{2-} \rightarrow NiC_2O_4^0$	2[c]	1
$CO_2 + H_2O \rightarrow H_2CO_3^0$	1	10
$HCO_3^- \rightarrow CO_2 + OH^-$	1	10^3

[a]Taken from data compiled in J. F. Pankow and J. J. Morgan, Kinetics for the aquatic environment, *Environ. Sci. Technol.* **15**:1155(1981). All reactants and products are aqueous species.

[b]Sum of the orders with respect to reactant species other than water.

[c]Initial concentrations of 10 mmol m^{-3} assumed for both reactants in second-order reactions.

The fundamental pathway for the formation of strong metal complexes with ligands that are not polymeric species is the *Eigen-Wilkins-Werner mechanism*,[8] which is illustrated in Eqs. 1.6 and 1.7 for the case of Mn^{2+} complexation by Cl^-. This mechanism involves a two-step process if the ligand is monodentate: the rapid formation of an outer-sphere complex followed by the slower formation of an inner-sphere complex, usually preceded by ejection of a water molecule from the first solvation layer of the metal cation. Thus the water-exchange rate in the metal cation solvation complex reaction in Eq. 2.3 is related closely to the rate of metal complexation by unidentate ligands. Figure 2.3 shows that this conclusion implies that rates of metal complexation, in turn, should be correlated negatively with metal cation ionic potentials. Given the HSAB principle, this correlation should be strongest for complexes formed between hard Lewis acids and bases.

A reaction rate law for the Eigen-Wilkins-Werner mechanism is developed in Section 1.5 (Eqs. 1.50, 1.52, 1.54a, 1.54c). If inner-sphere complex formation is rate limiting and the concentration of water remains constant, the rate of inner-sphere complex formation is (cf. Eq. 1.57)

$$\frac{dc_{ML}}{dt} = k_f \, K_{osc} \, c_M^{(eq)} \, c_L^{(eq)} \tag{2.4}$$

where M and L represent metal and ligand, respectively, K_{osc} is the conditional stability constant for the outer-sphere complex that is in equilibrium with the "free" metal and ligand, and k_f is a first-order rate coefficient for the outer-sphere → inner-sphere transformation. Values of K_{osc} are often estimated by using a simple electrostatic model equation for the formation of an ion pair from two point-charge species in aqueous solution.[8] Within the limitations of this semiquantitative approach, the measured ratio of the apparent second-order rate coefficient for metal complex formation to the calculated K_{osc} often approximates the measured value of k_{wex} reasonably well (i.e., within an order of magnitude).[8] Thus $k_f \approx k_{wex}$.

In a soil solution (or any natural water system), the Eigen-Wilkins-Werner mechanism occurs in a milieu of competition. For a given metal cation, different species of a given ligand may compete to form a complex and once a complex has formed, other ligands may compete to destabilize it and form new complexes with the metal cation. (The same applies, of course, to a given ligand for which different metal species or different metals compete to form a complex.) The final outcome of this competition can be predicted if the relevant thermodynamic stability constants for all possible complexes are known (see Section 2.4). Before equilibrium is achieved, however, kinetic species will appear whose concentrations will depend on the dynamics of a system of complex formation and dissociation reactions that reflects a kind of "chemical ecology" based on competition. If the metastability of these kinetic species is significant on a time scale of interest in a soil solution, their behavior becomes more important than that of the equilibrium species.

The flavor of the effects of competitive reactions on metal complex formation can be revealed by considering the example of the hard acid–hard base adduct, AlF^{2+}, at low pH (pH < 3.5).[9] Under this condition, two fluoride species, HF^0 and F^-, compete for the metal species $Al^{3+}(H_2O)_6$ to form the inner-sphere complex, $AlF^{2+}(H_2O)_5$:

$$Al^{3+}(H_2O)_6\,(aq) + F^-\,(aq) \overset{K_{os}^{12}}{=} Al^{3+}(H_2O)_6F^-\,(aq)$$
$$\overset{k_{12}}{\rightarrow} AlF^{2+}(H_2O)_5\,(aq) + H_2O(\ell) \qquad (2.5a)$$

$$Al^{3+}(H_2O)_6\,(aq) + HF^0(aq) \overset{K_{os}^{13}}{=} Al^{3+}(H_2O)_6HF^0(aq)$$
$$\overset{k_{13}}{\rightarrow} AlF^{2+}(H_2O)_5\,(aq) + H^+(aq) + H_2O(\ell) \qquad (2.5b)$$

where the Eigen-Wilkins-Werner mechanism has been invoked, as described above, and the superscripts, 1, 2, and 3, refer to solvated Al^{3+}, F^-, and HF^0, respectively. Given the low pH value assumed, hydrolytic species of Al have been neglected. The rate of proton transfer between the two F species is assumed to be much higher than the rate of inner-sphere complex formation of either species with Al^{3+}.

The rate laws for the reactions in Eq. 2.5 each have the mathematical form of Eq. 2.4, and the overall rate law for these two *parallel* reactions is simply the *sum* of the individual rate laws:[10]

$$\frac{d[AlF^{2+}]}{dt} = k_{12}\,K_{osc}^{12}\,[Al^{3+}]\,[F^-] + k_{13}\,K_{osc}^{13}\,[Al^{3+}]\,[HF^0] \qquad (2.6)$$

where the subscript designation for equilibrium of Al^{3+} and F^- with $Al^{3+}(H_2O)_6F^-$ or $Al^{3+}(H_2O)_6HF^0$ has been dropped to avoid confusion with the equilibrium condition for the inner-sphere complex, AlF^{2+}. (Recall that Al^{3+} and F^- are assumed to be in equilibrium only with the *outer-sphere* complex species.) Since the reaction

$$H^+(aq) + F^-(aq) \overset{K_{HF}}{=} HF^0(aq) \qquad (2.7)$$

is assumed to equilibrate much more quickly than the two reactions in Eq. 2.5, Eq. 2.6 can be rewritten in the compact form

$$\frac{d[AlF^{2+}]}{dt} = K_f([H^+])\,[Al^{3+}]\,[F^-] \qquad (2.8)$$

where

$$K_f([H^+]) \equiv k_{12} K_{osc}^{12} + k_{13} K_{osc}^{13} K_{HFc} [H^+] \tag{2.9}$$

is a pH-dependent, overall second-order rate coefficient, and

$$K_{HFc} = [HF^0]_e / [H^+]_e [F^-]_e \tag{2.10}$$

is a conditional stability constant for the rapid formation of HF^0 according to the reaction in Eq. 2.7. (Note that $[\ ]_e$ in Eq. 2.10 refers *only* to the equilibrium of this latter reaction.) Equation 2.8 can be tested by measuring the *initial* rate of increase in the concentration of AlF^{2+} to avoid interference from the (slower) backward reaction in which the outer-sphere complex reappears. One way to do this[9] is to measure $[F^-]$ electrochemically and then convert it to a total F concentration with the expression

$$F_T \equiv [F^-]_e + [HF^0]_e = [F^-]_e (1 + K_{HFc} [H^+]_e) \tag{2.11}$$

which applies to the *initial* condition; that is, the rate of decrease of F_T is *initially* equal and opposite to the rate of formation of AlF^{2+}. With fixed pH, a graph of F_T against time can be extrapolated to "time zero" to permit an estimate of its initial slope, which is perforce equal in absolute value to the left side of Eq. 2.8. Data generated by repeating this operation for varying initial $[Al^{3+}]$, $[F^-]$, and $[H^+]$ can then be fit to Eq. 2.8 using nonlinear least-square methods.[9] The resulting overall rate coefficient K_f can be plotted against $[H^+]$ to determine whether a linear relationship exists, as stipulated in Eq. 2.9. The slope and y-intercept of this line yield the apparent second-order rate coefficients for the reactions in Eqs. 2.5b and 2.5a, respectively, given that the value of K_{HFc} is available. At 298 K, and for an ionic strength of 100 mol m^{-3} (NaCl background), these two rate coefficients have the values 1.4 and 37 $dm^3\ mol^{-1}\ s^{-1}$, respectively,[9] leading to a time scale[7] on the order of 10^3 s for the (initial) formation of AlF^{2+} at pH 3 and an initial Al concentration, $[Al^{3+}]_0$ = 10 mmol m^{-3} [time scale $\approx (K_f[Al^{3+}]_0)^{-1}$]. Thus the reactions in Eq. 2.5 are relatively slow in the context of the kinetics data in Table 2.3. This fact, implicit in Fig. 2.3, may have important ecotoxicological implications because of the expected greater toxicity of Al^{3+} as compared to AlF^{2+} in natural waters.[9]

Competition for Al^{3+} between F^- and a different ligand (e.g., oxalate, $C_2O_4^{2-}$) can be pictured in terms of the *associative* ligand-exchange reaction:[8,11]

$$AlC_2O_4^+(aq) + F^-(aq) \underset{k_b}{\overset{k_f}{\rightleftarrows}} AlC_2O_4F^0(aq) \xrightarrow{k_f'} AlF^{2+}(aq) + C_2O_4^{2-}(aq) \qquad (2.12)$$

Other pathways than this sequential reaction can be considered (e.g., dissociation of $AlC_2O_4^+$ prior to the formation of AlF^{2+}, termed *dissociative* ligand exchange; or formation of AlF^{2+} and exchange with $C_2O_4^{2-}$ as a means of competition), but the present example suffices to show how competition affects the rate of AlF^{2+} formation. The reaction in Eq. 2.12 is a special case of the abstract reaction scheme in Eq. 1.52, with A = $AlC_2O_4^+$, B = F^-, C = $AlC_2O_4F^0$, D = AlF^{2+}, and E = $C_2O_4^{2-}$. Therefore the set of rate laws in Eq. 1.54 (with $k'_b \equiv 0$) can be applied to describe the kinetics of AlF^{2+} formation by associative ligand exchange. Under the assumption that $AlC_2O_4F^0$ ("species C") equilibrates much more rapidly with $AlC_2O_4^+$ and F^- than it decomposes into AlF^{2+} and $C_2O_4^{2-}$, the left side of Eq. 1.54b can be set equal to zero in order to derive the expression:

$$[AlC_2O_4F^0] = \frac{k_f}{k_f' + k_b}[AlC_2O_4^+][F^-] \qquad (2.13)$$

on a time scale that is small relative to $1/k_f'$. Equation 2.13 can be introduced into the rate law for the increase in concentration of AlF^{2+} ("species D" in Eq. 1.54c) to yield the equation

$$\frac{d[AlF^{2+}]}{dt} = \frac{k_f' k_f}{k_f' + k_b}[AlC_2O_4^+][F^-] \qquad (2.14)$$

This result can be transformed further under the assumption that $AlC_2O_4^+$ equilibrates much more rapidly with Al^{3+} and $C_2O_4^{2-}$ than does AlF^{2+} with F^- (otherwise it would not be useful to consider displacement of $C_2O_4^{2-}$ by F^-, but rather the converse). Then the equilibrium constraint

$$K_{AlOxc} \equiv [AlC_2O_4^+]_e / [Al^{3+}]_e [C_2O_4^{2-}]_e \qquad (2.15)$$

can be imposed on Eq. 2.14 to produce the rate law

$$\frac{d[AlF^{2+}]}{dt} = \frac{k_f' k_f}{k_f' + k_b} K_{AlOxc} [C_2O_4^{2-}][Al^{3+}][F^-]$$

$$\equiv K_f'([C_2O_4^{2-}])[Al^{3+}][F^-] \tag{2.16}$$

where

$$K_f' \equiv \frac{k_f' k_f}{k_f' + k_b} K_{AlOxc} [C_2O_4^{2-}] \tag{2.17}$$

is an overall second-order rate coefficient. Equation 2.16 has the same mathematical form as Eq. 2.8; that is, both equations are second-order rate laws depending on the concentrations of "free" Al and F. The relative importance of K_f' as compared to K_f is a matter for experiment to determine, but scenarios can be imagined[12] in which oxalate *enhances* the rate of formation of AlF^{2+} over that produced by the direct Eigen-Wilkins-Werner mechanism.

2.2 Oxidation–Reduction Reactions

In an oxidation–reduction reaction, or *redox reaction*, one or more electrons are transferred completely from one species to another. Every redox reaction may be separated into a reduction half-reaction and an oxidation half-reaction. A reduction half-reaction, in which a chemical species accepts electrons, may usually be written in the form

$$mA_{ox} + nH^+(aq) + e^-(aq) = pA_{red} + qH_2O(\ell) \tag{2.18}$$

where A represents a chemical species (e.g., $NO_3^-(aq)$) and "ox" and "red" denote oxidized and reduced states, respectively. The parameters m, n, p, and q are stoichiometric coefficients, whereas H^+ and e^- refer to the proton and the electron in aqueous solution. A complete redox reaction is a combination of Eq. 2.18 and an oxidation half-reaction, in which a chemical species gives up electrons. This combination yields a chemical equation that does not exhibit the electron as a reacting species.

In soil solutions the most important chemical elements that undergo redox reactions are C, N, O, S, Mn, and Fe. For contaminated soils the elements As, Se, Cr, Hg, and Pb could be added. Table 2.4 lists reduction half-reactions (most of which are heterogeneous) and their equilibrium constants at 298.15 K under 1 atm pressure for the six principal elements involved in soil redox phenomena. Although the reactions listed in the table are not full redox reactions, their equilibrium constants have thermodynamic significance and may be calculated with the help of Standard-State chemical potentials in the manner

Table 2.4 Some Important Reduction Half-Reactions in Soil Solutions

Reduction half-reaction	log K_{298}
$1/4\ O_2(g) + H^+(aq) + e^-(aq) = 1/2\ H_2O(\ell)$	20.8
$H^+(aq) + e^-(aq) = 1/2\ H_2(g)$	0.0
$1/2\ NO_3^-(aq) + H^+(aq) + e^-(aq) = 1/2\ NO_2^-(aq) + 1/2\ H_2O(\ell)$	14.1
$1/4\ NO_3^-(aq) + 5/4\ H^+(aq) + e^-(aq) = 1/8\ N_2O(g) + 5/8\ H_2O(\ell)$	18.9
$1/5\ NO_3^-(aq) + 6/5\ H^+(aq) + e^-(aq) = 1/10\ N_2(g) + 3/5\ H_2O(\ell)$	21.1
$1/8\ NO_3^-(aq) + 5/4\ H^+(aq) + e^-(aq) = 1/8\ NH_4^+(aq) + 3/8\ H_2O(\ell)$	14.9
$1/2\ MnO_2(s) + 2\ H^+(aq) + e^-(aq) = 1/2\ Mn^{2+}(aq) + H_2O(\ell)$	20.7
$1/2\ MnO_2(s) + 1/2\ HCO_3^-(aq) + 3/2H^+(aq) + e^-(aq) = 1/2\ MnCO_3(s) + H_2O(\ell)$	20.2
$Fe(OH)_3(s) + 3\ H^+(aq) + e^-(aq) = Fe^{2+}(aq) + 3H_2O(\ell)$	16.4
$FeOOH(s) + 3\ H^+(aq) + e^-(aq) = Fe^{2+}(aq) + 2H_2O(\ell)$	13.5
$1/2\ Fe_3O_4(s) + 4\ H^+(aq) + e^-(aq) = 3/2\ Fe^{2+}(aq) + 2H_2O(\ell)$	14.9
$1/2\ Fe_2O_3(s) + 3\ H^+(aq) + e^-(aq) = Fe^{2+}(aq) + 3/2\ H_2O(\ell)$	12.3
$1/4\ SO_4^{2-} + 5/4\ H^+ + e^-(aq) = 1/8\ S_2O_3^{2-}(aq) + 5/8\ H_2O(\ell)$	4.9
$1/8\ SO_4^{2-}(aq) + 9/8\ H^+(aq) + e^-(aq) = 1/8\ HS^-(aq) + 1/2\ H_2O(\ell)$	4.3
$1/8\ SO_4^{2-}(aq) + 5/4\ H^+(aq) + e^-(aq) = 1/8\ H_2S(aq) + 1/2\ H_2O(\ell)$	5.1
$1/2\ CO_2(g) + 1/2\ H^+(aq) + e^-(aq) = 1/2\ CHO_2^-(aq)$	-5.2
$1/4\ CO_2(g) + 7/8\ H^+(aq) + e^-(aq) = 1/8\ C_2H_3O_2^-(aq) + 1/4\ H_2O(\ell)$	1.2
$1/4\ CO_2(g) + 1/12\ NH_4^+(aq) + 11/12\ H^+(aq) + e^-(aq)$ $= 1/12\ C_3H_4O_2NH_3(aq) + 1/3\ H_2O(\ell)$	0.8
$1/4\ CO_2(g) + H^+(aq) + e^-(aq) = 1/24\ C_6H_{12}O_6(aq) + 1/4\ H_2O(\ell)$	-0.2
$1/8\ CO_2(g) + H^+(aq) + e^-(aq) = 1/8\ CH_4(g) + 1/4\ H_2O(\ell)$	2.9

described in Special Topic 1 (Chapter 1). Two important points are worth mentioning in this connection. First, the fact that the reactions in Table 2.4 describe electrochemical equilibria has no direct bearing on the method by which the equilibrium constants were determined. The Standard-State chemical potentials employed to compute a value of log K_R may be the results of solubility, themochemical, spectroscopic, or electrochemical experiments. No particular experimental method is implied just because the reaction of interest is a redox reaction. Second, the convention that $\mu^0 \equiv 0$ for $H^+(aq)$, $e^-(aq)$, and any element in its most stable phase (see Special Topic 1) often simplifies the calculation of log K_R for a reduction reaction. For example, log K_R for the reduction of $O_2(g)$ at 298.15 K and 1 atm pressure is calculated with the equation

$$\log K_R = -\left[\frac{1}{2}\mu^0(H_2O) - \frac{1}{4}\mu^0(O_2) - \mu^0(H^+) - \mu^0(e^-)\right] / 5.708$$

$$= -\mu^0(H_2O) / 11.416 \text{ kJ mol}^{1-} = 20.8$$

Because of the convention for μ^0, the value of log K for the oxidation of $H_2(g)$ is equal to zero:

$$\frac{1}{2}H_2(g) = H^+(aq) + e^-(aq)$$

$$\log K = -[\mu^0(H^+) + \mu^0(e^-) - \frac{1}{2}\mu^0(H_2)]/5.708 \equiv 0$$

It follows that log K_R for any reduction half-reaction may be considered equally as log K for a complete redox reaction in which electrons are transferred from $H_2(g)$ to the reduced species in the reduction reaction. In the case of Eq. 2.18, for example, log K_R for the reduction half-reaction is the same as log K for the redox reaction:

$$mA_{ox} + (n-1)H^+ + \frac{1}{2}H_2(g) = pA_{red} + qH_2O(\ell) \tag{2.19}$$

Taking the reduction of NO_3^- (denitrification) as a concrete example from Table 2.4, one has

$$\frac{1}{5}NO_3^-(aq) + \frac{1}{5}H^+(aq) + \frac{1}{2}H_2(g) = \frac{1}{10}N_2(g) + \frac{3}{5}H_2O(\ell) \tag{2.20}$$

$$\log K = 21.16$$

The firm thermodynamic status of log K_R for reduction half-reactions permits the use of these parameters in the normal way (see Section 1.2 and Special Topic 1) to evaluate equilibrium activities of oxidized and reduced species and to compare the stabilities of reactants and products in redox reactions. As an example of a stability comparison, consider the possible reduction of N(V) to N(0) through the oxidation of C(0) to C(IV) in a soil solution.[13] The reduction half-reaction for denitrification is implicit in Eq. 2.20; that for C oxidation is

$$\frac{1}{24}C_6H_{12}O_6(aq) + \frac{1}{4}H_2O(\ell) = \frac{1}{4}CO_2(g) + H^+(aq) + e^-(aq) \tag{2.21}$$

with log K = 0.2 according to Table 2.4. The combination of the half-reactions produces the redox reaction

$$NO_3^-(aq) + \frac{1}{24}C_6H_{12}O_6(aq) + \frac{1}{5}H^+(aq)$$

$$= \frac{1}{10}N_2(g) + \frac{1}{4}CO_2(g) + \frac{7}{20}H_2O(\ell) \qquad \log K = 2: \tag{2.22}$$

Since log K for this reaction is positive, $\Delta_r G^0$ is negative, and the products are more stable than the reactants in the Standard State. The situation in a state other than the Standard State may be evaluated, as usual, by a consideration of the equilibrium constant itself. In the present case (assuming unit water activity)

$$\log K = 21.3 = \frac{1}{10} \log p_{N_2} + \frac{1}{4} \log p_{CO_2} + \frac{1}{5} \mathrm{pH} - \frac{1}{24} \log(\mathbf{C_6H_{12}O_6}) - \frac{1}{5} \log(\mathbf{NO_3^-}) \tag{2.23}$$

If a soil solution is in equilibrium with $p_{CO_2} = 10^{-3.52}$ atm and $p_{N_2} = 0.78$ atm at pH 5.6 and has $(\mathbf{C_6H_{12}O_6}) = 10^{-6}$, the preceding expression for log K leads to $\log(\mathbf{NO_3^-}) = -104$, which shows that nitrate reduction is highly favored thermodynamically.

The chemical potentials of the "free" proton and electron in aqueous solution,[14] introduced in Special Topic 1, are used in a conventional thermodynamic way to compute log K_R values. The same is true for the aqueous activities of these two species. These kinds of formal manipulations of single-ion activities are discussed from a fundamental perspective in Special Topic 2 at the end of this chapter. Given that a suitable meaning can be assigned to the more familiar proton activity, it is instructive to develop the analogous concepts for the electron activity. Brønsted acidity is measured by the negative base 10 logarithm of the free proton activity (i.e., the pH value). As is well known, large positive values of pH strongly favor the existence of Brønsted bases (free proton acceptors), whereas small values of pH strongly favor Brønsted acids (free proton donors). This role of pH in Brønsted acid–base reactions (proton-transfer reactions) may be understood directly in terms of the fact that at a given temperature, pH is proportional to the chemical potential of $H^+(aq)$. The larger the pH value, the smaller is the value of $\mu(H^+)$ in solution and the greater is the tendency of a Brønsted acid to lose its transferable protons. Statements exactly analogous with these can be made about the free electron in aqueous solutions. The Jørgensen oxidizability is measured by the negative base 10 logarithm of the free electron activity, the *pE value*. Large positive values of pE strongly favor the existence of oxidized species (free electron acceptors), whereas small values of pE strongly favor reduced species (free electron donors). The role of the pE value in redox reactions (electron-transfer reactions) may be understood in terms of the proportionality of the chemical potential of $e^-(aq)$ to pE at a given temperature. The larger the pE value, the smaller is the value of $\mu(e^-)$ and the greater is the tendency of a reduced species to lose its transferable electrons.[15]

The range of pE values possible in soil solutions can be estimated by considering the decomposition of liquid water. The *upper* extreme of pE will occur when $H_2O(\ell)$ decomposes to form protons and $O_2(g)$ according to the first half-reaction in Table 2.4:

$$\frac{1}{2}\ H_2O(\ell) = \frac{1}{4}\ O_2(g) + H^+(aq) + e^-(aq) \qquad \log K = -20.8 \qquad (2.24)$$

If liquid water is in its Standard State, the pE value corresponding to this reaction at equilibrium is

$$pE \equiv -\log(\mathbf{e}^-) = -\log K - pH + \frac{1}{4}\log p_{O_2} = 20.6 - pH \qquad (2.25)$$

where $p_{O_2} = 0.21$ atm, the atmospheric partial pressure of $O_2(g)$, has been introduced in the last step along with the value of log K. The pE value will be greatest when the pH value is least. With pH = 4 taken as a typical lower bound for soil solutions, the estimated upper extreme value of pE is +16.6. The *lower* extreme of pE will occur when $H_2O(\ell)$ decomposes to form hydroxide ions and $H_2(g)$ according to the reaction

$$H_2O(\ell) + e^-(aq) = \frac{1}{2}\ H_2(g) + OH^-(aq) \qquad \log K = -14 \qquad (2.26)$$

After the water ionization reaction (Eq. 1.4) is subtracted from this reaction in order to introduce H^+ into the products in place of OH^-, the half-reaction

$$H^+(aq) + e^-(aq) = \frac{1}{2}\ H_2(g) \qquad \log K_R = 0 \qquad (2.27)$$

is the result. In this case, the pE value is given by the expression

$$pE \equiv -\log(\mathbf{e}^-) = \log K_R - pH - \frac{1}{2}\log p_{H_2}$$

$$= -pH - \frac{1}{2}\log p_{H_2} \qquad (2.28)$$

If $H_2(g)$ is in its Standard State (to simplify calculation) and pH = 9 is taken as a typical upper bound for soil solutions, the estimated lower extreme value of pE is −9.0, although a larger value would result if a less extreme value of p_{H_2} were used instead of 1.0 atm. Thus the range of pE values expected theoretically in soil solutions is $-9.0 < pE < +16.6$, which corresponds to a change in electron activity by 26 orders of magnitude. In actuality, pE values in soil solutions tend to remain between −6 and +13, regardless of the pH value.[16] At pH 7.0, *oxic* soils have $+7 < pE$; *suboxic* soils have $+2 < pE < +7$; and *anoxic* soils have $pE < +2$. Suboxic soils differ from oxic soils in having enough "free" electrons available to reduce $O_2(g)$, but not enough to reduce $SO_4^{2-}(aq)$.

Two or more aqueous chemical species that contain the same element, but in different oxidation states, are termed a *redox couple*.[17] Their concentrations in a soil solution can be described as a function of pE and pH in precisely the same way as the concentrations of carbonate species were described as a function of pH alone in Section 1.2. As a typical example of the calculation of the concentrations of aqueous redox species, consider the speciation of dissolved ionic forms of nitrogen in the absence of denitrification. The species to be considered are NO_3^-, NO_2^-, and NH_4^+. These three ions are linked through the nitrate reduction reactions (Table 2.4):

$$NO_3^-(aq) + 2H^+(aq) + 2e^-(aq) = NO_2^-(aq) + H_2O(\ell) \qquad \log K_1 = 28.2 \tag{2.29a}$$

$$NO_3^-(aq) + 10H^+(aq) + 8e^-(aq) = NH_4^+(aq) + 3H_2O(\ell) \qquad \log K_2 = 119.2 \tag{2.29b}$$

These two reactions, in turn, must be incorporated into the equation of mole balance:

$$N_T = [NO_3^-]_e + [NO_2^-]_e + [NH_4^+]_e \tag{2.30}$$

This expression can be solved for the *ligand distribution coefficient*:

$$\beta_{NO_3} \equiv \frac{[NO_3^-]_e}{N_T} \tag{2.31a}$$

in terms of pE, pH, and the equilibrium constants K_1 and K_2. After factoring the nitrate concentration from the left side of Eq. 2.30, one obtains

$$\begin{aligned} N_T &= [NO_3^-]_e \{1 + ([NO_2^-]_e / [NO_3^-]_e) + ([NH_4^+]_e / [NO_3^-]_e)\} \\ &= [NO_3^-]_e \{1 + K_1(H^+)^2(e)^2 + K_2(H^+)^{10}(e)^8\} \\ &= [NO_3^-]_e \{1 + 10^{2(14.1 - pH - pE)} + 10^{8(14.91 - 1.25pH - pE)}\} \end{aligned} \tag{2.32}$$

where it has been assumed that the Davies equation (Eq. 1.21) can be used to express the activity coefficient of each ion, thereby making all concentration ratios equal to activity ratios. It follows from Eqs. 2.31a and 2.32 that

$$\beta_{NO_3} = \{1 + 10^{2(14.1 - H - pE)} + 10^{8(14.91 - 1.25pH - pE)}\}^{-1} \tag{2.31b}$$

The other two distribution coefficients of interest now may be found immediately:

$$\beta_{NO_2} \equiv \frac{[NO_2^-]_e}{N_T} = \frac{[NO_2^-]_e}{[NO_3^-]_e}\beta_{NO_3} = 10^{2(14.1 - pH - pE)}\beta_{NO_3} \tag{2.31c}$$

$$\beta_{NH_4} \equiv \frac{[NH_4^+]_e}{N_T} = \frac{[NH_4^+]_e}{[NO_3^-]_e}\beta_{NO_3} = 10^{8(14.91 - 1.25pH - pE)}\beta_{NO_3} \tag{2.31d}$$

Equations 2.31 provide the complete solution to the speciation problem in this example. For a given pH value of a soil solution, the values of the distribution coefficients can be calculated as functions of pE, or vice versa. Figure 2.4 shows graphs of the three distribution coefficients for $4 \leq pH \leq 8$ and $3 \leq pE \leq 12$. The most important features of the graphs are the increasing range of pE over which nitrate predominates as the pH value is increased and the very narrow pE range over which predominance shifts from nitrate to nitrite to ammonium at any fixed pH value. This latter property suggests that the three aqueous nitrogen species will exhibit a great deal of variability in their concentrations in soils, where the degree of aeration fluctuates significantly.

Of course, redox reactions do not occur in isolation but are coupled through complexation reactions to other species. For example, Eq. 2.30 could include terms for nitrate and amine complexes, in addition to those for free nitrate, nitrite, and ammonium ions, if a typical soil solution were under consideration. The calculation of nitrogen speciation then would proceed just as described above. Indeed, redox reactions introduce no new mathematical elements into a speciation computation, any more than would the consideration of, for example, CO_2 reactions. The only new item brought in is an additional variable, the pE value, which, like the partial pressure of $CO_2(g)$, must be specified in order to solve mole balance equations for distribution coefficients.

It is very important to understand that this kind of speciation calculation indicates that certain redox reactions *can* occur in soils, but not that they *will* occur: a chemical reaction that is favored by a large value of log K is not necessarily favored *kinetically*. This fact is especially applicable to redox reactions because they are often extremely slow, and because reduction and oxidation half-reactions often do not couple well to each other. For example, the coupling of the half-reaction for $O_2(g)$ reduction with that for $N_2(g)$ oxidation leads to log K = −0.3 for the overall redox reaction:

$$\frac{1}{4}O_2(g) + \frac{1}{10}N_2(g) + \frac{1}{10}H_2O(\ell) = \frac{1}{5}NO_3^-(aq) + \frac{1}{5}H^+(aq) \tag{2.33}$$

For a soil solution that is in equilibrium with the atmosphere (p_{O_2} = 0.21 atm, p_{N_2} = 0.78 atm), the preceding value of log K predicts complete oxidation of

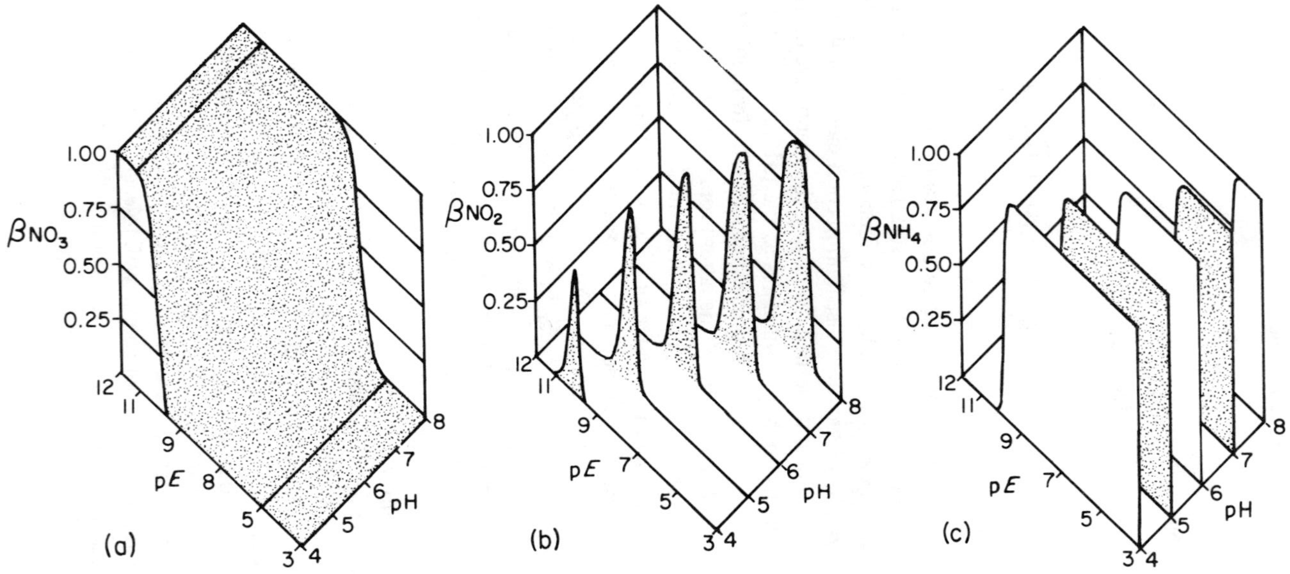

FIG. 2.4. Graphs of the distribution coefficients (Eq. 2.31) for redox nitrogen species as functions of pE and pH.

nitrogen from N(0) to N(V) at any pH > 2.5. But this prediction is contradicted dramatically by the observed persistence of N_2 in the atmosphere under surface terrestrial conditions.

The typical lack of effective coupling and slowness of redox reactions mean that *catalysis* is required if equilibrium is ever to come about. In soil solutions, and in other natural water systems, the catalysis of redox reactions is mediated by microbial organisms. In the presence of the appropriate microbial species, a redox reaction can proceed quickly enough in soil to produce activity values of the reactants and products that agree with thermodynamic predictions. Of course, this possibility is dependent entirely on the growth and ecological behavior of the soil microbial population and the degree to which the products of the attendant biochemical reactions can diffuse and mix in the soil solution. This perspective is essential to a complete interpretation of the terms *oxic, suboxic,* and *anoxic,* introduced above. Below pE 5, oxygen is not stable in neutral soils. Above pE 5, it is consumed in the respiration processes of aerobic microorganisms. As pE decreases below 8, electrons become available to reduce NO_3^-. This reduction is catalyzed by nitrate respiration (i.e., NO_3^- serving as a biochemical electron acceptor like O_2) involving bacteria that ultimately excrete NO_2^-, N_2, N_2O, or NH_4^+.[13] As the pE value drops into the range 7–5, electrons become plentiful enough to support the reduction of Fe and Mn in solid phases. Iron reduction does not occur until O_2 and NO_3^- are depleted, but Mn reduction can be initiated in the presence of nitrate. *Thus Fe and Mn reduction are characteristic of a suboxic soil environment.* As the pE value decreases below +2, a soil becomes anoxic and, when pE < 0, electrons are available for sulfate reduction catalyzed by a variety of anaerobic bacteria. Typical products in aqueous solution are H_2S, bisulfide (HS^-), or thiosulfate ($S_2O_3^-$) ions, as indicated in Table 2.4. The chemical reaction sequence for the reduction of O, N, Mn, Fe, and S induced by changes in pE is also a microbial ecology sequence for the biological catalysts that mediate the reactions. Aerobic microorganisms that utilize O_2 to oxidize organic matter do not function below pE 5. Denitrifying bacteria thrive in the pE range between +10 and 0, for the most part. Sulfate-reducing bacteria do not grow at pE values above +2.[16]

Often, redox reactions will be controlled by the highly variable dynamics of an open biological system, with the result that redox products will at best correspond to local conditions of partial equilibrium in soil. In other cases, including the very important case of flooded soil, redox reactions will be controlled by the behavior of a closed, isobaric chemical system that is catalyzed effectively by microorganisms and for which a thermodynamic description is especially apt. Regardless of which of these extremes is more appropriate for characterizing the redox reactions in a soil, the role of organisms deals only with the reaction pathways aspect of redox.[17] *Soil organisms can affect the rate or extent of a redox reaction, but not its standard Gibbs energy change.* If a redox reaction is not favored thermodynamically, microbial intervention cannot change that fact.

Kinetic studies of redox reactions in the absence of microbial catalysis have produced only a few broad insights as to the mechanisms involved.[15,18] For redox reactions involving metals, one of which is coordinated to a ligand that has available lone-pair electrons, the other of which is in a solvation complex, an associative metal exchange process with electron transfer, the *Taube mechanism*, has been proposed (Fig 2.5).[18] This mechanism is formally similar to that described in Eq. 2.12, but with metal and ligand roles reversed:

$$ML^{(m-\ell)+}(aq) + N^{n+}(aq) \underset{k_b}{\overset{k_f}{\rightleftarrows}} MLN^{(m+n-\ell)}(aq)$$

$$\overset{k_f^{\prime}}{\rightarrow} M^{(m-1)+}(aq) + NL^{(n+1-\ell)+}(aq) \qquad (2.34)$$

where M and N are metals and L is the complexing ligand. This reaction sequence is a special case of the abstract reaction scheme in Eq. 1.52 ($A = ML^{(m-\ell)+}$, $B = N^{n+}$, etc.) and therefore the rate laws in Eq. 1.54 apply. Typical examples of M and N are transition metals, with L being a ligand containing an electron donor from Group VIA or VIIA of the Periodic Table. As is evident in Fig. 2.5, the overall second-order rate coefficient for the redox reaction (cf. Eq. 2.16) varies over many orders of magnitude. Equation 2.34 describes what is termed an *inner-sphere mechanism* (because ML and NL are inner-sphere complexes). If both M and N are in solvation complexes while electron transfer occurs (i.e., L is replaced by a water molecule), the mechanism is termed *outer-sphere* (the *Marcus mechanism*). This latter mechanism is typical of metals in the +II oxidation state, which do not tend to form hydrolytic species (i.e., L = OH; cf. Fig. 2.1).[18]

With less mechanistic information available, rate laws for redox reactions often are expressed in a *pseudo-first-order* form:[19]

$$-\frac{d[A]}{dt} = K(Env)\,[A] \qquad (2.35)$$

where A is an oxidized (reduced) species whose reduction (oxidation) is monitored with time and K(Env) is a "pseudo-first-order" rate coefficient that depends on one or more chemical "environmental factors," Env, which are held fixed as the redox reaction is monitored. Actually, Eq. 2.35 can be applied to a variety of reactions, including metal complexation.[19] For example, Eq. 2.8 can be written in the form of Eq. 2.35 after noting that the rate of decrease in $[Al^{3+}]$ equals the rate of increase in $[AlF^{2+}]$ and setting $K(Env) \equiv K([H^+], [F^-]) \equiv K_f([H^+])\,[F^-]$. In the case of redox reactions, Env refers typically to the concentrations of oxidants, reductants, protons, and hydroxide ions. As a concrete example, let A = Fe(II) and Env include p_{O_2} and $[OH^-]$. Then Eq. 2.35 describes Fe(II) oxidation and, for pH > 4.5, takes the form[15]

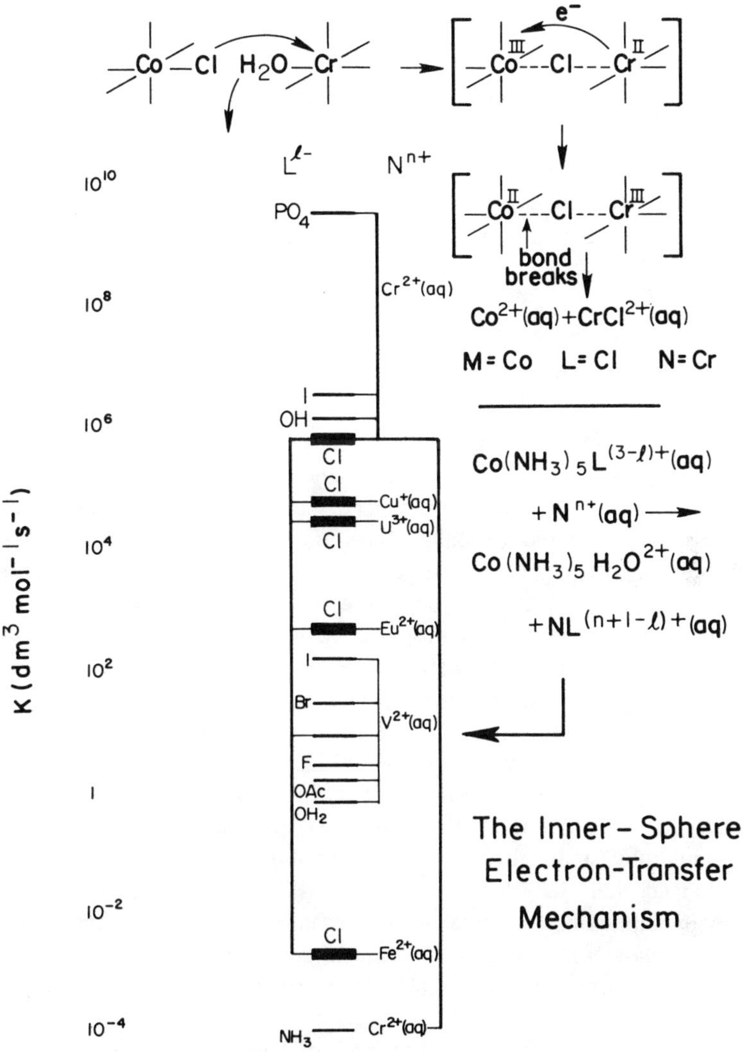

FIG. 2.5. The associative inner-sphere reaction mechanism for Co reduction by Cr or other metals via Cl (thick bars) or other ligands. The overall second-order rate coefficient for the redox reaction (K) varies over 14 orders of magnitude.

$$-\frac{d[Fe(II)]}{dt} = K(p_{O_2}, [OH^-])[Fe(II)] \tag{2.36}$$

where

$$K(p_{O_2}, [OH^-]) = k p_{O_2} [OH^-]^2 \tag{2.37}$$

is the pseudo-first-order rate coefficient for the case where p_{O_2} and $[OH^-]$ are constants and $k \approx 2.5 \times 10^{11}$ dm^6 mol^{-2} atm^{-1} s^{-1} at 293 K. Given $p_{O_2} = 0.21$ atm and pH 6, the time scale[7] for Fe(II) oxidation according to Eq. 2.37 is approximately 10^5 s (just about three days). Although Eq. 2.35 is successful as an empirical rate law, it should be evident that no specific reaction pathway is implied by this success.

2.3 Polymeric Species

Chemical species involving molecular units in a repetitive structure to form a polymer are ubiquitous in soils. Multinuclear hydrolytic complexes, such as $Al_2(OH)_2^{4+}$ or $Fe_2(OH)_2^{4+}$, and biopolymers, such as proteins or polysaccharides, come to mind as familiar examples. The complexation reactions of hydrolytic and biological polymers are investigated in much the same way as described in Section 2.1, the principal issues being the characterization of the stoichiometry of the polymers, their metastability as aqueous species, and the effects of the close proximity among their functional groups on their reactivity.[20]

The situation is different for aqueous species of *humic substances*, the organic matter in soil that is not identifiable as unaltered or partially altered biomass or as conventional biomolecules.[21] Humic substances comprise organic compounds that are not synthesized directly to sustain the life cycles of the soil biomass. More specifically, they comprise polymeric molecules produced through microbial action that differ from biopolymers because of their molecular structure and their long-term persistence in soil. This definition of humic substances implies no particular set of organic compounds, range of relative molecular mass, or mode of chemical reactivity. What is essential is dissimilarity to conventional biomolecular structures and biologically refractory behavior.

The principal structural characteristics of humic substances that influence their chemical reactivity are[22] (1) *polyfunctionality*, the existence of a variety of different functional groups with a broad range of functional group reactivity, typical of mixtures of interacting heteropolymers; (2) *macromolecular charge*, the development of (typically anionic) polyelectrolyte character with the resultant effect on functional group reactivity and molecular confirmation; (3) *hydrophilicity*, the tendency to form strong hydrogen bonds with water molecules that solvate polar functional groups, and (4) *structural lability*, the capacity to associate intramolecularly and to change molecular conformation in response to variations in pH value, pE value, electrolyte concentration, and functional group binding. These four properties of humic substances, of course, are common also to biomolecules, but for humic substances they reflect the behavior of a heterogeneous mixture of polymeric molecules instead of the behavior of a structurally well defined, single macromolecule.

Given that humic substances typically bear an overall negative charge in the pH range of soils,[23] the principal complexation reactions they engage in are those with protons and metal cations, that is, with Brønsted and Lewis acids. These reactions, at the simplest level of characterization, require chemical composition data for which the moles of metal, for example, complexed by a dissolved humic substance must be determined by a method that separates the metal–humate product from the reactant humic substance and metal species, and from all other chemical forms of the metal in aqueous solution. The method of separation and the procedures for quantitating the metal–humate product, however, are generally much more difficult to design than for conventional metal complexation reactions like those in Eq. 2.5. For example, if an electroactive metal such as Cu(II) is complexed by a humic substance such as fulvic acid[23] and if the moles of metal complex are determined by an electrodeposition technique (e.g., anodic stripping voltammetry) as the difference between the total moles of metal present and the moles of metal deposited, careful consideration must be given to the effects of imposed conditions (the electrodeposition potential, the pH value, the extent of Cu saturation of the binding capacity, etc.) and of electrode system–aqueous system interactions (humic substance adsorption, kinetic currents, etc.) on the determination of noncomplexed metal. Usually it is necessary to evaluate these effects quantitatively through ancillary experiments and then apply empirical corrections to the data of the metal complexation reaction. Since the structures and the bonding characteristics of metal–humate complexes are known poorly,[23] no alternative to careful supplementary experimentation exists for estimating the perturbing influence of the methods used to determine metal complexation. This situation is illustrated in Fig. 2.6, which lists eight methods of quantitating Cu(II) complexation by fulvic acid. Each method is associated with a characteristic sensitivity to detect $Cu^{2+}(aq)$ concentrations, time scale of detection, and Cu saturation of the binding capacity of fulvic acid (equivalent to decreasing stability of the Cu–fulvate complex). This last qualifying property of the methodologies is unique to heterogeneous, polymeric ligands.

The prospect of interpreting the complexation reactions of humic substances at the level of detail typical of well-characterized biomolecules is dimmed by the need to consider many competing elementary reactions involving interacting polyelectrolyte fragments.[22] Even if the molecular structure of each component organic complex species were somehow established definitively, the quantitative characterization of their collective reaction behavior would necessitate the determination of a very large number of molecular parameters. Although relatively simple relationships sometimes can be found between complex stability and molecular structure, typically the data available on a complexation reaction will be inadequate for an equilibrium or kinetics description with more than a few model parameters. For this reason, and because the intended use of the complexation data is nearly always in the context of chemical speciation, *the complexation reactions involving humic substances are interpreted in some*

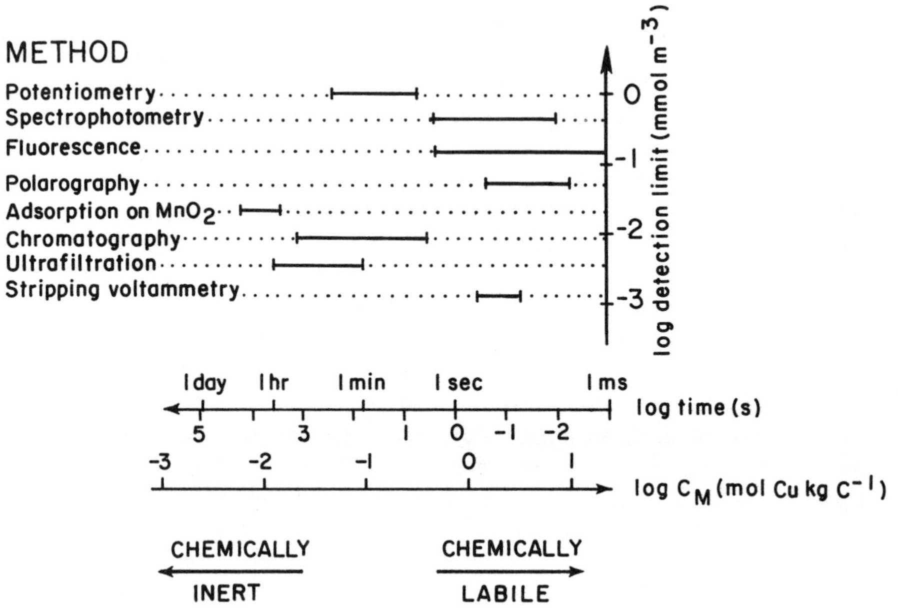

FIG. 2.6. Methods of quantitating Cu complexation by fulvic acid, their sensitivity, time scales, and range of complexation capacity (C_M). (Adapted with permission from J. Buffle.[21] Copyright © 1988 by Ellis Horwood Ltd., Chichester, England.)

average sense: The humic substance reactant and product are taken to represent averages over a heterogeneous organic mixture whose overall reaction with a proton or a metal species can be described by analogy with what is known to apply to complexation reactions with well-defined ligands.

This perspective on the complexation reactions of humic substances leads to a generic model approach termed a *quasiparticle description*.[24] A quasiparticle description is a mathematical model of an aqueous solution containing a humic substance in which the actual heterogeneous, interacting, organic ligand mixture is replaced conceptually by a set of hypothetical, *non*interacting macromolecular ligands whose chemical behavior mimics closely that of the actual mixture in respect to complexation reactions observed under given experimental conditions. The hypothetical macromolecules are called *quasiparticles* because they are only theoretical constructs, not real molecules. Their use is justified in a model sense because they permit a mathematical description of complexation reactions that is sufficiently accurate under given experimental conditions, while avoiding the greatly complicated modeling that would be required for a molecular-level description of an actual humic substance. However, it must always be borne in mind that the parameters generated by quasiparticle models only reflect average properties and therefore are more useful for speciation calculations than for understanding reaction mechanisms. Moreover, quasiparticle models are not valid for predictions under any experimental conditions except those for which the data they are based on were collected.[24]

Logical relationships among several quasiparticle models of humic substance complexation reactions are shown as a flowchart in Fig. 2.7.[24] The descending series of prompts on the left side of the diagram serves to define the quasiparticle functional groups involved in complexation. Each model indicated on the right side of the diagram does not assume the functional group property queried to its immediate left, but does assume the properties queried either above it or below it. For example, the "polyelectrolyte model" does not assume independent functional groups, but does assume a discrete "affinity spectrum" and 1:1 complexes exclusively. The Scatchard model is the result of a positive response to all of the quasiparticle functional group properties queried in Fig. 2.7.

The "Henderson-Hasselbalch," "Perdue-Lytle," and "Scatchard" models are special cases of an "affinity spectrum" model, defined by[23, 24]

$$[\text{metal complex}]_e = \int_{-\infty}^{\infty} p(\ln K) \frac{K[\text{metal}]_e}{1 + K[\text{metal}]_e} \, d\ln K \qquad (2.38)$$

for metal complexation, where $[\ \ ]_e$ is an equilibrium concentration in moles per cubic decimeter and $p(\ln K)d\ln K$ is the relative probability that the natural logarithm of the conditional stability constant for the complex is between $\ln K$ and $\ln K + d\ln K$, subject to the constraint

$$\lim_{[\text{metal}]\uparrow\infty} [\text{metal complex}]_e = \int_{-\infty}^{\infty} p(\ln K) \, d\ln K \equiv C_M \qquad (2.39)$$

which usually is termed the *complexation capacity* of a humic substance quasiparticle.

The Scatchard model[23, 24] is perhaps the best-known quasiparticle description of complexation by humic substances. The quasiparticles in this case are hypothetical polymeric molecules bearing one class of functional group that forms a 1:1 complex with a cation. The conditional stability constant for the complexation of 1 mol of a metal by 1 mol of a given class of Scatchard quasiparticles can be expressed as

$$K_{ic}^{qp} \equiv \frac{[\text{metal complex}]_{ie}}{[\text{metal}]_e \, [\text{quasiparticle}]_{ie}} \qquad (i = 1, 2, \ldots) \qquad (2.40)$$

where $[\text{quasiparticle}]_{ie}$ is the equilibrium concentration of quasiparticles of class i not complexed with metal. The Scatchard model then results from Eq. 2.38 after introduction of the delta-"function" affinity spectrum[23, 24]

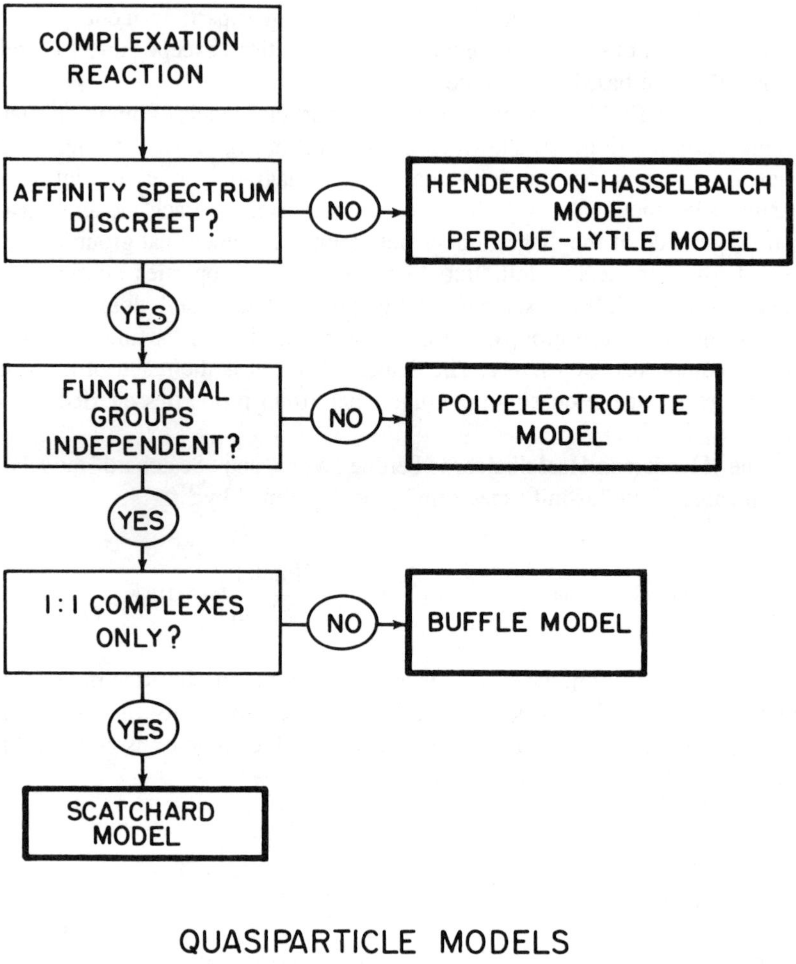

FIG. 2.7. Logical flowchart relating several quasiparticle models of metal complexation reactions.

$$p(\ln K) = \sum_{i=1}^{N} b_i \delta (\ln K - \ln K_{ic}^{qp}) \tag{2.41}$$

and the defining relationships

$$\int_{-\infty}^{\infty} f(x)\ \delta(x - a)\, dx \equiv f(a)$$

and

$$C_M \equiv \sum_{i=1}^{N} b_i \qquad (2.42)$$

to obtain

$$[\text{metal complex}]_e = \sum_{i=1}^{N} b_i \frac{K_{ic}^{qp}[\text{metal}]_e}{1 + K_{ic}^{qp}[\text{metal}]_e} \qquad (2.43)$$

where b_i is the complexation capacity of the class i quasiparticles. Thus if the "affinity spectrum" comprises a row of very sharp "spikes" centered at $\ln K = \ln K_{ic}^{qp}$ (i = 1, 2, . . . , N), the Scatchard model is generated; often, in applications, $N \leq 3$.[23, 24]

An obvious generalization of this model would be to determine the "affinity spectrum," p(ln K), by numerical inversion of Eq. 2.38, given precise experimental data on $[\text{metal complex}]_e$ as a function of $[\text{metal}]_e$.[25] The clear conceptual advantage of a general "affinity spectrum" model is its avoidance of an assumed probability function, p(ln K); but its concomitant disadvantage is the absence of any model parameters, such as K_{ic}^{qp} with which to facilitate comparison among the results of different metal complexation studies. The use of an experimentally determined p(ln K) does not affect the quasiparticle aspect of "affinity spectrum" models. The quasiparticles in this more general situation are still noninteracting, hypothetical molecules, but they exhibit a measured continuum of (noninteracting) functional group classes, with metal complexation in each class governed by a Scatchard-model expression (any one term on the right side of Eq. 2.43).[23, 24]

In addition to its relative simplicity, the quasiparticle approach has the advantage of a formal mathematical structure that is analogous to that described in Section 2.1 for complexes involving nonpolymeric ligands, such as F^- or $C_2O_4^{2-}$. Thus, for example, the complexation reaction between Al^{3+} and a humic substance Scatchard quasiparticle, L^-, can be written by analogy with Eq. 2.5a:

$$Al^{3+}(aq) + L^-(aq) \rightleftarrows AlL^{2+}(aq) \qquad (2.44)$$

where solvation water molecules are omitted, as they were in Eq. 1.8 depicting the formation of $MnCl^+(aq)$. The measured concentration of the complex, AlL^{2+}, can be represented by Eq. 2.43 in the Scatchard model and nonlinear least-squares methods can be applied to estimate, under given conditions of pH, ionic strength, and so on, the model parameters, b_i, K_{ic}^{qp} (i = 1, . . . , N), from the observed dependence of $[AlL^{2+}]_e$ on $[Al^{3+}]_e$ or Al_T.[23]

The same considerations apply to rate laws developed to describe the kinetics of Al^{3+} complexation by humic substances. This fact can be illustrated by the associative metal-exchange reaction:[11, 25]

$$AlL^{2+}(aq) + Cu^{2+}(aq) \underset{k_b}{\overset{k_f}{\rightleftarrows}} AlLCu^{4+}(aq)$$

$$\overset{k_f'}{\rightarrow} CuL^{+}(aq) + Al^{3+}(aq) \qquad (2.45)$$

in which Cu^{2+} replaces Al^{3+} in a humic substance complex via an intermediate binuclear metal complex, $AlLCu^{4+}$. This reaction, like that in Eq. 2.12, is a special case of the abstract reaction scheme in Eq. 1.52, with A = AlL^{2+}, B = Cu^{2+}, C = $AlLCu^{4+}$, D = CuL^{+}, and E = Al^{3+}. The set of rate laws in Eq. 1.54 can be applied under the condition $k'_b = 0$ and, if $AlLCu^0$ equilibrates with AlL^{2+} and Cu^{2+} much more rapidly than it decomposes to form CuL^{+} and Al^{3+}, the left side of Eq. 1.54b can be set equal to zero in order to derive the equation

$$[AlLCu^{4+}] = \frac{k_f}{k_f' + k_b}[AlL^{2+}][Cu^{2+}] \qquad (2.46)$$

which applies on a time scale that is very much smaller than $1/k'_f$. The combination of Eqs. 2.46 and 1.54a (with "species B" selected for the left side) leads to the expression

$$\frac{d[Cu^{2+}]}{dt} = -\frac{k_f k_f'}{k_f' + k_b}[AlL^{2+}][Cu^{2+}]$$

$$\equiv -K_{Cu}^{AlL}[AlL^{2+}][Cu^{2+}] \qquad (2.47)$$

for the rate of Cu^{2+} consumption in the exchange reaction. Under experimental conditions in which $[AlL^{2+}]$ remains constant, Eq. 2.47 can be written in the pseudo-first-order form of Eq. 2.35:[19, 25]

$$-\frac{d[Cu^{2+}]}{dt} = K([AlL^{2+}])[Cu^{2+}] \qquad (2.48)$$

on a time scale defined by[7] 0.693/K. For example, if $K_{Cu}^{AlL} \approx 10\ dm^3\ mol^{-1}\ s^{-1}$ and $[AlL^{2+}] \approx 10\ mmol\ m^{-3}$, then the time scale for Cu^{2+} consumption (and Al^{3+} release) is about 7×10^3 s, or 2 h, which would imply a significant kinetics effect on the redistribution of the humic substance ligands among Al and Cu species.

2.4 Multispecies Equilibria

The distribution of a chemical element among aqueous species in a soil solution can be described in a straightforward manner if the time scale of interest is either very long or very short as compared to those for the chemical reactions of importance. Although data on the composition of soil solutions represent only transitory states, characteristic of open systems, they can be analyzed in a chemical thermodynamic framework so long as the time scale of experimental observation is typically incommensurable with the time scales of transformation among aqueous species of differing stability.[26] The application of chemical thermodynamics to aqueous speciation thus is possible following an analysis of the natural time scales over which key complexation reactions take place, coupled with intuition about how to make the "free-body cut," the choice of a *closed* model equilibrium system whose behavior is to mimic an investigated *open* system in nature.

Speciation calculations based on chemical equilibrium are commonplace in soil and water chemistry.[27] Before the methodology is described in detail, it is well to remember the inherent limitations of this approach:[28]

1. The representation of chemical composition data in terms of "free" ionic and complex aqueous species is neither unique nor necessary to a thermodynamic description of the data (cf. Section 1.2).
2. Full recognition must be given and full account taken of the fact that few chemical transformations of importance in soil solutions go to completion exclusively outside the time domain of their observation at laboratory or field scales. A critical implication of this fact, noted in Section 1.3, is that one must always distinguish carefully between *thermodynamic* chemical species, sufficient in number and variety to represent the stoichiometry of a chemical transformation between stable states, and *kinetic* chemical species, required to depict completely the mechanisms of the transformation.
3. Pertinent chemical equilibria (e.g., oxidation–reduction reactions) may be ignored unintentionally in formulating mass balance, or important species in a soil solution may not have been considered.
4. Conditional stability constants for the assumed species may be incorrect or inadequate in some other way for soils.
5. Analytical methods for certain constituents in the soil solution may be inadequate to distinguish between various chemical forms (e.g., dissolved versus suspended, oxidized versus reduced, monomeric versus polymeric).
6. Temperature and pressure variations may need to be considered. Significant temperature and pressure gradients exist at times in nearly all natural soils. Adequate data on the temperature and pressure

dependence of the relevant conditional stability constants often are not available. Usually the stability constant data will refer to 298 K and 1 atm pressure.

The way in which conditional stability constants are used to calculate the distribution of chemical species can be illustrated by consideration of the forms of dissolved Cu(II) in a dilute, acidic soil solution. Suppose that the pH of a soil solution is 6.0 and that the total concentration of Cu is 0.1 mmol m^{-3}. The concentrations of the complex-forming ligands sulfate and fulvic acid have the values 50 and 10 mmol m^{-3}, respectively. The important complexes between these ligands and Cu are $CuSO_4^0$ and CuL^+, where L refers to fulvic acid ligands (see Section 2.3). These illustrative complexes are not the only ones formed among Cu, SO_4, or L, nor are the three ligands the only ones that form Cu complexes in soil solution.[29] Under the conditions assumed, the equation of mole balance for Cu is (cf. Eqs. 2.11 and 2.30)

$$Cu_T = [Cu^{2+}]_e + [CuSO_4^0]_e + [CuL^+]_e \tag{2.49}$$

Thermodynamic data now are necessary in order to proceed. The relevant chemical reactions are

$$Cu^{2+}(aq) + SO_4^{2-}(aq) = CuSO_4^0(aq) \tag{2.50a}$$

$$Cu^{2+}(aq) + L^-(aq) = CuL^+(aq) \tag{2.51a}$$

Each of the complexes in Eq. 2.49 can be characterized by a conditional stability constant:

$$K_{sc} = \frac{[CuSO_4^0]_e}{[Cu^{2+}]_e\,[SO_4^{2-}]_e} \approx 10^{2.4}\ mol^{-1}\ dm^3 \tag{2.50b}$$

$$K_{sc}^{qp} = \frac{[CuL^+]_e}{[Cu^{2+}]_e\,[L^-]_e} \approx 10^{6}\ mol^{-1}\ dm^3 \tag{2.51b}$$

where the second equation refers to a Scatchard quasiparticle stability constant (Eq. 2.40), with just one class of quasiparticle assumed. The combination of Eqs. 2.49, 2.50b, and 2.51b produces the mole-balance expression:

$$Cu_T = [Cu^{2+}]_e \left\{ 1 + \frac{[CuSO_4^0]_e}{[Cu^{2+}]_e} + \frac{[CuL^+]_e}{[Cu^{2+}]_e} \right\}$$

$$= [Cu^{2+}]_e \left\{ 1 + K_{sc}[SO_4^{2-}]_e + K_{sc}^{qp}[L^-]_e \right\} \quad (2.52a)$$

The quantity of interest here is the *metal distribution coefficient* (cf. Eq. 2.31):

$$\alpha_{Cu} \equiv \frac{[Cu^{2+}]_e}{Cu_T}$$

$$= \{1 + K_{sc}[SO_4^{2-}]_e + K_{sc}^{qp}[L^-]_e\}^{-1} \quad (2.53a)$$

which evidently can be calculated if the concentrations of the free ligand species are known.

Given the large value of K_{sc}^{qp} relative to K_{sc}, it is reasonable to assume that $[CuL^+]_e \approx Cu_T$. Since $Cu_T << L_T$ in the present example, $[L^-]_e \approx L_T$ in a first approximation, and Eq. 2.51b can be written in the form

$$K_{sc}^{qp} = \alpha_{CuL} / \alpha_{Cu}\beta_L L_T \approx (\alpha_{Cu}L_T)^{-1} \quad (2.54a)$$

where

$$\alpha_{CuL} \equiv \frac{[CuL^+]_e}{Cu_T} \qquad \beta_L \equiv \frac{[L^-]_e}{L_T} \quad (2.53b)$$

are the metal and ligand distribution coefficients for Cu^+ and L^-, respectively, and L_T is the total fulvic acid ligand concentration. It follows from Eq. 2.54a that $\alpha_{Cu} \approx (K_{sc}^{qp} L_T)^{-1} = 0.1$ in the first approximation, implying that only about 10% of Cu_T is the species Cu^{2+}. This approximate result can be used to estimate the distribution coefficient for $CuSO_4^0$, given the assumption $[SO_4^{2-}]_e \approx SO_{4T}$:

$$\alpha_{CuSO_4} \equiv \frac{[CuSO_4^0]_e}{Cu_T} = \alpha_{Cu}\frac{[CuSO_4^0]_e}{[Cu^{2+}]_e} = \alpha_{Cu}K_{sc}[SO_4^{2-}]_e$$

$$\approx \alpha_{Cu}K_{sc}SO_{4T} \approx 10^{-3} \quad (2.53c)$$

The estimates of α_{CuL}, α_{Cu}, and α_{CuSO_4} can be refined by considering the ligand speciation in more detail:

$$SO_{4T} = [SO_4^{2-}]_e + [CuSO_4^0]_e = [SO_4^{2-}]_e\{1 + K_{sc}[Cu^{2+}]_e\} \quad (2.52b)$$

$$L_T = [L^-]_e + [HL^0]_e + [CuL^+]_e$$

$$= [L^-]_e \{1 + K_{Hc}^{qp} [H^+]_e + K_{sc}^{qp} [Cu^{2+}]_e\} \tag{2.52c}$$

where

$$K_{Hc}^{qp} = \frac{[HL^0]_e}{[H^+]_e [L^-]_e} \approx 10^4 \text{ mol}^{-1} \text{ dm}^3 \tag{2.55}$$

is a quasiparticle protonation constant analogous to K_{sc}^{qp}. Given the first estimate, $[Cu^{2+}]_e = \alpha_{Cu}Cu_T \approx 10^{-8}$ mol dm^{-3}, the ligand distribution coefficients are

$$\beta_{SO_4} \equiv \frac{[SO_4^{2-}]_e}{SO_{4T}} = \{1 + K_{sc} [Cu^{2+}]_e\}^{-1} \approx 1.0 \tag{2.53d}$$

$$\beta_L = \{1 + K_{Hc}^{qp} [H^+]_e + K_{sc}^{qp} [Cu^{2+}]_e\}^{-1} \approx 0.96 \tag{2.53e}$$

for $[H^+]_e \approx 10^{-6}$ mol dm^{-3}. The refined estimate of α_{Cu} that results from Eqs. 2.53e and 2.54a is found by rearranging the latter equation,

$$K_{sc}^{qp} \approx (1 - \alpha_{Cu})/\alpha_{Cu}\beta_L L_T \tag{2.54b}$$

to read

$$\alpha_{Cu} = (1 + K_{sc}^{qp}\beta_L L_T)^{-1} \approx 0.094 \tag{2.54c}$$

where the refined mole-balance constraint, $\alpha_{Cu} + \alpha_{CuL} \approx 1$, has been used to derive Eq. 2.54b from Eq. 2.54a. The improved estimate of α_{Cu} should be accurate, given its closeness to the first estimate and the smallness of α_{CuSO_4}. Thus about 90% of Cu_T is organically complexed at pH 6, with the remainder being in free-ionic form, in agreement with Table 2.1. This example, despite its "back-of-the-envelope" nature, illustrates all the salient features of a more exact chemical speciation calculation: *mole balance* (Eq. 2.52), *conditional stability constants* (Eqs. 2.50b, 2.51b, and 2.55), *distribution coefficients* (Eq. 2.53), and the *iterative refinement of distribution coefficients* through additional mole balance on the ligands.

The appropriateness of this speciation calculation from the kinetics standpoint can be assessed by considering the associative ligand-exchange reaction (cf. Eq. 2.12):

$$CuSO_4^0 + L^-(aq) \underset{k_b}{\overset{k_f}{\rightleftarrows}} CuSO_4L^-(aq) \overset{k_f'}{\rightarrow} CuL^+(aq) + SO_4^{2-}(aq) \quad (2.56)$$

Following the steps used to derive Eq. 2.14, one finds the rate law:

$$\frac{d[CuL^+]}{dt} = \frac{k_f' k_f}{k_f' + k_b} [CuSO_4^0][L^-] \quad (2.57)$$

for the rate of CuL^+ formation by the pathway of $CuSO_4^0$ destruction, where $[CuSO_4^0]$ can be estimated with Eq. 2.50b (i.e., $CuSO_4^0$ equilibrates much more rapidly with Cu^{2+} and SO_4^{2-} than CuL^+ does with L^-). If a scenario is imagined with Cu_T and SO_{4T} as the given input to the speciation calculation, then Eq. 2.50b leads to the result $[CuSO_4^0] \approx 10^{-9}$ mol dm^{-3} in a solution without fulvic acid. For an initial value of added free ligand, $[L^-] \approx 10^{-5}$ mol dm^{-3}, and under the conditions $k_f' >> k_b$, $k_f \approx k_{wex}K_{osc} \approx 10^{10}$ dm^3 mol^{-1} s^{-1}, the rate of $[CuL^+]$ formation would be about 100 mmol m^{-3} s^{-1}, which is large enough to justify an equilibrium calculation of speciation.

Chemical speciation in soil solutions and other natural waters can be calculated routinely with a number of software products offered in a variety of computational media.[27, 30] Five examples of these products are listed in Table 2.5. They differ principally in the method of solving the chemical equilibrium problem numerically, or in the chemical species and equilibrium constants considered, or in the model used to estimate single-species activity coefficients. Irrespective of these differences, all the examples follow a similar algorithm:

1. Mole-balance equations are established for each metal and each ligand in terms of the concentrations of all species assumed to exist in solution.
2. Thermodynamic stability constants for the formation of each complex are compiled.
3. The mole-balance equations are expressed in terms of conditional stability constants and the concentrations of the species chosen as components[31] (e.g., free metals and ligands) to provide a set of coupled, nonlinear algebraic equations in the component concentrations.
4. The set of coupled equations is solved numerically to a prescribed degree of approximation in order to obtain the concentrations or, equivalently, the distribution coefficients for these species. The concentrations so calculated may be used, in turn, to compute those of all other species.

Table 2.5 Representative Chemical Speciation Programs[32]

Property	PHREEQE	SOILCHEM	EQ3/6	MINTEQ	WATEQ4F
Chemical elements	19	47	47	31	32
Aqueous species	120	1853	686	373	245
Gases	3	3	11	3	2
Redox elements	3	11	25	8	7
Activity coefficient models[a]	Davies, Pitzer, and special models	Davies and special models	Davies, Pitzer, and special models	Davies and special models	Special models and Davies
Temperature (°C)	0–100	25	0–300	25	0–100
Pressure (bar)	1	1	1 and saturation pressure above 100°C	1	1

[a]"Davies" = Eq. 1.21; "Pitzer" = Eq. 1.13 and correlation equations based on Young's rules; "special models" = expressions like Eq. 1.23.

For example, in the computer program SOILCHEM,[32] "free" metal cations, the proton, and "free" ligands are selected as components (Fig. 2.8). The data input to the program are the total concentrations of all metals and ligands, as well as the pH value. If desired, the soil solution may be treated as an open system, with the partial pressures of CO_2, O_2, and/or N_2 specified in addition to total concentrations. A mole-balance equation is developed for each metal and ligand in terms of the free-ionic concentrations of the proton and all metals and ligands through the use of conditional stability constants (at 298.15 K), as was done, for example, in Eq. 2.52. Values of the conditional stability constants are estimated with the help of equilibrium constants stored in the computer program, and single-species activity coefficients are calculated according to Eqs. 1.21–1.24 with an input guess of the effective ionic strength. The set of coupled, nonlinear algebraic equations represented by the mole-balance equations is solved numerically, using the Newton-Raphson method, to find the free-ionic concentrations and to compute the ionic strength. The results are substituted into the mole-balance equations to examine them for self-consistency, with the conditional stability constants used now being those estimated at the newly computed ionic strength. The calculation is repeated until self-consistency is achieved within a selected tolerance. Then the concentrations of all the complexes and free-ionic species are tabulated. An example of this kind of speciation calculation is shown in Table 2.6 for a soil solution at pH 5.6 containing four metal and three ligand components. Even in this very simple case of complexation involving only inorganic ligands, about 40 aqueous species were considered.

From the chemical point of view, the heart of an equilibrium speciation calculation is the set of conditional stability constants used. These parameters, in turn, require equilibrium constants and single-species activity coefficients (cf. Eq. 1.26). The accuracy and consistency of equilibrium constants are major concerns in chemical thermodynamics, as discussed in Special Topic 1 (Chapter

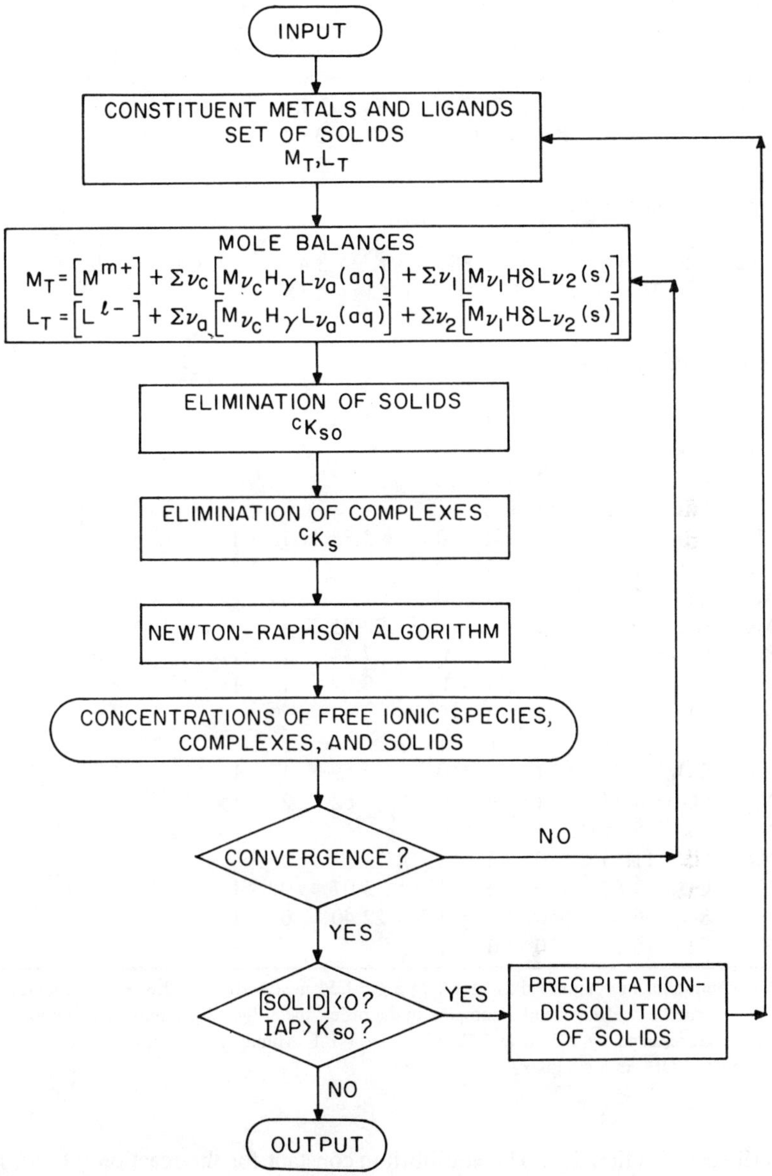

FIG. 2.8. Flowchart of a chemical speciation calculation based on "free" ionic species (M^{m+}, $L^{\ell-}$) as components that form complexes and solids.

1). If Standard-State chemical potentials are used to calculate equilibrium constants (with Eqs. s1.6 and s1.8), these concerns become paramount.[33, 34] Consider, for example, the complex, AlF^{2+}, whose kinetics of formation was

Table 2.6 Results of a Chemical Speciation Calculation[a]

Total Concentrations:

Na_T = 2.55	Mg_T = 2.82	K_T = 3.40	Ca_T = 3.15
CO_{3T} = 3.00	SO_{4T} = 2.65	Cl_T = 2.68	

Component concentrations

Na^+ = 2.55	Mg^{2+} = 2.88	K^+ = 3.41	Ca^{2+} = 3.22
CO_3^{2-} = 8.38	SO_4^{2-} = 2.71	Cl^- = 2.68	

Complex concentrations

Ca	CO_3	8.76	1	1	0[b]	6.23	1	1	1			
Ca	SO_4	3.99	1	1	0							
Ca	Cl	5.78	1	1	0							
Ca	OH	10.35	1	0	-1							
Mg	CO_3	8.72	1	1	0	5.88	1	1	1			
Mg	SO_4	3.75	1	1	0							
Mg	Cl	5.54	1	1	0							
Mg	OH	9.21	1	0	-1	28.81	4	0	-4			
K	CO_3	11.07	1	1	0	7.34	1	1	1	14.69	2	1	0
K	SO_4	5.40	1	1	0	9.97	1	1	1	8.22	2	1	0
K	Cl	6.21	1	1	0							
K	OH	12.28	1	0	1							
Na	CO_3	9.82	1	1	0	6.59	1	1	1	14.39	2	1	0
Na	SO_4	4.75	1	1	0	7.62	2	1	0			
Na	Cl	5.75	1	1	0							
Na	OH	11.13	0	1	-1							
H	CO_3	3.82	0	1	1	3.07	0	1	2			
H	SO_4	6.45	0	1	1	22.40	0	1	2			
H	Cl	16.73	0	1	1							

[a]Concentrations are expressed as $-\log [\ ]_e$; e.g., 2.55 means $10^{-2.55} = 2.8 \times 10^{-3}$ mol dm^{-3}.
[b]Stoichiometric coefficients of a complex in the order: metal-ligand-proton/hydroxide ion; e.g., $CaCO_3^0$ is represented by 1 1 0, and $CaHCO_3^+$ by 1 1 1. Minus signs denote the stoichiometric coefficients of OH^- in a complex.

described in Section 2.1. The equilibrium constant for the reaction (at 298 K)

$$Al^{3+}(aq) + F^-(aq) = AlF^{2+}(aq) \tag{2.58}$$

can be calculated by using the equation

$$\log K_s = -[\mu^0(AlF^{2+}) - \mu^0(F^-) - \mu^0(Al^{3+})] / 5.708 \tag{2.59}$$

Values of the Standard-State chemical potentials in Eq. 2.58 are available from the published compilation sources described in Special Topic 1. One source provides the following set of values:

$$\mu^0(AlF^{2+}) = -803 \text{ kJ mol}^{-1}$$
$$\mu^0(F^-) = -278.8 \text{ kJ mol}^{-1}$$
$$\mu^0(Al^{3+}) = -485 \text{ kJ mol}^{-1}$$

leading to log K_s = 6.87 via Eq. 2.59. Another source provides the two values:

$$\mu^0(Al^{3+}) = -489.8 \text{ kJ mol}^{-1} \qquad \mu^0(F^-) = -281.7 \text{ kJ mol}^{-1}$$

without data for the AlF^{2+} species. If the second set of μ^0-values is combined with $\mu^0(AlF^{2+})$ from the first set, log K_s = 5.52 results. These very different log K_s estimates can be compared with a recommended, directly measured value,[35] log K_s = 7.0 ± 0.1. Evidently, either the imprecision or the inaccuracy of the first set of μ^0-values is near 1.0 kJ mol^{-1}, such that the resulting log K_s is uncertain by about 0.1–0.2 (see Eq. s1.12 in Special Topic 1). Also, the value of $\mu^0(AlF^{2+})$ in the first set of data is not consistent with those of $\mu^0(F^-)$ and $\mu^0(Al^{3+})$ in the second set, such that log K_s from the latter is underestimated by the large error of 1.5 (i.e., $K_s = 3.3 \times 10^5$ instead of $K_s = 10^7$, an error of nearly two orders of magnitude).

Equally critical to the accuracy of the conditional stability constants is the estimation of single-species activity coefficients. As is pointed out in Section 1.2, these parameters have thermodynamic significance only to the extent that products containing them conform to Eqs. 1.15 and 1.20. Figure 2.8 implies that, computationally, single-species activity coefficients and species concentrations are connected closely. Almost arbitrary, yet mutually consistent, values of the two parameters are possible in speciation calculations that otherwise may meet general thermodynamic and mole-balance constraints. This ambiguous feature of the calculation of single-species activity coefficients is made apparent in Table 2.5 by the variety of different mathematical models indicated for them in chemical speciation programs whose predictive reliability has been demonstrated, at least in part.[32] Several programs even use combinations of the two kinds of activity coefficient in Eqs. 1.13 and 1.15, despite the fundamental conceptual difference between them.[36] This rather empirical approach to defining speciation algorithms only serves to emphasize the relative nature of the calculations that result: Their essential purpose should not really be to predict, but instead to reveal important trends in speciation that assist in the design of improved methods to *measure* species concentrations.[37]

Special Topic 2: Electrochemical Potentials

The introduction of ionic species into the framework of a thermodynamic analysis of chemical phenomena in soil solutions is principally a convenient mathematical device, since an ionic species cannot be strictly described in thermodynamic terms (see Special Topic 1). For example, in Section 2.2 the chemical potential of an electron is accorded a formal thermodynamic significance and manipulated in the same way as the chemical potential of a macroscopic quantity of matter, but with care taken always to meet the condition of overall electroneutrality. In this fashion the chemical potentials of ionic species are incorporated usefully into discussions of complexation, precipitation–dissolution, and redox phenomena to derive results of thermodynamic importance in Chapter 1, as well as in the present chapter. However, in none of these applications was it necessary to consider the process of interphase *charge transfer,* in which, by definition, a single ionic species passes alone from one phase to another. This kind of process occurs at the interface between an electrode and a soil solution (or suspension). Therefore charge-transfer processes are of paramount concern in the study of the electrochemistry of aqueous media. By its very definition, a charge-transfer process is not subject to the direct imposition of the electroneutrality condition. This fact necessitates a more careful examination of the concept of the chemical potential of an ionic species than was carried out in Special Topic 1.

The distribution of a charged species between two bulk phases, regardless of their composition, is determined by the *electrochemical potential*[38] of the species, $\tilde{\mu}$. The gradient of the electrochemical potential of a species drives its diffusive transfer between phases, and equilibrium with respect to this transfer is described by the equality:

$$\tilde{\mu}_\lambda(\mathrm{i}) = \tilde{\mu}_\sigma(\mathrm{i}) \qquad \text{(s2.1)}$$

where λ and σ denote two different phases that contain the charged chemical species i. The electrochemical potential, like the chemical potential, has the SI units joules per mole.[39] The operational meaning of $\tilde{\mu}$ can be illustrated by considering the following soil suspension–soil solution system into which two identical electrodes are immersed:

L		W		R
Ag | AgCl |	NaX(s), NaCl(aq)	▪▪▪▪	NaCl(aq) solution	| AgCl | Ag

The larger vertical line refers to the interface between a silver–silver chloride electrode and either a soil suspension (left) or a soil solution (right). The heavy, dashed vertical line marked W refers to a membrane that is impermeable to the soil colloidal anion, X^{-1}, but permeable to dissolved ions and water.[39] There may be ions other than Na^+ and Cl^- in the suspension or in the solution, but they are assumed not to interfere with the reversible behavior of the silver–silver chloride electrode toward Cl^-. At the left electrode (L), by convention,[39] the oxidation reaction

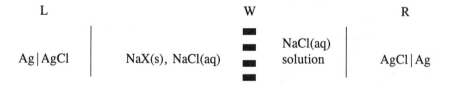

$$Ag(s) + Cl^-(L, aq) \rightarrow AgCl(s) + e^-$$

takes place, where L refers to a point inside the left electrode assembly. At the right electrode (R), the reduction reaction

$$AgCl(s) + e^- \rightarrow Ag(s) + Cl^-(R, aq)$$

takes place, where R refers to a point inside the right electrode assembly. The overall electrode-pair reaction, obtained by combining the two half-reactions, is simply

$$Cl^- (L, aq) \rightleftarrows Cl^- (R, aq) \tag{s2.2}$$

According to thermodynamic conventions for the assignment of emf values to electrode assemblies, summarized in Table s2.1, the emf across the silver–silver chloride electrode pair is given by the following equation:

$$\tilde{\mu}^{R}(Cl^-) - \tilde{\mu}^{L}(Cl^-) = -FE \tag{s2.3}$$

where F is the Faraday constant and E is the emf in volts. If equilibrium exists between $Cl^-(L, aq)$ and Cl^- in the soil suspension, as well as between $Cl^-(R, aq)$ and Cl^- in the solution, by Eq. s2.1,

$$\tilde{\mu}^{R}(Cl^-) = \tilde{\mu}_{so}(Cl^-) \quad \text{and} \quad \tilde{\mu}^{L}(Cl^-) = \tilde{\mu}_{su}(Cl^-) \tag{s2.4}$$

Table s2.1 Conventions in the Assignment of emf to Galvanic Cells[39]

1. The cell reaction is written as if oxidation occurs spontaneously at the left electrode and reduction at the right electrode.

2. If the overall reaction is

   $$aA + bB + \cdots = xX + yY + \cdots$$

   The cell emf (in volts) may be defined by the relation:

   $$x\tilde{\mu}(X) + y\tilde{\mu}(Y) + \cdots - a\tilde{\mu}(A) - b\tilde{\mu}(B) - \cdots \equiv -FE$$

   where F is the Faraday constant and E is the cell emf.

3. In these conventions it is assumed that all reduction or oxidation half-reactions are written in terms of the transfer of 1 mol of electrons.

where *so* refers to the solution and *su* to the suspension. The combination of Eqs. s2.3 and s2.4 leads to an expression for the difference between the electrochemical potential of the chloride ion in the suspension and that in the aqueous solution:

$$\tilde{\mu}_{so}(Cl^-) - \tilde{\mu}_{su}(Cl^-) = -FE \qquad (s2.5)$$

Equation s2.5 illustrates the general rule that *differences in electrochemical potential for a charged species can be measured by determining the emf developed across a pair of electrodes that behave reversibly toward the charged species*. The derivation of Eq. s2.5 can be carried through for any charged species in two different aqueous systems by invoking electrodes that behave reversibly toward the species. The corresponding general result for the interphase electrochemical potential difference is

$$\Delta_\sigma^\lambda \tilde{\mu}_i \equiv \tilde{\mu}_\lambda(i) - \tilde{\mu}_\sigma(i) = Z_i FE \qquad (s2.6)$$

where λ denotes the phase containing an electrode (reversible to charged species i) at which a reduction occurs and σ denotes the phase containing the electrode at which an oxidation occurs. The parameter Z_i is the valence of species i. If equilibrium exists with respect to the transfer of species i between the phase λ and σ, then Eq. s2.1 applies and E = 0 in Eq. s2.6. Thus *the absence of an emf across a pair of electrodes that behave reversibly toward a charged species can be used to indicate equilibrium with respect to the transfer of the species between two phases.*

The electrochemical potential of a charged species can be envisioned as the potential difference involved with the transfer of 1 mol of a charged species from a point at charge-free infinity to a point inside a material phase.[38] It is evident from this conceptualization that $\tilde{\mu}_\lambda(i)$ depends on the purely chemical nature of the species i as well as on the purely electrostatic interactions that can occur between i and all other charged species in the phase λ during the transfer process. For example, in the case of the chloride ion, discussed above, $\tilde{\mu}_{su}(Cl^-)$ should depend on the chemical properties of chloride in the suspension as well as on the electrostatic interactions between Cl^- and all the other ions or charged solid surfaces in the suspension. Similarly, $\tilde{\mu}_{so}(Cl^-)$ should depend on the chemical properties of chloride in the soil solution and on the electrostatic interactions between Cl^- and all the other dissolved ions. This dual characteristic of the electrochemical potential suggests that it is worthwhile to inquire as to the significance of the *formal* definition:

$$\tilde{\mu}(i) \equiv \xi_0 + RT \ln(\mathbf{i}) + Z_i F\phi \qquad (s2.7)$$

where ξ_0 is some Reference-State function of temperature and pressure, as well as of the "purely chemical" nature of the species whose activity is **(i)**; R is the molar gas constant, T is absolute temperature, and ϕ epitomizes the electrostatic interactions in the form of a *macroscopic* electric potential to which i is subjected. Evidently, Eq. s2.7 separates the electrochemical potential into a "purely chemical" part—the first two terms on the right side—and a "purely electrostatic" part containing the macroscopic electric potential, ϕ. For an electrically neutral species, $Z_i = 0$ and Eq. s2.7 becomes the same as Eq. s1.4, assuring consistency with the case in which no charge transfer occurs.

Suppose that Eq. s2.7 is applied to Na^+ in two solutions of NaCl, one of concentration 10 mol m^{-3} and the other of concentration 1 mol m^{-3}. The difference in electrochemical potential of $Na^+(aq)$ between the two solutions, denoted by 10 and 1, is

$$\tilde{\mu}(Na^+)_{10} - \tilde{\mu}(Na^+)_1 = RT \ln [(\mathbf{Na^+})_{10} / (\mathbf{Na^+})_1] + F(\phi_{10} - \phi_1) \quad (s2.8)$$

The left side of Eq. s2.8 is always measurable, in principle, by means of a voltmeter and a pair of suitable electrodes. The ratio of Na^+ *concentrations* implicit in the ratio of activities on the right side of Eq. s2.8 is also measurable, in principle. Therefore the extent to which the macroscopic electric potential difference, $(\phi_{10} - \phi_1)$, *which can never be measured directly*,[38] has chemical meaning depending entirely on whether the ratio of single-ion activity coefficients, γ_{10}/γ_1 implicit in the ratio of Na^+ activities, can be evaluated unambiguously. If, for example, this ratio could be calculated accurately by using the Davies equation (Eq. 1.21), the electrical potential difference would become well defined. If, on the other hand, the ratio of activity coefficients could not be represented accurately with the help of some model expression, or, in general, if the activity ratio for Na^+ in the two solutions could not be measured, there would be *no chemical significance* to separating the electrochemical potential into "purely electrical" and "purely chemical" parts.

Consider now Eq. s2.7 applied to Cl^- in the soil suspension–soil solution system diagram that follows Eq. s2.1. The left side of Eq. s2.5 can be expressed:

$$\Delta_{su}^{so}\tilde{\mu}_{Cl^-} = \xi_{0,so} - \xi_{0,su} + RT \ln\left[\frac{(\mathbf{Cl}^-)_{so}}{(\mathbf{Cl}^-)_{su}}\right] - F(\phi_{so} - \phi_{su}) \quad (s2.9)$$

The left side of Eq. s2.9 is always measurable, as indicated in Eq. s2.5. The right side of Eq. s2.9 contains the difference between ξ_0 for Cl^- in the two aqueous systems, the ratio of the activities of Cl^-, and the *membrane potential difference*, $\phi_{so} - \phi_{su}$. These three quantities have operational meaning only if it is possible to measure any two of them unambiguously, that is, to measure them without making unverifiable assumptions about the *thermodynamic* nature of the two aqueous systems. Unfortunately, no experimental method exists that can determine even ratios of the single-ion activities without making extrathermodynamic assumptions.[38] Moreover, no experimental technique exists for an unambiguous measurement of the difference in ξ_0 values for two phases of differing chemical composition, and no electrode assembly can measure a membrane potential difference without the data requiring interpretation through extrathermodynamic assumptions.[38] It follows that in this case the formal partitioning of the right side of Eq. s2.9 has absolutely no chemical significance.

The macroscopic electric potential on the right side of Eq. s2.7 is known as an *inner electric potential*, and the potential difference on the right side of Eq. s2.9 is an example of a *Galvani potential difference*[39]

$$\Delta_\sigma^\lambda \phi \equiv \phi_\lambda - \phi_\sigma \qquad \text{(s2.10)}$$

between two phases denoted λ and σ. The discussion presented in this section is intended to illustrate the general principle[38] that *Galvani potential differences between two phases of different chemical composition cannot be measured.* The root problem is that the inner potential is the *electric* potential difference between a point at charge-free infinity and a point inside a material phase. Although this potential is a well-defined entity in classical electrostatics (the theory of a hypothetical charged fluid), it is not well defined in thermodynamics (the macroscopic theory of matter in stable states) because the interactions that produce the chemical properties of charged species cannot be divided unambiguously into electrostatic and nonelectrostatic categories, even at a macroscopic level. However, if a charged species occurs in two phases with identical chemical composition, this difficulty is obviated by the fact that the purely chemical part of the electrochemical potential of a charged species must be the same in the two phases. It follows from Eq. s2.8 that in this case

$$\Delta_\sigma^\lambda \tilde{\mu}_i = Z_i F \Delta_\sigma^\lambda \phi \qquad \text{(s2.11)}$$

and therefore that *Galvani potential differences between two phases of identical chemical composition are measurable.* Examples of the application of Eq. s2.11 include Galvani potential differences between two identical metal wires or between two identical aqueous solutions.

Now consider a soil suspension into which is immersed an electrode pair consisting of a glass electrode reversible to Na^+ and a silver–silver chloride electrode:

L		R
Na(gl)	NaX(s), NaCl(aq)	AgCl | Ag

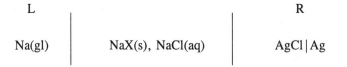

Other colloidal solids than NaX(s) and other electrolytes than NaCl may be present in the soil suspension, but they are assumed not to interfere with the performance of the electrode pair. The development of electrical potential at the sodium electrode is a somewhat complicated process that includes the creation of a difference in $\tilde{\mu}(Na^+)$ across the glass membrane separating the soil suspension from the inner solution of the electrode, the slow diffusion of Na^+ inside the glass membrane, and an oxidation reaction at the inner electrode. Since a detailed account of these processes is not required, it will be sufficient to describe the electrode reactions formally by the expression:

$$\mathrm{Na(gl) + AgCl(s) = Na^+(L, aq) + Cl^-(R, aq) + Ag(s)} \qquad \text{(s2.12)}$$

According to Table s2.1, the emf for this reaction is

$$-FE \equiv \tilde{\mu}^L(\mathrm{Na^+}) + \tilde{\mu}^R(\mathrm{Cl^-}) + \mu(\mathrm{Ag}) - \tilde{\mu}(\mathrm{Na^+(gl)}) - \mu\mathrm{Ag(Cl(s))} \qquad \text{(s2.13)}$$

If chemical equilibrium exists with respect to Na(aq) and Cl^-(aq), then

$$\tilde{\mu}^L(\mathrm{Na^+}) = \tilde{\mu}_{su}(\mathrm{Na^+}) \qquad \tilde{\mu}^R(\mathrm{Cl^-}) = \tilde{\mu}_{su}(\mathrm{Cl^-}) \qquad \text{(s2.14)}$$

and Eq. s2.13 can be written in the form:

$$-FE = [\mu(\mathrm{Ag}) - \tilde{\mu}(\mathrm{Na^+(gl)}) - \mu\mathrm{Ag(Cl(s))}] + \tilde{\mu}_{su}(\mathrm{Na^+}) + \tilde{\mu}_{su}(\mathrm{Cl^-}) \qquad \text{(s2.15)}$$

But, as in any single phase,

$$\tilde{\mu}_{su}(\mathrm{Na^+}) + \tilde{\mu}_{su}(\mathrm{Cl^-}) \equiv \mu_{su}(\mathrm{NaCl}) = \mu^0(\mathrm{NaCl}) + RT\ \ln(\mathbf{NaCl})_{su} \qquad \text{(s2.16)}$$

according to Eqs. s1.4, 1.12, and 1.15, where $(\ )_{su}$ refers to an activity in the suspension. Therefore

$$E = A - \frac{RT}{F}\ln(\mathbf{NaCl})_{su} \qquad \text{(s2.17)}$$

where

$$A \equiv (1/F)\,[\mu(\mathrm{AgCl(s)}) + \tilde{\mu}(\mathrm{Na^+(gl)}) - \mu(\mathrm{Ag}) - \mu^0(\mathrm{NaCl})] \qquad \text{(s2.18)}$$

Equation s2.17 demonstrates that *the emf of a reversible electrode pair can be employed to measure the activity of NaCl in a soil suspension*. Equation s2.17, on the other hand, *cannot* provide direct information about the electrical nature of colloid–electrolyte interactions in soil suspensions. Whatever contributions electrical potentials may make to $\tilde{\mu}_{su}(Na^+)$ and $\tilde{\mu}_{su}(Cl^-)$, they are completely canceled when these two electrochemical potentials are added, as in Eq. s2.16.

Suppose next that the silver–silver chloride electrode is replaced by a saturated calomel electrode:

L				R
Na(gl) |	NaX(s), NaCl(aq)	‖	KCl(satd.) |	$H_2Cl_2(s)$ | Hg(l)

The dashed vertical line refers to a liquid–liquid junction between the suspension and the saturated KCl salt bridge (as well as to an interface through which soil colloidal material cannot pass).[39] The overall reaction for this system can be written out formally in a manner similar to Eq. s2.12:

$$\mathrm{Na(gl)} + \tfrac{1}{2}\,\mathrm{Hg_2Cl_2(s)} = \mathrm{Na^+(L,\ aq)} + \mathrm{Cl^-(R,\ aq)} + \mathrm{Hg(l)} \qquad \text{(s2.19)}$$

The expression for the emf developed in this case differs from that in Eq. s2.13 because of an electric potential difference, E_J, which arises across the liquid–liquid junction:

$$-F(E - E_J) = \tilde{\mu}^{L}(\mathrm{Na^+}) + \tilde{\mu}^{R}(\mathrm{Cl^-}) + \mu(\mathrm{Hg(l)})$$

$$-\tilde{\mu}(\mathrm{Na(gl)}) - \tfrac{1}{2}\,\tilde{\mu}(\mathrm{Hg_2Cl_2(s)}) \qquad \text{(s2.20)}$$

The liquid junction potential occurs because of macroscopic variations in chemical composition in passing from the suspension into the KCl salt bridge.[40] This variation in composition induces mixing processes across the liquid junction. However, because of the way a saturated calomel electrode is constructed, these mixing processes are prevented from bringing about equilibrium with respect to Cl^-(aq) transfer. Thus the steady-state transport process that ensues does not permit the equality $\tilde{\mu}^{R}(Cl^-) = \tilde{\mu}_{su}(Cl^-)$ to be applied in this case. Instead, permanent differences in the electrochemical potentials of Cl^-(aq) and other charged species in the system across the junction combine to produce an electric potential difference, E_J. The calculation of E_J, accordingly, cannot be done by thermodynamic methods, but instead requires information relating to macroscopic steady-state ionic transport processes.

Since the only variable quantity on the right side of Eq. s2.20 is the first term, and since $\tilde{\mu}^{L}(Na^+) = \tilde{\mu}_{su}(Na^+)$, the emf developed may be expressed by the following equation:

$$E = E_J + B - (1/F)\tilde{\mu}_{su}(\mathrm{Na^+}) \qquad \text{(s2.21)}$$

where B equals $(-1/F)$ times the sum of all the constant electrochemical and chemical potentials in Eq. s2.20. Suppose that the electrochemical potential of Na^+ in the suspension is separated arbitrarily into the two terms:

$$\tilde{\mu}_{su}(\mathrm{Na^+}) \equiv \lambda_{su}^{Na} + RT\ln(\mathbf{Na^+})_{su} \qquad \text{(s2.22)}$$

where λ is a parameter that incorporates both the electrical state of the suspension and the Standard-State properties of $Na^+(aq)$. The combination of Eqs. s2.21 and s2.22 produces the formal relationship

$$-\ln(\mathbf{Na}^+)_{su} = \frac{F}{RT}(E - E_J - B') \qquad \text{(s2.23)}$$

where

$$B' = B - (\lambda_{Na}^{su}/F) \qquad \text{(s2.24)}$$

Equation s2.23 can be used to calculate the activity of Na^+ in the suspension *provided that the parameters E_J and B′ can be determined experimentally or eliminated by some method of calibration.*

If the soil suspension were instead an aqueous solution, a scale of activity values for Na^+ could be defined in terms of emf data obtained for standard reference solutions of prescribed $(\mathbf{Na}^+)$, in exactly the same way as the scale of $(\mathbf{H}^+)$ values (the operational pH scale) is defined (arbitrarily) in terms of emf data for standard buffer solutions.[39,40] However, the success of this extrathermodynamic calibration technique depends entirely on the extent to which E_J and B′ in the standard reference solutions are the same as E_J and B′ in the solution of interest. For the case of a soil suspension, the presence of colloidal material may cause these two parameters to differ very much from what they would be in a reference aqueous solution. If the difference is indeed large, the value of $(\mathbf{Na}^+)$, $(\mathbf{H}^+)$, or any other ionic activity estimated with the help of standard solutions and an equation like Eq. s2.23 would be of *no chemical significance.*

This point can be illustrated in a concrete manner by a consideration of the double electrode-pair system:

L R

$$Hg(l)\,|\,Hg_2Cl_s\,\Big|\,KCl(satd.)\,\Big\|\,NaX(s),\,NaCl(aq)\,\Big|\,Na(gl)\,\Big\|\,Na(gl)\,\Big|\,NaCl(aq)\,\Big\|\,KCl(satd.)\,\Big|\,Hg_2Cl_s(s)\,|\,Hg(l)$$

The emf developed by this double system is the algebraic sum of those developed by two calomel–sodium electrode pairs:

$$E^L = E_J^L - B + (1/F)\tilde{\mu}_{su}(Na^+) \qquad \text{(s2.25a)}$$

$$E^R = E_J^R + B - (1/F)\tilde{\mu}_{so}(Na^+) \qquad \text{(s2.25b)}$$

according to Eq. s2.21; therefore

$$E = E^R + E^L = (1/F)\,[\tilde{\mu}_{su}(Na^+) - \tilde{\mu}_{so}(Na^+)] + E_J^R - E_J^L \qquad \text{(s2.26)}$$

If equilibrium exists with respect to $Na^+(aq)$ between the suspension and the solution, then $\tilde{\mu}_{su}(Na^+) = \tilde{\mu}_{so}(Na^+)$, and Eq. s2.26 reduces to the simple expression

$$E = E_J^R - E_J^L \qquad (s2.27)$$

Thus, under equilibrium conditions, the emf of the double electrode-pair system is determined *solely* by electric potential differences developed at the two liquid junctions that involve KCl salt bridges. The two E_J may differ because of the effect of soil colloids. Thus the fact that this emf can develop is known as the *suspension effect*.[40] Only ionic transport processes across the liquid junctions need be taken into account in order to evaluate E. *Ionic transport processes across the semipermeable membrane between the suspension and the solution are not germane*. Moreover, since neither E_J^R nor E_J^L can be calculated by strictly thermodynamic methods, the interpretation of E must be made in terms of specific models of ionic transport across salt bridges contacting suspensions and solutions. Thus the relation between E and the behavior of ions in soil suspensions is not direct.

Finally, the equality $\tilde{\mu}_{su}(Na^+) = \tilde{\mu}_{so}(Na^+)$, provided by thermodynamics (Eq. s2.1), is only a statement of electrochemical equilibrium, not a conclusion relating to the details of ionic interactions within soil suspensions. This equality says nothing, for example, about the electric potentials experienced by $Na^+(aq)$ in either the suspension or the solution, nor can it do so, because the division of each electrochemical potential into electrical and chemical parts would be, in this case, a completely arbitrary step without chemical significance.

NOTES

1. A fine introduction to complexes and complexation reactions is in Chap. 2 of F. A. Cotton and G. Wilkinson, *Advanced Inorganic Chemistry*, Wiley, New York, 1988. Note that "ligand" is defined exclusively as an electron donor (Lewis base) in this book.

2. W. B. Jensen, *The Lewis Acid–Base Concepts*, Wiley, New York, 1980. See also Chap. 12 in G. N. Lewis, *Valence and the Structure of Atoms and Molecules*, Chemical Catalog Co., New York, 1923.

3. M. Misono, E. Ochiai, Y. Saito, and Y. Yoneda, A new dual parameter scale for the strength of Lewis acids and bases with evaluation of their softness, *J. Inorg. Nucl. Chem.* **29**:2685 (1967).

4. R. D. Shannon and C. T. Prewitt, Effective ionic radii in oxides and fluorides, *Acta Crystallogr. Sec. B* **25**:925 (1969); Revised values of effective ionic radii, *Acta Crystallogr. Sec. B* **26**:1046(1970). The fact that crystallographic ionic radii are the same as unsolvated ionic radii in aqueous solution is shown by G. Sposito, Distribution of potentially hazardous trace metals, *Metal Ions Biol. Systems* **20**:1 (1986).

5. See, for example, Chap. 6 in J. Burgess, *Ions in Solution*, Wiley, New York, 1988.

6. J. G. Hering and F. M. M. Morel, The kinetics of trace metal complexation: Implications for metal reactivity in natural waters, pp. 145–171 in *Aquatic Chemical Kinetics*, ed. by W. Stumm, Wiley, New York, 1990.

7. The "time scale" in Table 2.3 is strictly the *half-life* of a reaction, which can be obtained by graphical analysis of rate laws cast into the mathematical form:

$$-\left(\frac{d[A]}{dt}\right) = K\,[A]^b$$

where A is a reactant species and b is the order of the reaction (see Section 1.3). The parameter K, which is always positive, may equal the product of a rate coefficient with the concentration of a reactant species maintained constant during a reaction (e.g., H_2O in the second reaction in Table 2.3). Depending on the value of b, a suitable choice of plotting variables will produce a straight-line graph that can be used to calculate the reaction half-life:

Reaction order (b)	Plotting variables	Slope	Intercept[a]	Half-life
0	[A] vs. time	-K	$[A]_0$	$[A]_0/2K$
1	ln[A] vs. time	-K	$\ln[A]_0$	0.693/K
2	1/[A] vs. time	+K	$1/[A]_0$	$1/K[A]_0$

[a]$[A]_0$ is equal to the initial concentration of species A.

A complete discussion of reaction half-life is given in Chap. 2 of K. J. Laidler, *Chemical Kinetics*, Harper & Row, New York, 1987.

8. See, for example, A. McAuley and J. Hill, Kinetics and mechanism of metal-ion complex formation in solution, *Quart. Rev.* **23**:18(1969); Chap. 12 in J. Burgess, *Metal Ions in Solution*, Wiley, New York, 1978; G. W. Neilson and J. E. Enderby, The coordination of metal aquaions, *Adv. Inorg. Chem.* **34**:195(1989); Chap. 4 in R. G. Wilkins, *Kinetics and Mechanism of Reactions of Transition Metal Complexes*, VCH Publishers, New York, 1991.

9. B. J. Plankey, H. H. Patterson, and C. S. Cronan, Kinetics of aluminum fluoride complexation in acidic waters, *Environ. Sci. Technol.* **20**:160 (1986).

10. See, for example, Z. G. Szabó, Kinetic characterization of complex reaction systems, pp. 1–80 in *The Theory of Kinetics*, ed. by C. H. Bamford and C. F. H. Tipper, Elsevier, Amsterdam, 1969.

11. Dissociative and associative pathways in metal- or ligand-exchange reactions are discussed in the context of natural water chemistry by J. G. Hering and F. M. M. Morel, op. cit.[6]

12. M. A. Anderson and P. M. Bertsch, Dynamics of aluminum complexation in multiple ligand systems, *Soil Sci. Soc. Am. J.* **52**:1597 (1988).

13. The pathways of denitrification in soil (Eq. 2.22 is merely an overall reaction) are described in Chap. 4 of F. J. Stevenson, *Cycles of Soil*, Wiley, New York, 1986.

14. Solvated electrons are evanescent aqueous species that exist on time scales of

milliseconds (see, e.g., Chap. 13 in J. G. Burgess, op. cit.[8]) and therefore are not to be associated with the "free" electron species, $e^-(aq)$ in Eq. 2.18, which has only a formal, thermodynamic significance (see Special Topic 2).

15. These concepts are developed in greater detail in Chap. 7 of W. Stumm and J. J. Morgan, *Aquatic Chemistry*, Wiley, New York, 1981. As with any single-ion activity, it is always possible to relate the electron activity to a chemical potential formally (see, e.g., Eq. s2.22) and thus define an equivalent electrochemical emf (see Eq. s2.23). Accordingly, the (electro)chemical potential of an electron in aqueous solution is related to the pE value by the equation (see Special Topic 2):

$$\tilde{\mu}(e^-) = -RT(\ln 10)\ pE$$

since $\mu^0(e^-) \equiv 0$. It is always possible to express $\tilde{\mu}(e^-)$ in units of energy per unit charge, instead of energy per mole, by dividing both sides of the equation by the Faraday constant to give the chemical potential of the electron the units of joules per coulomb, or volts. Then the equation may be written in the form (T = 298.15 K):

$$E_H \equiv \frac{RT \ln 10}{F}\ pE = 0.059155\ pE$$

where $E_H \equiv -\tilde{\mu}(e^-)/F$ is the *electrode potential.* (The second step results from substituting 5708 J mol^{-1} for RT ln 10 at 298 K.) The electrode potential, with the convention that its sign be the same as that of pE, is just an alternative formal way to express the chemical potential of the electron. Its use in the thermodynamic description of redox reactions is convenient only if $\tilde{\mu}(e^-)$ can be determined with an electrochemical cell. In soil solutions and other natural waters, it is often not possible to make an accurate electrochemical measurements of the electrode potential. When this situation prevails, there is no particular advantage in employing E_H instead of the more fundamental pE value.

16. See the classic paper, L. G. M. Baas Becking, I. R. Kaplan, and D. Moore, Limits of the natural environment in terms of pH and oxidation-reduction potentials, *J. Geol.* **68**:243 (1960).

17. The catalytic role of microorganisms in redox reactions is discussed in exemplary fashion in Chap. 7 of F. M. M. Morel, *Principles of Aquatic Chemistry*, Wiley, New York, 1983.

18. See Chap. 13 in J. G. Burgess, op. cit.[8]

19. The application of Eq. 2.35 to chemical reactions in natural waters is described by J. Hoigné, Formulation and calibration of environmental reaction kinetics; oxidations by aqueous photo oxidants as an example, pp. 43–70 in W. Stumm, op. cit.[6]

20. See, for example, J. Steinhardt and J. A. Reynolds, *Multiple Equilibria in Proteins*, Academic Press, 1969; K. S. Murray, Binuclear oxo-bridged iron(III) complexes, *Coord. Chem. Rev.* **12**:1 (1974); W. Schneider and B. Schwyn, The hydrolysis of iron in synthetic, biological, and aquatic media, pp. 167–196, in *Aquatic Surface Chemistry*, ed. by W. Stumm, Wiley, New York, 1987; and P. M. Bertsch, Aqueous polynuclear aluminum species, pp. 87–115 in *The Environmental Chemistry of Aluminum*, ed. by G. Sposito, CRC Press, Boca Raton, FL, 1989.

21. See, for example, Chap. 3 in J. Buffle, *Complexation Reactions in Aquatic*

Systems, Wiley, New York, 1988.

22. J. Buffle, Natural oganic matter and metal-organic interactions in aquatic systems, *Metal Ions Biol. Systems* **18**:165 (1984).

23. See Chap. 5 in J. Buffle, op. cit.[21]

24. G. Sposito, Sorption of trace metals by humic materials in soils and natural waters, *Crit. Rev. Environ. Control* **16**:193 (1986).

25. J. G. Hering and F. M. M. Morel, Kinetics of trace metal complexation: Role of alkaline-earth metals, *Environ. Sci. Technol.* **22**:1469 (1988).

26. See, for example, Chap. 2 in G. N. Lewis and M. Randall, *Thermodynamics*, McGraw-Hill, New York, 1961.

27. D. C. Melchior and R. L. Bassett, *Chemical Modeling of Aqueous Systems II*, American Chemical Society, Washington, DC, 1990.

28. J. J. Morgan, Applications and limitations of chemical thermodynamics in natural water systems, pp. 1–29 in *Equilibrium Concepts in Natural Water Systems*, ed. by W. Stumm, American Chemical Society, Washington, DC, 1967.

29. For a listing of other complexes of Cu(II), see, for example, Chap. 14 in W. L. Lindsay, *Chemical Equilibria in Soils*, Wiley, New York, 1979. Figure 14.6 in this book suggests that $CuSO_4^0$ is the only important *inorganic* complex of Cu(II) in soil solutions at pH 6.

30. A. Martell and R. Motekaitis, *Determination and Use of Stability Constants*, VCH Publishers, New York, 1992.

31. Components are species whose concentrations can be varied independently (in a mathematical sense) and whose combination in reactions can produce all other species in an aqueous system. The number of components is unique to a given system, but not their identity, which is chosen for convenience in developing a thermodynamic (or kinetics) description. For an excellent discussion of equilibrium speciation calculations in terms of components, see Chap. 3 in F. M. M. Morel, *Principles of Aquatic Chemistry*, Wiley, New York, 1983.

32. These programs are discussed by D. C. Mangold and C.-F. Tsang, A summary of subsurface hydrological and hydrochemical models, *Rev. Geophys.* **29**:51 (1991). See also Chap. 1 in D. C. Melchior and R. L. Bassett, op. cit.[7]

33. D. K. Nordstrom et al., Revised chemical equilibrium data for major water-mineral reactions and their limitations, Chap. 31 in D. C. Melchior and R. L. Bassett, op. cit.[27]

34. See Section 4 in D. D. Wagman et al., The NBS tables of chemical thermodynamic properties, *J. Phys. Chem. Ref. Data* **11**:Suppl. No. 2 (1982).

35. D. K. Nordstrom and H. M. May, Aqueous equilibrium data for mononuclear aluminum species, pp. 29–53 in G. Sposito, op. cit.[20]

36. G. Sposito, The future of an illusion: Ion activities in soil solutions, *Soil Sci. Soc. Am. J.* **48**:531 (1984).

37. These methods are described, for solutions not containing humic substances, by F. R. Hartley, C. Burgess, and R. M. Alcock, *Solution Equilibria*, Wiley, New York, 1980. For solutions containing humic substances, a comprehensive discussion is given by J. Buffle, op. cit.[21]

38. The clearest introduction to the electrochemical potential that is given by its creator, in Chap. 8 of E. A. Guggenheim, *Thermodynamics*, North-Holland, Amsterdam, 1967. The issue of electric potentials near interfaces is discussed in detail in R. Parsons, Equilibrium properties of electrified interphases, *Modern Aspects Electrochem.* **1**:103 (1954).

39. The IUPAC nomenclature used in this section is described by I. Mills, T. Cvitaš, K. Homann, N. Kallay, and K. Kuchitsu, *Quantities, Units and Symbols in Physical Chemistry*, Blackwell, Oxford, 1988.

40. See, for example, R. G. Bates, The modern meaning of pH, *Crit. Rev. Anal. Chem.* **10**:247 (1981); K. L. Babcock and R. Overstreet, On the use of calomel half cells to measure Donnan potentials, *Science* **117**:686 (1953); and J. Th. G. Overbeek, Donnan-E.M.F. and suspension effect. *J. Colloid Sci.* **8**:593 (1953).

FOR FURTHER READING

Buffle, J., *Complexation Reactions in Aquatic Systems*, Wiley, New York, 1988. An extraordinary, encyclopedic discussion of the measurement and modeling of complexation reactions involving humic substances. Chapters 7–10 focus on methods of measurement.

Burgess, J., *Ions in Solution*, Wiley, New York, 1988. A very readable account of complexation reactions from the point of view of solution physical chemistry, emphasizing molecular mechanisms.

Hartley, F. R., C. Burgess, and R. M. Alcock, *Solution Equilibria*, Wiley, New York, 1980. A comprehensive survey of equilibrium chemical speciation in ionic solutions, including methods of measurement, data analysis, and stability constants.

Hering, J. G., and F. M. M. Morel, The kinetics of trace metal complexation: Implications for metal reactivity in natural waters, pp. 145–171 in *Aquatic Chemical Kinetics*, ed. by W. Stumm, Wiley, New York, 1990. A survey of metal complexation kinetics and mechanisms written with applications in mind; a fine complement to the book by Burgess cited above.

Margerum, D. W., G. R. Cayley, D. C. Weatherburn, and G. K. Pagenkopt, Kinetics and mechanisms of complex formation and ligand exchange, pp. 1–220 in *Coordination Chemistry*, Vol. 2, ed. by A. E. Martell, American Chemical Society, Washington, DC, 1978. *The* classic, rigorous survey of complexation reactions in aqueous solutions; a "must" for the dedicated student.

Martell, A. E., and R. J. Motekaitis, *Determination and Use of Stability Constants*, VCH Publishers, New York, 1992. A very useful description of how complex stability constants can be measured and applied, as well as advice on choosing values form published data from one of the modern masters of coordination chemistry.

Stumm, W., and J. J. Morgan, *Aquatic Chemistry*, Wiley, New York, 1981. Chapter 7 of this standard reference gives an insightful discussion of the problems in measuring pE in natural waters.

Wilkins, R. G., *Kinetics and Mechanism of Reactions of Transition Metal Complexes*, VCH Publishers, New York, 1991. This classic textbook, by one of the key figures in the study of complexation reactions, offers a wealth of detail on the experimental aspects of aqueous speciation.

PROBLEMS

1. The solvation complexes, $Fe(H_2O)_6^{3+}$ and $Hg(H_2O)_4^{2+}$, can form complexes with halide ions according to the scheme in Eq. 2.5a. Some thermodynamic data for the overall complexation reactions at 298 K follow. Use these data to calculate $\Delta_r H^0$ and then explain the differences among the values for the metal–ligand adducts. (*Hint*: See Eq. 2.4.)

Metal	Ligand	$\Delta_r G^0$ (kJ mol^{-1})	$\Delta_r S^0$ (J mol^{-1} K^{-1})
Fe(III)	F^-	-29.5	+131.5
Fe(III)	Cl^-	-2.6	+67.8
Hg(II)	Cl^-	-38.5	+52.0
Hg(II)	Br^-	-51.5	+29.5
Hg(II)	I^-	-73.2	-7.0

2. A solution of hydrofluoric and oxalic acids containing 20 mmol m^{-3} of each is mixed with a solution of 20 mmol m^{-3} $AlCl_3$ in 1 mol m^{-3} HCl. Given uniform mixing and the rate coefficients k_{wex} = 1.3 s^{-1}, K_f = 12 dm^3 mol^{-1} s^{-1}, K_f' = 24 dm^3 mol^{-1} s^{-1}, and $k_f K_{osc}$ = 100 dm^3 mol^{-1} s^{-1}, respectively, for the formation of the complexes $Al(H_2O)_6^{3+}$ (Eq. 2.3, with M^{m+} = Al^{3+}); AlF^{2+} (Eq. 2.5a); AlF^{2+} via associative exchange with oxalate (Eq. 2.12); and $AlC_2O_4^+$ (Eq. 2.4, with ML = $AlC_2O_4^+$), prepare a "time-scale diagram," like that in Fig. 2.6, showing the expected sequence of complex formation. (*Hint*: Use the table in Note 7 to calculate reaction time scales.)

3. At pH 3, the measured value of K_f in Eq. 2.8 is 12 dm^3 mol^{-1} s^{-1}, whereas that of K_f' in Eq. 2.16 is 23 dm^3 mol^{-1} s^{-1} in the presence of 18 mmol m^{-3} $C_2O_4^{2-}$. Given $k_f' >> k_b$ nd $K_{AlOxc} \approx 10^7$ dm^3 mol^{-1} in Eq. 2.17, estimate k_f. What value of $C_2O_4^{2-}$ concentration would lead to *no* effect of oxalate on AlF^{2+} kinetics? (*Answer*: $k_f \approx 0.13$ dm^3 mol^{-1} s^{-1} and $[C_2O_4^{2-}] \approx 10^{-6}$ mol dm^{-3}.)

4. Develop a pseudo-second-order rate law for the formation of Al–fulvic acid complexes according to the reaction in Eq. 2.44 and the parallel reaction:

$$Al^{3+}(aq) + HL^0(aq) \rightleftarrows AlL^{2+}(aq) + H^+(aq)$$

under the assumption that the Eigen-Wilkins-Werner mechanism applies to both reactions. Explain how the values of the rate coefficients for each contributing reaction can be obtained from experimental kinetics data. (*Hint*: See Eq. 2.5)

5. At pH < 4, the pseudo-first-order rate constant in Eq. 2.37 takes the pH-independent form, $K(p_{O_2}) = k_f\, p_{O_2}$, with $k_f \approx 10^{-8}\ atm^{-1}\ s^{-1}$. For pH ≈ 3, the oxidation of Fe^{2+} to Fe^{3+} should be followed quickly by the formation of the hydrolytic species $FeOH^{2+}$. A reaction scheme describing this possibility is

$$Fe^{2+} + O_2(g) \xrightarrow{k_f} Fe^{3+}(aq) + O_2^-(aq)$$

$$Fe^{3+}(aq) + H_2O(\ell) \overset{*K_1}{=} FeOH^{2+}(aq) + H^+(aq)$$

where $O_2^-(aq)$ is the superoxide anion and $*K_1 = 360$ is the equilibrium constant for the second reaction. Derive an equation for the time dependence of the concentration of Fe^{2+} at fixed pH and p_{O_2}. Estimate the time scale, at $p_{O_2} = 0.2$ atm and pH 3, over which $Fe^{2+}(aq)$ disappears to form $FeOH^{2+}(aq)$. (*Hint*: Adapt the reaction scheme in Eq. 1.51 under the assumptions of no back-reduction of Fe^{3+} and very rapid equilibration of Fe^{3+} with $FeOH^{2+}$. Time scale $\approx 0.693/k_f\, p_{O_2} \approx 10$ yr.)

6. An additional pathway to that considered in Problem 5 is the direct oxidation of $FeOH^+$:

$$FeOH^+(aq) + O_2(g) \xrightarrow{k_f'} FeOH^{2+}(aq) + O_2^-(aq)$$

where $k_f' \approx 0.032\ atm^{-1}\ s^{-1}$. Given that $*K_1' = 3.2 \times 10^{-10}$ for the hydrolysis reaction,

$$Fe^{2+}(aq) + H_2O(\ell) \overset{*K_1'}{=} FeOH^+(aq) + H^+(aq)$$

estimate the time scale over which Fe(II) disappears by the combination of the oxidation of $Fe^{2+}(aq)$ and $FeOH^+(aq)$ at $p_{O_2} = 0.2$ atm and pH 3. (*Hint*: The combined-pathway, pseudo-first-order rate coefficient is $(k_f + k_f'\ {*K_1'}/[H^+])p_{O_2}$ with neglect of ionic strength effects on $*K_1'$.)

7. Consider a solution containing initially $Fe(II)_T = 20\ mmol\ m^{-3}$ and $L_T = 2\ mol\ m^{-3}$ where L refers to a quasiparticle fulvic acid ligand. The solution pH = 7.5 and $p_{O_2} = 0.2$ atm. The species Fe^{2+} makes up all of $Fe(II)_T$ initially, but then oxidizes to Fe^{3+}, following the rate law in Eq. 2.36, or reacts with L^-

to form the organic complex FeL^+:

$$Fe^{2+}(aq) + L^-(aq) \xrightarrow{k_f} FeL^+(aq)$$

Show that the rate law for the change in Fe^{2+} concentration is pseudo-first-order and calculate the concentrations of the competing products, $Fe^{3+}(aq)$ and $FeL^+(aq)$, as functions of time. Take $k_f = 5\ dm^3\ mol^{-1}\ s^{-1}$ and $K_{sc}^{cp} \equiv [FeL^+]_e / [Fe^{2+}]_e[L^-]_e = 10^4\ dm^3\ mol^{-1}$. (*Hint*: With $[L^-] \approx L_T$, the product concentrations are (t in seconds):

$$[Fe^{3+}] = (10^{-4}/15)[1 - \exp(-0.015t)]$$

$$[FeL^+] = 2 \times (10^{-4}/15)[1 - \exp(-0.015t)]$$

Thus the time scale for Fe^{2+} oxidation is about 140 s, whereas that for the formation of FeL^+ is about 70 s.)

8. Given the generic relationship for a single-ion activity coefficient,

$$\log \gamma_J = -A_{DH} Z_J^2 \left[\frac{\sqrt{I_{ef}}}{1 + A_J \sqrt{I_{ef}}} + B_J I_{ef} \right]$$

derive an equation relating K_s to K_{sc} for the overall complexation reaction in Eq. 2.5a. For the Davies equation, $A_J = 1$, $B_J = -0.3$, whereas for the extended Debye-Hückel equation, $A_J = 2.97$ for Al^{3+}, 1.0 for F^-, 1.32 for AlF^{2+}, and $B_J = 0$. Calculate K_{sc} at 298 K for an effective ionic strength of 100 mol m^{-3} and compare your results, using both activity coefficient equations, with $K_{sc} = 2.7 \times 10^6\ dm^3\ mol^{-1}$ (experimental value).

$$\left(\textit{Answer}: \log K_s = \log K_{sc} + A_{DH} \left\{ 9 \left[\frac{\sqrt{I_{ef}}}{1 + A_{Al}\sqrt{I_{ef}}} + B_{Al} I_{ef} \right] \right. \right.$$

$$\left. \left. + \left[\frac{\sqrt{I_{ef}}}{1 + A_F \sqrt{I_{ef}}} + B_F I_{ef} \right] - 4 \left[\frac{\sqrt{I_{ef}}}{1 + A_{AlF}\sqrt{I_{ef}}} + B_{AlF} I_{ef} \right] \right\} . \right)$$

9. The consensus of experts is that pH cannot be measured more *accurately* than ±0.05 in natural waters. Two important limitations on accuracy are “activity coefficient corrections” and liquid junction potentials.

(a) Derive a mathematical relationship between pH $\equiv$ $-\log(\mathbf{H^+})$ and $[H^+]_e$, first using the Davies equation, then the extended Debye-Hückel equation:

$$\log \gamma_H = -A_{DH}\left[\frac{\sqrt{I_{ef}}}{1 + 2.97\sqrt{I_{ef}}}\right]$$

Calculate pH versus $-\log[H^+]_e$ at 298 K for $0 < I_{ef} < 0.5$ mol dm^{-3} with the two derived relationships and estimate the inaccuracy of pH that could result from "activity coefficient corrections."

(b) Given the following expression for the liquid junction potential at 298 K,

$$E_J = -50[H^+]_e / I_{ef} \qquad (E_J \text{ in mV, } I_{ef} \text{ in mol dm}^{-3})$$

calculate the error in pH (Eq. s2.23) that would result by neglecting the variation of E_J with $[H^+]_e$ or I_{ef}, assuming that calibration is done at pH 4 in a solution of ionic strength 0.01 mol dm^{-3}. Let $4 < \text{pH} < 7$ and $0.01 < I_e < 0.5$ mol dm^{-3}. (*Answer*: (a) ± 0.1 (b) $E_J = -0.5$ mV at calibration; at $[H^+]_e = 10^{-7}$ mol dm^{-3} and $I_{ef} = 0.5$ mol dm^{-3}; $E_J = -10^{-5}$ mV. The inaccuracy in pH $\approx$ $\Delta E_J\, F/RT = 0.5/25.7 \approx 0.02$.)

10. Consider a solution that has equilibrated according to the reactions in Eqs. 2.44 and 2.51a, with the species Cu^{2+}, Al^{3+}, L^-, CuL^+, and AlL^{2+}, where L^- is a fulvic acid quasiparticle ligand, with K_{sc}^{qp} values of 10^5 and 10^6 dm^3 mol^{-1}, respectively, for the two metal complexes. Given $Cu_T = 0.1$ mmol m^{-3}, $Al_T = 10$ mmol m^{-3}, and $L_T = 10$ mmol m^{-3}, estimate the distribution of fulvic acid between Al and Cu. (*Hint*: Develop a mole-balance equation for L_T analogous to Eq. 2.52c and justify neglecting the term in $[Cu^{2+}]_e$. Then develop a mole-balance equation for Al_T analogous to Eq. 2.52a and show that $\beta_L \equiv [L^-]_e/L_T \approx \alpha_{Al} \equiv [Al^{3+}]_e/Al_T$, under the conditions in the solution. The mole-balance equation for L_T then becomes a quadratic equation for β_L: $\beta_L^2 + \beta_L - 1 = 0$, with the solution $\beta_L = 0.618 \approx \alpha_{Al}$. This result is consistent with $\alpha_{Cu} \equiv [Cu^{2+}]_e/Cu_T \approx 0$. Thus $\beta_{AlL} \equiv [AlL^{2+}]_e/L_T \approx 0.382$ and $\beta_{CuL} \equiv [CuL^+]_e/L_T \approx Cu_T/L_T = 0.01$. Note that the larger value of K_{sc}^{qp} for CuL^+ is overwhelmed in this calculation by the smallness of Cu_T. As a general rule, the distribution of metals (M) among complexes with a given ligand is roughly proportional to the *product*, $K_{sc}M_T$, which considers both complex stability (K_{sc}) and the concentration of metal (M_T) available to react with the ligand.)

3

MINERAL SOLUBILITY

3.1 Dissolution–Precipitation Reactions

An *overall* reaction describing the dissolution or precipitation of a two-component solid, $M_aL_b(s)$, in contact with a soil solution, can be written as follows

$$M_aL_b(s) \underset{k_p}{\overset{k_{dis}}{\rightleftarrows}} aM^{m+}(aq) + bL^{\ell -}(aq) \tag{3.1}$$

where M is a metal and L is a ligand, while a and b are stoichiometric coefficients. An equilibrium constant for this heterogeneous reaction (taken as an overall dissolution process) is defined analogously to K in Eq. 1.11:

$$K_{dis} \equiv (\mathbf{M}^{m+})^a \, (\mathbf{L}^{\ell -})^b / (\mathbf{M}_a\mathbf{L}_b) \tag{3.2}$$

in which the activity of the solid phase, $(\mathbf{M}_a\mathbf{L}_b)$, is related to its chemical potential as in Eq. s1.10 (Special Topic 1 in Chapter 1). A thermodynamic parameter closely related to K_{dis} is the *solubility product constant* K_{so}:

$$K_{so} \equiv (\mathbf{M}_a\mathbf{L}_b)K_{dis} \tag{3.3}$$

Formally, K_{so} is equal to the numerator on the right side of Eq. 3.2 when the reaction in Eq. 3.1 is at equilibrium. If the solid phase is in its Standard State (see Table s1.1 in Chapter 1), then $K_{so} = K_{dis}$, as is commonly assumed.

Dissolution–precipitation reactions often exhibit characteristic time scales that are much larger than those for complexation reactions in aqueous solution. When this is true, the aqueous species in a soil solution will come to mutual equilibrium long before they equilibrate with the solid phase via the reaction in Eq. 3.1. It is possible under these circumstances to define two useful criteria for dissolution–precipitation equilibrium, the *ion activity product* (IAP):

$$\text{IAP} \equiv (\mathbf{M}^{m+})^a\,(\mathbf{L}^{\ell-})^b \tag{3.4}$$

and its ratio to K_{so}, the *relative saturation*:

$$\Omega \equiv \text{IAP}/K_{so} \tag{3.5}$$

Values of Ω can be monitored as time passes to determine whether equilibrium between a solid phase and a soil solution exists. If $\Omega < 1$, the soil solution is termed *undersaturated* with respect to the solid phase; if $\Omega > 1$, the soil solution is *supersaturated*. Measurements of Ω thus lead to one or more of the following conclusions about the reaction of Eq. 3.1 whenever $\Omega \neq 1$:

1. The reaction is not at equilibrium.
2. No solid phase corresponding to the reaction is present.
3. The reaction is at (possibly metastable) equilibrium, but the solid phase is not in the Standard State assumed in computing K_{so}.

As an example of the use of Ω to probe a dissolution reaction, consider the soil mineral gibbsite [$Al(OH)_3$] dissolving in an Oxisol:

$$Al(OH)_3(s) \rightarrow Al^{3+}(aq) + 3\ OH^-(aq) \tag{3.6}$$

The value of K_{dis} can be calculated with Standard-State chemical potentials according to the relationships in Eqs. s1.12–s1.14:[1]

$$\begin{aligned}\log K_{dis} &= [\mu^0(Al(OH)_3) - \mu^0(Al^{3+}) - 3\mu^0(OH^-)]/5.708 \\ &= [-1154.8 \pm 1.3 + 489.8 \pm 4.0 + 3(157.2)]/5.708 = -33.9 \pm 0.7\end{aligned}$$

at 298.15 K, where the μ^0 are expressed in units of kJ mol^{-1}. Figure 3.1 shows measured values of Ω based on Eq. 3.5 and the assumption that $(\mathbf{Al(OH)_3}) = 1.0$ (i.e., $K_{so} = K_{dis} = 1.3 \times 10^{-34}$). Ion activity products $[(\mathbf{Al^{3+}})(\mathbf{OH^-})^3]$ were determined in aliquots of the leachate from an Oxisol during slow elution. After about 40 days of elution, $\Omega \approx 1.0$, and thermodynamic equilibrium between the soil solution and dissolving gibbsite may be assumed to have been achieved. (The presence of gibbsite in the soil was confirmed by x-ray diffraction analysis.) Matters can become much more complicated, however, in the case of gibbsite *precipitation*, both because of the formation of metastable Al-hydroxy polymers that transform slowly in the aqueous solution phase and because of structural disorder in the growing solid phase.[1] Under these obfuscating conditions, the interpretation of measured values of Ω as an indicator of disequilibrium becomes problematic.

The quantitative role of Ω in dissolution–precipitation kinetics can be sharpened by a rate-law analysis of the reaction in Eq. 3.1. The rate of increase in the concentration of $M^{m+}(aq)$ can be postulated to be equal to the difference

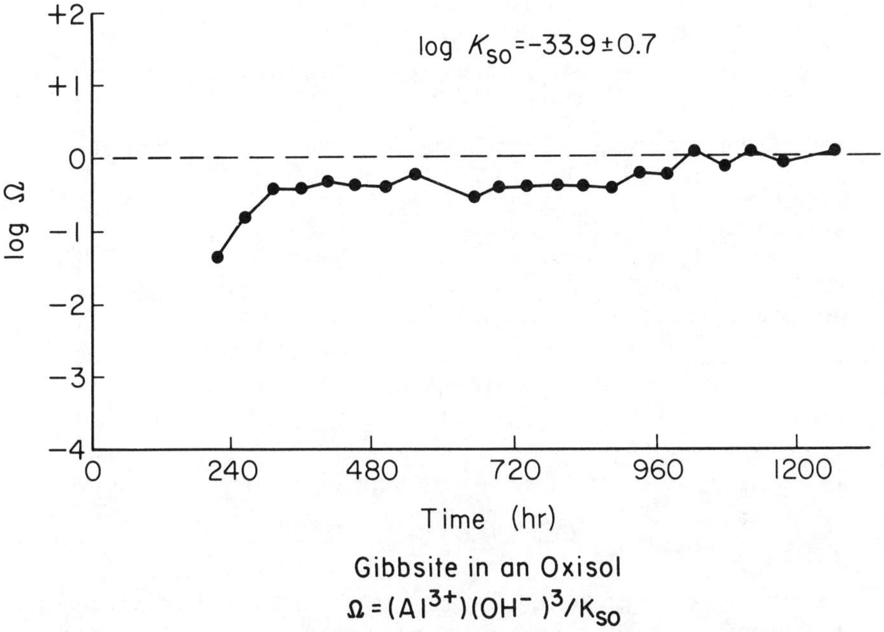

FIG. 3.1. Evolution of the relative saturation (Ω) for gibbsite toward equilibrium in an Oxisol. (Reprinted by permission of Oxford University Press from G. Sposito.[12])

of two functions, R_f and R_b, which depend, respectively, on powers of the concentrations of the reactant and products in Eq. 3.1. This line of reasoning is analogous to that associated with Eqs. 1.33 and 1.34. The *overall* rate law is then

$$\frac{d[M^{m+}]}{dt} = k_{dis}[M_aL_b]^{\delta} - k_p[M^{m+}]^{\alpha}[L^{\ell-}]^{\beta} \tag{3.7a}$$

where k_{dis} and k_p are rate coefficients, and α, β, δ are partial reaction orders. If the solid phase is in great excess during dissolution or precipitation, its concentration factor can be absorbed into the dissolution rate coefficient: $k_d \equiv k_{dis}[M_aL_b]^{\delta}$. If also it is assumed that $\alpha = a$, $\beta = b$, as done in connection with Eq. 1.35, then the overall rate law

$$\frac{d[M^{m+}]}{dt} = k_d - k_p[M^{m+}]^a[L^{\ell-}]^b \tag{3.7b}$$

can be *postulated*, where k_d and k_p depend on temperature, pressure, aqueous-solution composition, and nature of the solid phase. These rate coefficients, by analogy with Eq. 1.36, are related to a *conditional solubility product constant*:

$$K_{soc} \equiv [M^{m+}]_e^a \, [L^{\ell-}]_e^b = k_d/k_p \tag{3.8}$$

The combination of Eqs. 3.7 and 3.8 produces the model rate expression

$$\frac{d[M^{m+}]}{dt} = k_d\,(1 - \Omega) \tag{3.9}$$

where the relationship (cf. Eq. 1.15)

$$\begin{aligned} \Omega_c &\equiv [M^{m+}]^a [L^{\ell-}]^b / K_{soc} \\ &= \gamma_M^a \, \gamma_L^b \, [M^{m+}]^a [L^{\ell-}]^b / K_{so} \\ &= (M^{m+})^a (L^{\ell-})^b / K_{so} = \Omega \end{aligned} \tag{3.10}$$

has been applied. Equation 3.9 demonstrates the role of Ω as a discriminant in the kinetics of dissolution–precipitation reactions. As discussed in Section 1.3, the mathematical form of the right side of Eq. 3.9 is not unique unless the reaction in Eq. 3.1 is elementary. For example, a more general expression than Eq. 3.8 can be invoked to connect rate coefficients to equilibrium concentrations (cf. Eq. 3.7a):

$$k_d/k_p = [M^{m+}]_e^{\alpha} \, [L^{\ell-}]_e^{\beta} \tag{3.11}$$

where α and β are partial reaction orders, not necessarily stoichiometric coefficients. If the partial reaction orders are constrained as in Eq. 1.40b, then Eq. 3.9 will take the form

$$\frac{d[M^{m+}]}{dt} = k_d(1 - \Omega^n) \tag{3.12}$$

where $n \equiv \alpha/a = \beta/b$. On the other hand, if Eq. 1.37 is applied, say, by setting the arbitrary function $g(\cdot) \equiv k_d$, the rate law in Eq. 3.9 would become

$$\frac{d[M^{m+}]}{dt} = k_d\,(1 - \Omega)^n \tag{3.13}$$

where $n > 1$ is now an exponent unrelated to reaction order. The rate laws in Eqs. 3.9, 3.12, and 3.13 meet the thermodynamic requirements of exhibiting a positive reaction rate for dissolution processes far from equilibrium and a zero rate for processes at equilibrium. All three rate laws are in common use to describe the kinetics of heterogeneous reactions.[2]

The rate coefficients in Eq. 3.7 characterize an overall reaction and therefore can be expected to vary with system properties, insofar as these properties affect reaction mechanisms for dissolution–precipitation processes. The nonchemical influence of the aqueous phase volume is removed by the widespread practice of scaling the rate coefficients with the factor $a_s c_s$, where a_s is specific surface area and c_s is the concentration of solid in the aqueous phase.[2–4] This factor, the solid-phase surface area per unit volume of aqueous phase, normalizes rate coefficients to permit the effects of differing particle size and degree of solid-phase crystallinity to be determined. The rate coefficient k_d typically is observed also to vary significantly with pH, as can be seen in Fig. 3.2, which shows log–log plots of $(k_d/a_s c_s)$ versus $(\mathbf{H^+})$ for several silicates. This pH dependence, equivalent to a power-law dependence on $(\mathbf{H^+})$, suggests proton promotion of mineral dissolution far from equilibrium at low pH and hydroxide ion promotion at high pH.[3] Significant variation of k_d with the concentration of adsorptive ligands in the aqueous-solution phase also is observed.[3] Adsorption reactions leading to changes in solid surface reactivity thus can affect the values of k_d and k_p. For example, in the case of calcite ($CaCO_3$), there is an increase in k_d with p_{CO_2} that reflects the dissolution-enhancing influence of adsorbed CO_2,[4,5] and there is a decrease in k_p with fulvic acid concentration in the aqueous phase that reflects the precipitation-inhibiting influence of adsorbed organic ligands.[6]

Figure 3.3 shows an application of Eq. 3.7 to the dissolution–precipitation reaction for calcite:[7,8]

$$CaCO_3(s) \underset{k_p}{\overset{k_d}{\rightleftharpoons}} Ca^{2+}(aq) + CO_3^{2-}(aq) \tag{3.14}$$

The dissolution rate of calcite can be investigated in a fluidized bed reactor[9] whose operation is optimized to enhance solid–aqueous solution contact by turbulent mixing while minimizing solid-phase abrasion from particle contacts. The influence of pH on the rates of dissolution and precipitation can be seen in the graph on the left side of Fig. 3.3.[7] At pH $\approx$ 7, $\Omega << 1$ and $(k_d/a_s c_s) \approx 10^{-6}$ mol m^{-2} s^{-1}. At pH > 8, the precipitation rate of calcite at low p_{CO_2} can be investigated in supersaturated solutions inoculated with seed crystals.[8] The initial rate of precipitation then is measured by extrapolating the slope of a graph of $[Ca^{2+}]$ vs. time to zero elapsed time. A log–log plot of the initial rate versus initial $(\Omega - 1)$ is used to discriminate between Eqs. 3.9 and 3.13. For the data in the graph on the right side of Fig. 3.3, n = 1.2 ± 0.1, suggesting that Eq. 3.9 is satisfactory. The slope of the linear plot of *initial* rate versus *initial* $(\Omega - 1)$ leads to $(k_p/a_s c_s) \approx 1.5 \times 10^{-4}$ m^4 mol^{-1} s^{-1}. The ratio $k_d/k_p \approx 6.7 \times 10^{-3}$ mol^2 m^{-6} = 6.7×10^{-9} mol^2 dm^{-6}, in approximate agreement with K_{dis} for calcite

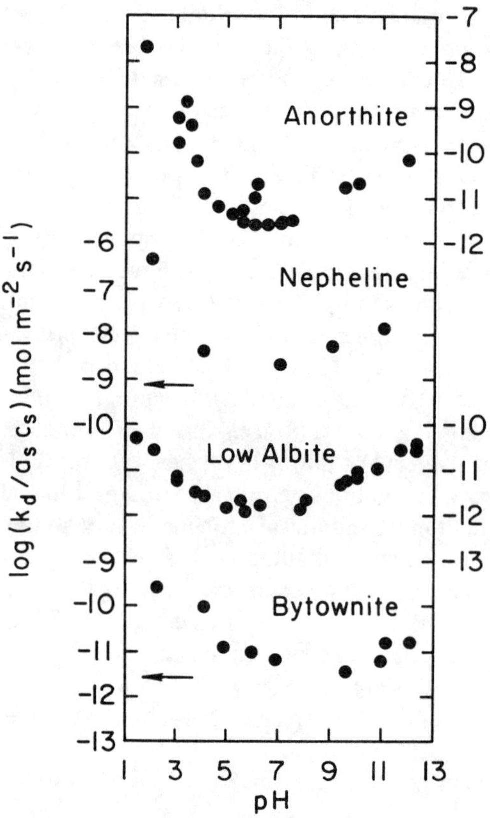

FIG. 3.2. Dependence of the dissolution rate coefficient for silicate minerals on pH. The data can be fit to a generic power-law expression: $(k_d/a_s c_s) = [k_1(H^+)^a + k_2 + k_3(H^+)^{-b}]$, where k_1, k_2, k_3, a, and b are positive parameters. (Adapted with permission from P. V. Brady and J. V. Walther.[39] Copyright © by Pergamon Press Ltd., Headington Hill Hall, Oxford OX3 0BW, England.)

(Eqs. s1.12 and s.1.13 in Chapter 1):

$$\log K_{dis} = [\mu^0(CaCO_3) - \mu^0(Ca^{2+}) - \mu^0(CO_3^{2-})]/5.708$$

$$= [-1128.8 \pm 1.4 + 553.5 \pm 1.2 + 527.9 \pm 0.1]/5.708$$

$$= -8.3 \pm 0.3$$

resulting in $2.5 \times 10^{-9} < K_{dis} < 10^{-8}$. Direct measurement[7,8] of IAP for calcite at precipitation–dissolution equilibrium leads to $K_{so} = 3.4 \times 10^{-9}$.

The reaction in Eq. 3.1 is an overall reaction that can, therefore, be the resultant of several parallel or sequential elementary reactions. For example, the

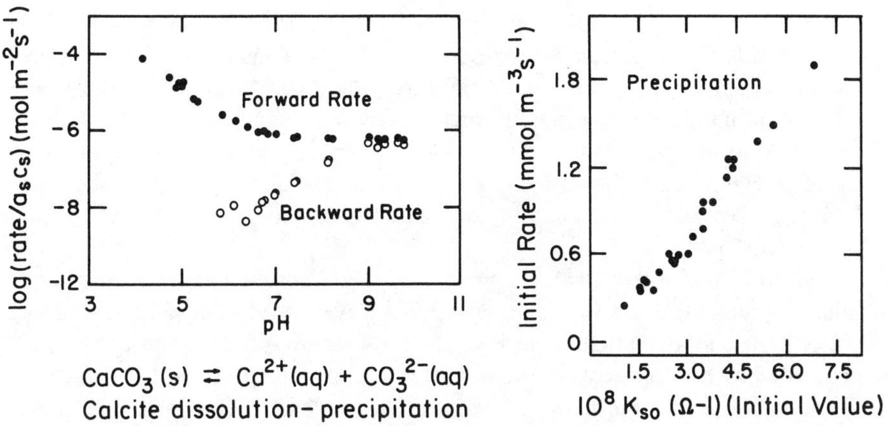

FIG. 3.3. Influence of pH on the rates of calcite dissolution and precipitation (left, adapted from Chou et al.[7] with the permission of Elsevier Science Publishers), and an initial-rate plot for calcite precipitation indicating the applicability of Eq. 3.9 (right, data from W. P. Inskeep and P. R. Bloom,[8] adapted with permission. Copyright © by Pergamon Press Ltd., Headington Hill Hall, Oxford OX3 0BW, England.)

calcite dissolution–precipitation reaction in Eq. 3.14 could be interpreted as the sum of the elementary protonation reaction

$$CaCO_3(s) + H^+(aq) \rightleftharpoons Ca^{2+}(aq) + HCO_3^-(aq) \tag{3.15}$$

combined with the dissociation–complexation reaction:

$$HCO_3^-(aq) \rightleftharpoons H^+(aq) + CO_3^{2-}(aq) \tag{3.16}$$

Another possibility is to combine the reaction

$$CaCO_3(s) + H_2CO_3^0(aq) \rightleftharpoons Ca^{2+}(aq) + 2HCO_3^-(aq) \tag{3.17}$$

with those in Eqs. 1.2, 1.4, and 3.16 to create the overall reaction in Eq. 3.14. For this second reaction pathway, the kinetics of $H_2CO_3^0$ formation, discussed in Section 1.5, and that of $CO_2(g)$ coming into solution can be important to the time scale of a dissolution rate law based on Eq. 3.17.[5]

These simple examples illustrate the complexity implicit in heterogeneous overall reactions, even those as straightforward to express as Eq. 3.1. On the other hand, the rate law in Eq. 3.9 (as opposed to those in Eqs. 3.12 and 3.13) is characteristic of the kinetics of an elementary heterogeneous reaction and, to the extent it proves to be accurate, suggests the possibility of interpreting k_d

mechanistically. Figure 3.4 shows a useful correlation in this spirit, a log–log plot of (k_d/a_sc_s), measured for several orthosilicates containing bivalent metal cations, against k_{wex} for the metal cations (see Eq. 2.3).[10] The excellent positive correlation implies more generally that the ionic potential of a metal cation is important to the pathways of dissolution of simple ionic solids containing the metal (see Fig. 2.3). Thus, for example, solids containing metal cations with high ionic potentials should exhibit low rates of dissolution (as described by Eq. 3.9 when $\Omega \ll 1$).

Correlations of the kind that appear in Fig. 3.4 must be tempered, however, with the reminder that Eqs. 3.1 and 3.7 always represent *hypotheses* about dissolution and precipitation processes. If the rates of these processes are controlled by how quickly aqueous-solution species can approach the surface of the solid phase (*transport control*), then a rate law based solely on an assumed chemical reaction at the surface (*reaction control*) is quite irrelevant. This issue cannot be decided simply by fitting rate data to models like that in Eq. 3.7, but instead must be resolved through direct experimentation (e.g., by comparing the temperature dependence of the reaction with that for aqueous species transport,

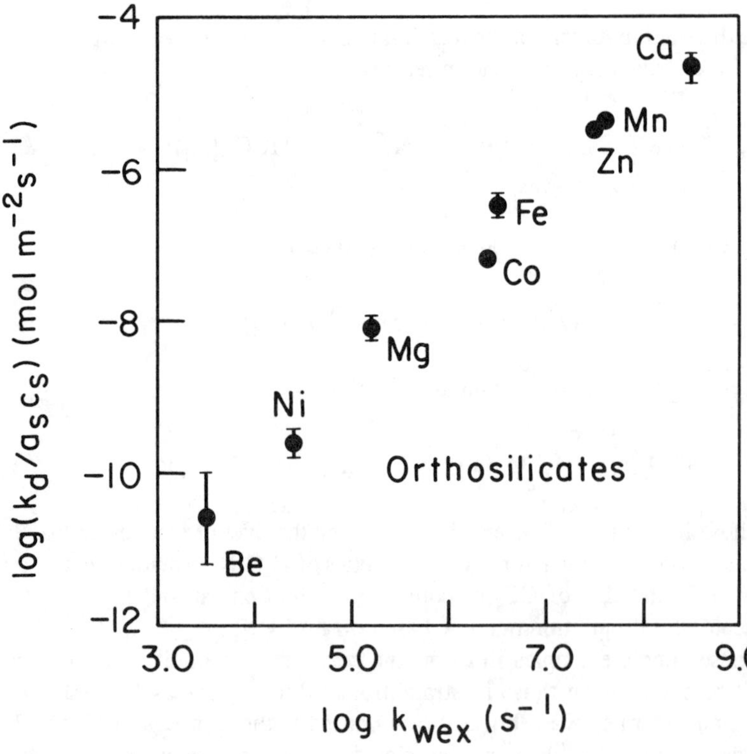

FIG. 3.4. Positive correlation between k_d and k_{wex} for orthosilicates containing bivalent metal cations.[10]

or by varying the rate of aqueous-phase mixing to determine its effect on the rate of dissolution).[7,11] In the same vein, it must be remembered that rate laws like those in Eq. 3.7 assume *constant* stoichiometry (see Section 1.3). If the reaction pathway changes significantly over the time scale on which a dissolution–precipitation process is investigated, the rate law assumed initially may become irrelevant. This can occur, for example, if a secondary solid phase is precipitated during dissolution (*incongruent dissolution*), such that the right side of the overall dissolution–precipitation reaction must exhibit a new solid phase containing one or more components of the original, dissolving solid phase.[11] These caveats serve to emphasize the fundamentally empirical nature of rate laws for overall heterogeneous reactions in natural systems like soils.

The situation becomes much less problematic if the reaction in Eq. 3.1 is considered only at equilibrium. Equilibrium states are steady states in which the net reaction fluxes to produce products or reactants, regardless of the reaction pathway, are equal and opposite. From the perspective of chemical thermodynamics, equilibrium states are unique and independent of thermodynamic path. Thus these states can be described by a unique set of chemical species irrespective of the intermediate steps of their formation. Dissolution– precipitation reactions and the chemical species that affect them at equilibrium can be described by an extension of the methodology discussed in Section 2.4 (cf. Fig. 2.8). The model balance equation for each metal and ligand (e.g., Eqs. 2.49 and 2.52) is augmented to include formally the concentration of each possible solid phase. By choosing an appropriate linear combination of these equations, it is always possible to eliminate the concentrations of the solid phases from the set of equations to be solved numerically. Moreover, some of the free ionic concentrations of the metals and ligands also can be eliminated from the equations because of the constraints imposed by K_{so} on their activities (combine Eqs. 3.2 and 3.3), which holds for each solid phase formed. The final set of nonlinear algebraic equations obtained from this elimination process will involve only independent free ionic concentrations, as well as conditional stability and solubility product constants. The numerical solution of these equations then proceeds much like the iteration scheme outlined in Section 2.4 for the case where only complexation reactions were considered, with the exception of an added requirement of self-consistency, that the calculated concentration of each solid formed be a positive number and that IAP not be greater than K_{so} (see Fig. 2.8).

An example of a speciation calculation involving calcite formation is shown in Table 3.1 for a soil solution containing the same metals and ligands as in the example in Table 2.6, but at pH 7.9 instead of 5.6. The nominal total concentration of Ca (5.25 mol m^{-3}) is predicted to be partitioned as 56% calcite and 44% aqueous species at equilibrium. Thus the solubility of Ca is predicted to be 2.3 mol m^{-3} (= 0.44 × Ca_T) under the conditions of the calculation. Note that about 12% of this solubility is contributed by metal–ligand complexes. As an additional bit of analysis, the IAP for calcite, $(Ca^{2+})(CO_3^{2-})$, in the soil solution can be calculated, given the values of the concentrations of Ca^{2+}

and CO_3^{2-} in Table 3.1 and the single-ion activity coefficients, computed at I_{ef} = 30 mol m^{-3} with the Davies equation (Eq. 1.21). Therefore IAP = $\gamma_{Ca}[Ca^{2+}]_e \gamma_{CO_3}[CO_3^{2-}]_e = (0.52)(1.69 \times 10^{-3})(0.52)(6.96 \times 10^{-6}) = 3.18 \times 10^{-9}$. This result may be compared with the equilibrium IAP = 3.4×10^{-9}, measured directly for calcite.[7,8]

3.2 Activity-Ratio and Predominance Diagrams

Graphical methods based on dissolution equilibria offer a simple and direct approach to the qualitative interpretation of mineral solubility data. The two most common of these methods are the *activity-ratio diagram* and the *predominance diagram.*[12] Although both methods ultimately tell the same story, each has visual features appealing uniquely to different aspects of the patterns of solubility in soils. Fundamentally, they respond to the question: Does a dissolving solid phase control the concentration of a given chemical element in a soil solution under given conditions, and, if so, which solid phase is it likely to be? This question, facile in appearance, turns out to be complex in application, thus preserving the need for qualitative, pictorial analyses despite the increasing sophistication of quantitative speciation calculations.[13]

The construction of an *activity-ratio diagram* can be summarized in the following four steps:

1. Identify a set of solid phases that contain a chemical element of interest and may be controlling its solubility. Write a dissolution reaction for each solid, with the *free* ionic species of the element as one of the products. Be sure that the stoichiometric coefficient of the free ion (metal or ligand) is 1.0.
2. Compile values of K_{dis} for the solid phases. Write an algebraic equation for each log K_{dis} in terms of log[activity] variables for the products and reactants in the corresponding dissolution reaction. Rearrange the equation to have log[**(solid phase)/(free ion)**]—the log *activity ratio*—on the left side, with all other log[activity] variables on the right side.
3. Choose an *independent* log[activity] variable against which log[**(solid)/(free ion)**] can be plotted for each solid phase. A typical example is pH = $-\log(\mathbf{H^+})$.
4. Select fixed values for all other log[activity] variables, corresponding to an assumed set of soil conditions. Use these values and that of log K_{dis} to develop a linear relation between log[**(solid)/(free ion)**] and the independent log[activity] variable for each solid phase considered. Plot all of these equations on the same graph.

For a chosen value of the log[activity] parameter taken as the independent variable, and under the assumption that all solid phases are in their Standard States, the solid that produces the largest value of the logarithm of the activity

Table 3.1 Results of a Chemical Speciation Calculation Involving Calcite Formation[a]

Total Concentrations

Na_T = 2.00	Mg_T = 2.64	K_T = 2.89	Ca_T = 2.28
CO_{3T} = 2.39	SO_{4T} = 2.12	Cl_T = 2.03	

Component Concentrations

Na^+ = 2.01	Mg^{2+} = 2.75	K^+ = 2.90	Ca^{2+} = 2.77
CO_3^{2-} = 5.16	SO_4^{2-} = 2.20	Cl^- = 2.03	

Species Distribution

Ca		CO_3	
as a free metal	32.1%	as a free ligand	0.2%
complexed with CO_3	0.3	complexed with Ca	0.3
in solid form with CO_3	56.1	in solid form with Ca	72.3
complexed with SO_4	11.2	complexed with Mg	0.3
complexed with Cl	0.3	complexed with Na	0.2
		complexed with H	26.7
Mg		SO_4	
as a free metal	77.5	as a free ligand	83.3
complexed with CO_3	0.5	complexed with Ca	7.7
complexed with SO_4	21.4	complexed with Mg	6.5
complexed with Cl	0.6	complexed with K	0.4
		complexed with Na	2.1
K		Cl	
as a free metal	96.7	as a free ligand	99.3
complexed with SO_4	2.5	complexed with Ca	0.2
complexed with Cl	0.7	complexed with Mg	0.1
		complexed with Na	0.1
Na			
as a free metal	98.0		
complexed with SO_4	1.6		
complexed with Cl	0.3		

[a]Concentrations are expressed as $-\log [\ ]_c$; for example, 2.28 means 5.25×10^{-3} mol dm^{-3}.

ratio is the one that is most stable and, therefore, the one that will be present at equilibrium. This conclusion follows directly from the fact that a solid phase that produces the smallest soil solution activity of a free ionic species will control that species. This, in turn, is true because the chemical potential of an ion in an aqueous phase will be smallest wherever the activity of the ion is least (cf. Eq. s1.10 in Chapter 1). The tendency of an ion, if several solids containing it were to be present initially, would be to diffuse ultimately to the region of the aqueous solution phase where its chemical potential will be least. Therefore the solid phase finally capable of producing the smallest soil solution activity of an ion also will be the most stable compound forming that contains the ion.

As an example of these ideas, an activity-ratio diagram for control of Al(III) solubility by secondary minerals in an acidic soil solution will be constructed. The Jackson–Sherman weathering scenario[14] indicates that when soil profiles are leached free of silica with fresh water, 2:1 layer-type clay minerals are replaced by 1:1 layer-type clay minerals, and ultimately these are replaced by metal oxyhydroxides. This sequence of clay mineral transformations can be represented by the successive dissolution reactions of smectite, kaolinite, and gibbsite:

$$\begin{aligned} &Mg_{0.208}[Si_{3.82}\,Al_{0.18}](Al_{1.29}\,Fe(III)_{0.335}\,Mg_{0.445})O_{10}(OH)_2(s) \\ &\quad + 3.28H_2O(\ell) + 6.72H^+(aq) = 1.47Al^{3+}(aq) + 0.335Fe^{3+}(aq) \\ &\quad + 0.653Mg^{2+}(aq) + 3.82H_4SiO_4^0(aq) \qquad \log K_{dis} = 3.2 \end{aligned} \tag{3.18a}$$

$$Al_2Si_2O_5(OH)_4(s) + 6H^+(aq) = 2Al^{3+}(aq) + 2H_4SiO_4^0(aq) + H_2O(\ell) \qquad \log K_{dis} = 7.43 \tag{3.18b}$$

$$Al(OH)_3(s) + 3H^+(aq) = Al^{3+}(aq) + 3H_2O(\ell) \qquad \log K_{dis} = 8.11 \tag{3.18c}$$

The solid-phase reactant in Eq. 3.18a is a montmorillonite, with Mg^{2+} as the interlayer exchangeable cation. Its dissolution reaction is a generalization of that in Eq. 3.1 to the case of a multicomponent solid. The value of K_{dis} for the dissolution of kaolinite (Eq. 3.18b) reflects a well-crystallized solid phase. Poorly crystallized kaolinite—typical of intensive soil weathering conditions—would yield $\log K_{so} \approx 10.5$. The reaction for gibbsite dissolution differs from that in Eq. 3.6 by the subtraction of the water ionization reaction (Eq. 1.4), with a corresponding change in the value of $\log K_{dis}$. Like kaolinite, gibbsite is assumed to be well crystallized; poorly crystallized gibbsite would yield $\log K_{so} \approx 9.35$.[14]

Equation 3.18 can be used to construct an activity-ratio diagram in respect to Al solubility as influenced by the leaching of silicic acid ($H_4SiO_4^0$). The equations for $\log[(\mathbf{solid})/(Al^{3+})]$ are as follows:

$$\begin{aligned} &\log[(\mathbf{montmorillonite})/(\mathbf{Al}^{3+})] = -2.18 + 0.228\log(Fe^{3+}) \\ &\quad + 0.444\log(\mathbf{Mg}^{2+}) + 4.57\,pH + 2.6\log(\mathbf{H_4SiO_4^0}) \\ &\quad - 2.23\log(\mathbf{H_2O}) \end{aligned} \tag{3.19a}$$

$$\log[(\mathbf{kaolinite})/(\mathbf{Al}^{3+})] = -3.72 + 3\,pH + \log(\mathbf{H_4SiO_4^0}) + \frac{1}{2}\log(\mathbf{H_2O}) \tag{3.19b}$$

$$\log[(\mathbf{gibbsite})/(\mathbf{Al}^{3+})] = -8.11 + 3\,\mathrm{pH} + 3\log(\mathbf{H_2O}) \qquad (3.19c)$$

Note that Eq. 3.18a and its log K_{dis} value must be divided by 1.47, and that Eq. 3.18b and its log K_{dis} value must be divided by 2, before Eq. 3.19 can be derived. If $(\mathbf{H_4SiO_4^0})$ is to be the "master variable," then pH, $(\mathbf{H_2O})$, and the activities of Fe^{3+} and Mg^{2+} in the soil solution must be prescribed. Representative values are pH = 5, $(\mathbf{H_2O})$ = 1.0, $(\mathbf{Fe^{3+}}) = 10^{-13}$, and $(\mathbf{Mg^{2+}}) = 6 \times 10^{-3}$. The resulting linear activity-ratio equations are then

$$\log[(\mathbf{montmorillonite})/(\mathbf{Al}^{3+})] = 16.72 + 2.6\log(\mathbf{H_4SiO_4^0}) \qquad (3.20a)$$

$$\log[(\mathbf{kaolinite})/(\mathbf{Al}^{3+})] = 11.28 + \log(\mathbf{H_4SiO_4^0}) \qquad (3.20b)$$

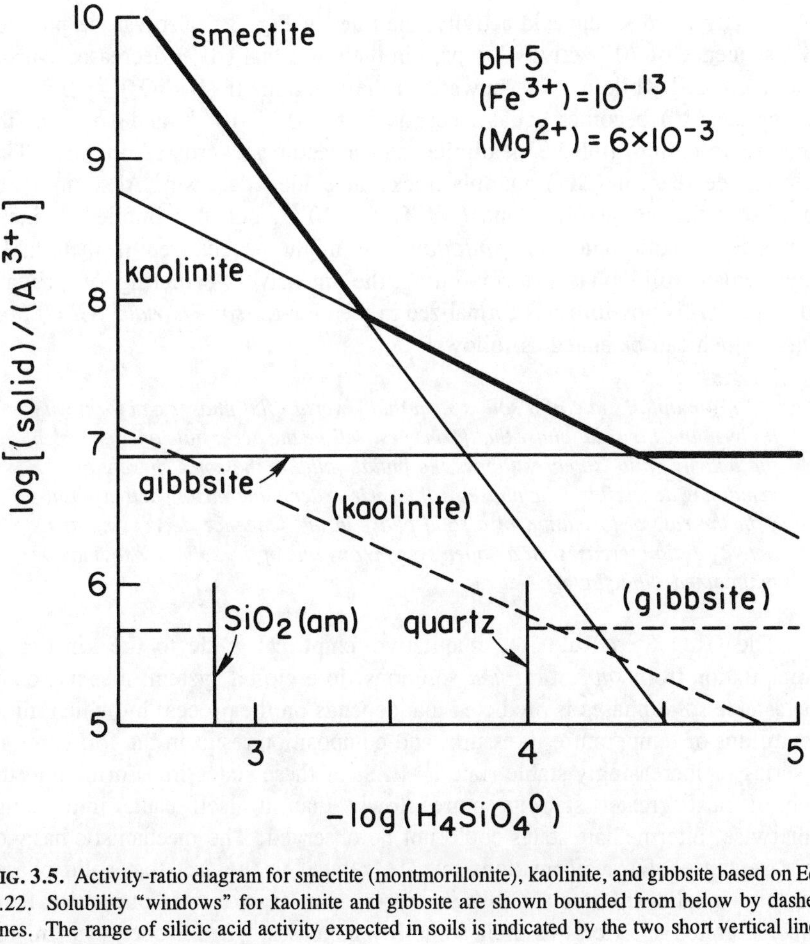

FIG. 3.5. Activity-ratio diagram for smectite (montmorillonite), kaolinite, and gibbsite based on Eq. 3.22. Solubility "windows" for kaolinite and gibbsite are shown bounded from below by dashed lines. The range of silicic acid activity expected in soils is indicated by the two short vertical lines labeled "SiO_2(am)" and "quartz."

$$\log[(\mathbf{gibbsite})/(\mathbf{Al}^{3+})] = 6.89 \tag{3.20c}$$

The activity-ratio diagram resulting from Eq. 3.20 is shown in Fig. 3.5.

The effect of soil profile leaching at fixed pH is represented in the activity-ratio diagram by moving from left to right along its x-axis. Amorphous silica supports $(\mathbf{H_4SiO_4^0}) \approx 10^{-2.7}$. Under this condition, which can reflect the intensive weathering of primary silicates in an acidic soil, Fig. 3.5 indicates that smectite is the most stable solid phase with respect to solubility controls on Al. As leaching and loss of silica proceed, the silicic acid activity will decrease, and when $(\mathbf{H_2SiO_4^0})$ falls well below 10^{-4}, the silicic acid activity supported by quartz, gibbsite becomes the most stable Al-bearing solid phase. This progression agrees with field observations as summarized in the Jackson– Sherman weathering stages.

At any fixed silicic acid activity, the lines in Fig. 3.5 also can be pictured as a sequence of Al^{3+} activity "steps," in the sense that $(\mathbf{Al^{3+}})$ decreases as each line is crossed while moving upward in the diagram. If $(\mathbf{H_2SiO_4^0}) = 10^{-2.7}$, for example, $(\mathbf{Al^{3+}})$ becomes equal successively to $10^{-6.9}$, $10^{-8.6}$, and $10^{-9.7}$, as the lines representing gibbsite, kaolinite, and smectite are crossed in turn. This stepwise decrease in $(\mathbf{Al^{3+}})$ not only tracks the evident deceasing Al solubility of the three minerals at pH 5 and $(\mathbf{H_2SiO_4^0}) = 10^{-2.7}$, but also implies a kinetic sequence of solid-phase *precipitation* that might be realized if metastable, intermediate solid phases form during the intensive weathering of primary silicates.[15] This possibility is formalized in the *Gay-Lussac-Ostwald (GLO) Step Rule*, which can be stated as follows:

> *If the initial state of a soil is such that several solid phases can form with a given ion, the solid phase that forms first will be the accessible one for which the activity ratio is nearest above the initial value in the soil. Thereafter, the remaining accessible solid phases will form in order of increasing activity ratio, with the rate of formation of a solid phase in the sequence decreasing as its activity ratio increases. In an open system, any one of the solid phases may be maintained "indefinitely."*

The GLO Step Rule is a qualitative, empirical guide to the kinetics of precipitation from *supersaturated* solutions. In a closed system, a sequence of metastable solid phases is predicted that depends on the process by which initial conditions of temperature, pressure, and composition result in the formation of a series of increasingly stable states.[15] Each of these states transforms into the one of next greater stability more slowly than it itself came into being; otherwise, intermediate states could not be observed. The mechanistic basis of this sequence of transformations may be related to the fact that solid phases exhibit a smaller interfacial Gibbs energy—and a larger rate of nucleation from supersaturated solution—as their solubility increases.[15] In an open system, the input of matter can be arranged to maintain the initial composition fixed, with the result that the solid phase of least stability consistent with that composition and the possible reaction pathways will also be preserved, despite its expected decay (at a lower rate than its formation) into more stable phases.[16]

Applied to the activity-ratio diagram in Fig. 3.5, the GLO Step Rule implies that, for example, if $(Al^{3+}) > 10^{-7}$ at pH 5 and $(H_4SiO_4^0) = 10^{-2.7}$ in a closed system, the least stable phase, gibbsite, could precipitate before the most stable phase, smectite, is formed. Behavior consistent with this prediction is seen in Fig. 3.6.[17] The activity-ratio diagrams in this figure represent the same mineral system as in Fig. 3.5 at pH 5 and 6. (Note that, as is common practice,[12] the scaling of the x- and y-axes is the reverse of that in Fig. 3.5, with *downward* movement reflecting increasing stability.) The filled circles are the respective initial states of the aqueous phase, with $(Al^{3+}) = 10^{-6.93}$ (pH 5) or $(Al^{3+}) = 10^{-9.27}$ (pH 6) and $(H_4SiO_4^0) = 10^{-2.8}$, 10^{-3}, $10^{-3.2}$, $10^{-3.5}$, 10^{-4}, or 0. The "ohs," connected by dotted lines to their corresponding initial states, are the states of the aqueous phase after 419 days of equilibration in the presence of 10 kg m^{-3} smectite in suspension and $p_{O_2} = 1$ atm. Movement toward the horizontal gibbsite line in each diagram is apparent at pH 6, whereas at pH 5, with (Al^{3+}) initially so close to (and just below) the gibbsite value (cf. Eq. 3.20c), little movement in the diagram occurs, except for the $(H_4SiO_4^0) = 0$ initial state. Evidently, a yet slower movement of the data points toward the kaolinite line would occur in due course, with coalescence on the smectite line achieved ultimately for all states supporting sufficiently high values of silicic acid activity.

The application of the GLO Step Rule and, for that matter, the interpretation of activity-ratio diagrams in general are influenced by the existence of varying degrees of crystallinity or of particle size in soil minerals, with a corresponding variation in their solubility.[15] For example, in the case of Fig. 3.5, very poorly crystallized forms of gibbsite and kaolinite, alluded to above, would require

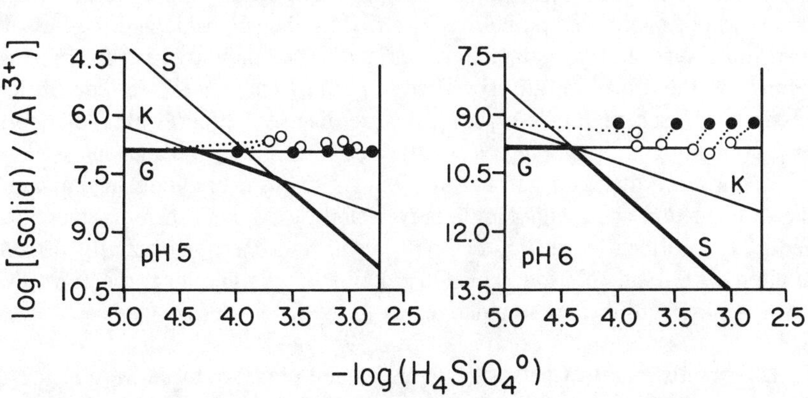

FIG. 3.6. Activity-ratio diagrams at pH 5 and 6 for the same set of solid phases as described in Fig. 3.5. Open circles represent initial states, whereas filled circles represent the subsequent states after 419 days of equilibration (dotted "reaction paths" link these two kinds of data point). (Data from May et al.[17])

replacing K_{dis} by K_{so} and making a consequent change in the constant terms on the right sides of Eqs. 3.19b and 3.19c to the values −5.25 and −10.8, respectively. Thus, the horizontal line in Fig. 3.5 would be plotted 1.24 units lower, and the kaolinite line would be shifted downward by 1.53 units (see the dashed lines in Fig. 3.5). The effects of these changes are to create "windows" of gibbsite and kaolinite stability,[18] instead of precise lines in the diagram, and to enlarge the range of silicic acid activity over which smectite remains the most stable solid phase, within the variability permitted by the windows. This kind of variability and the typical $(\mathbf{H_2SiO_4^0}) \approx 8 \times 10^{-4}$ in acid soils suggests that smectite, kaolinite, and gibbsite will coexist in active weathering environments.

Similar effects arise from varying particle size. The transformation of a macroscopic crystalline solid phase with negligible specific surface area to a finely divided powder whose particles have a significant specific surface area produces an additional term, proportional to the specific surface area, in the expression for $\Delta_r G^0$ describing the reaction in Eq. 3.1.[15] This term increases the value of the Standard-State chemical potential of the solid phase but does not alter K_{dis}, such that the equilibrium IAP increases also. The effect on Eq. 3.3 is to increase K_{so} according to the following equation:[15]

$$\log K_{so}(a_s) = \log K_{dis} + Ba_s \tag{3.21}$$

where a_s is specific surface area and B is a positive-valued parameter that depends on temperature, the interfacial Gibbs energy of the particles, and their geometric shape characteristics (surface-volume relationship). Equation 3.20 can be applied to calculate an appropriate value of K_{so} for use in constructing an activity-ratio diagram.

The kind of solubility information epitomized in Eq. 3.21 also can be represented in a convenient fashion through a *predominance diagram*. A predominance diagram is a two-dimensional field consisting of well-defined regions whose coordinate points are specified by the pH value and the base 10 logarithm of a second relevant activity variable. The boundary lines that enclose regions in the diagram are specified by equations based on equilibrium constants. In each region of a predominance diagram, either a particular solid phase containing an ion of interest or the free ion itself in the aqueous solution phase contacting the solid will be predominant. Thus a predominance diagram gives information about changing relative stabilities, at equilibrium, among the solid phases formed by an ion as the pH value and one other activity variable are altered in a soil solution. The construction of this graphical representation of solid-phase equilibria is summarized as follows:

1. Establish a set of solid-phase species and obtain values of log K for all reactions between the species.
2. Unless other information is available, set the activities of liquid water and all solid phases equal to 1.0. Set all fixed gas-phase pressures at values appropriate to soil conditions.

3. Develop each expression for log K into a relation between a dependent log[activity] variable and pH. In any relation involving aqueous species, choose values for the activities of the aqueous species.
4. Plot all the expressions resulting from step 3 as phase boundary lines on the same graph.

These steps will be illustrated with the dissolution reactions in Eq. 3.18. A corresponding set of chemical reactions that relates the solid-phase species to one another is

$$Al_2Si_2O_5(OH)_4(s) + 5H_2O(\ell) = 2Al(OH)_3(s) + 2H_4SiO_4^0(aq) \qquad \log K = -8.79 \qquad (3.22a)$$

$$Mg_{0.208}[Si_{3.82}Al_{0.18}](Al_{1.29}Fe(III)_{0.335}Mg_{0.445})O_{10}(OH)_2(s) + 7.69\,H_2O(\ell) + 2.31\,H^+(aq) = 1.47\,Al(OH)_3(s) + 3.82H_4SiO_4^0(aq) + 0.653\,Mg^{2+}(aq) + 0.335\,Fe^{3+}(aq) \qquad \log K = -8.72 \qquad (3.22b)$$

$$Mg_{0.208}[Si_{3.82}Al_{0.18}](Al_{1.29}Fe(III)_{0.335}Mg_{0.445})O_{10}(OH)_2(s) + 4.02\,H_2O(\ell) + 2.13\,H^+(aq) = 1.47\,Al_2Si_2O_s(OH)_4(s) + 2.35\,H_4SiO_4^0(aq) + 0.653\,Mg^{2+}(aq) + 0.335\,Fe^{3+}(aq) \qquad \log K = -2.26 \qquad (3.22c)$$

These three equations are merely algebraic combinations of Eq. 3.18. Inspection suggests that the activities of H_2O, Mg^{2+}, Fe^{3+}, and $H_4SiO_4^0$ are candidates for the dependent aqueous-phase variable in a predominance diagram. To preserve comparability with Fig. 3.5, $(\mathbf{H_4SiO_4^0})$ will be chosen, with the other three activities fixed as before. These choices reduce the general log K equations

$$-8.79 = 2\log(\mathbf{H_4SiO_4^0}) - 5\log(\mathbf{H_2O}) \qquad (3.23a)$$

$$-8.72 = 3.82\log(\mathbf{H_4SiO_4^0}) + 0.653\log(\mathbf{Mg^{2+}}) + 0.335\log(\mathbf{Fe^{3+}}) + 2.31\,\mathrm{pH} - 7.69\log(\mathbf{H_2O}) \qquad (3.23b)$$

$$-2.26 = 2.35\log(\mathbf{H_4SiO_4^0}) + 0.653\log(\mathbf{Mg^{2+}}) + 0.335\log(\mathbf{Fe^{3+}}) + 2.31\,\text{pH} - 4.02\log(\mathbf{H_2O}) \quad (3.23c)$$

to the boundary-line equations

$$\log(\mathbf{H_4SiO_4^0}) = -4.40 \quad \text{(kaolinite-gibbsite)} \quad (3.24a)$$

$$\log(\mathbf{H_4SiO_4^0}) = -0.763 - 0.605\,\text{pH} \quad \text{(smectite-gibbsite)} \quad (3.24b)$$

$$\log(\mathbf{H_4SiO_4^0}) = 1.51 - 0.983\,\text{pH} \quad \text{(smectite-kaolinite)} \quad (3.24c)$$

These equations are plotted as boundary lines in Fig. 3.7 for the range of pH values common in acid soils. At pH 5, the sequence of predominant solid phases that is predicted to occur as the activity of silicic acid changes is in agreement with the sequence predicted in Fig. 3.5. Note that if quartz controls the activity of silicic acid, there is a shift from kaolinite to smectite at pH 5.6 (water equilibrated with atmospheric CO_2). If the kaolinite and gibbsite solubility "windows" are incorporated, it is necessary to reconsider Eq. 3.22 with log K =-8.2, -10.5, and -4.5, respectively. The net effect of these changes will be to enlarge the field of stability of smectite at the expense of kaolinite and gibbsite. Thus, poor crystallinity in these two latter minerals makes the persistence of smectite possible in actively weathering soil profiles.

A pE–pH diagram is a predominance diagram in which the electron activity is the dependent activity variable chosen to plot against pH. Thus the pE value plays the same role as the value of log $(\mathbf{H_2SiO_4^0})$ in Fig. 3.7. The construction of a pE–pH diagram is, accordingly, another example of the construction of a predominance diagram. Differences come because of redox reactions involving only aqueous species and because of the interpretation of the diagram, which is in terms of redox species instead of solids alone. The steps in constructing a pE–pH diagram follow:

1. Establish a set of redox species and obtain values of log K for all reactions between the species.
2. Unless other information is available, set the activities of liquid water and all solid phases equal to 1.0. Set all fixed gas-phase pressures at values appropriate to soil conditions.
3. Develop each expression for log K into a pE–pH relation. In *one* relation involving an aqueous species and a solid phase wherein a change in oxidation number is involved, choose a value for the activity of the aqueous species.
4. In each reaction involving two aqueous species, set the activities of the two species equal.

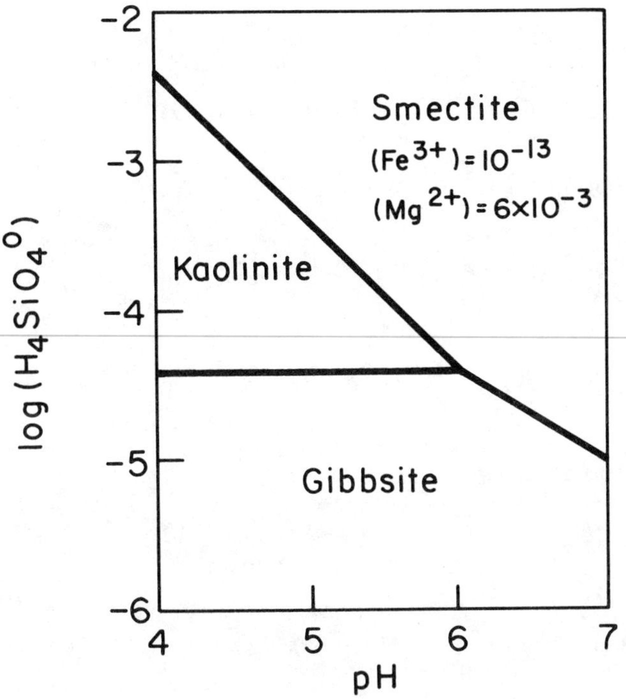

FIG. 3.7. Predominance diagram for the same set of solid phases as in Fig. 3.5. At pH 5, the sequence of stability with increasing silicic acid activity is gibbsite → kaolinite → smectite.

To illustrate the construction of a pE–pH diagram, the Fe redox species, Fe^{3+}(aq), Fe^{2+}(aq), amorphous ferric hydroxide [$Fe(OH)_3$], ferrihydrite ($Fe_{10}O_{15}\bullet 9H_2O$), and hydromagnetite ($Fe_3(OH)_8$) will be considered. Other Fe oxyhydroxides [e.g., goethite, α-FeOOH] are more stable thermodynamically than the three solids just listed, but short-term kinetics data (and the GLO Step Rule) suggest the metastable existence of hydrated, poorly crystalline Fe oxyhydroxide phases in soil systems.[19] The reduction reactions that will be used to define the boundary lines in the pE–pH diagram are the following (cf. Table 2.4):

$$Fe^{3+}(aq) + e^-(aq) = Fe^{2+}(aq) \qquad \log K_R = 13$$

$$pE = 13.0 - \log[(\mathbf{Fe^{3+}})/(\mathbf{Fe^{2+}})] \tag{3.25a}$$

$$Fe(OH)_3(s) + 3H^+(aq) + e^-(aq) = Fe^{2+}(aq) + 3H_2O(\ell) \qquad \log K_R = 16.4$$

$$pE = 16.4 - 3\log(\mathbf{H_2O}) - \log(\mathbf{Fe^{2+}}) - 3\ pH \tag{3.25b}$$

$$\frac{1}{10}\ Fe_{10}O_{15} \cdot 9H_2O(s) + 3H^+(aq) + e^-(aq) = Fe^{2+}(aq) + 2.4H_2O(\ell)$$

$$\log K_R = 17$$

$$pE = 17 - 2.4 \log \mathbf{(H_2O)} - \log \mathbf{(Fe^{2+})} - 3\,pH \tag{3.25c}$$

$$\frac{1}{2}\ Fe_3(OH)_8(s) + 4H^+(aq) + e^-(aq) = \frac{3}{2}\ Fe^{2+}(aq) + 4H_2O(\ell)$$

$$\log K_R = 21.9$$

$$pE = 21.9 - 4\ \log \mathbf{(H_2O)} - 1.5\ \log \mathbf{(Fe^{2+})} - 4\,pH \tag{3.25d}$$

$$Fe(OH)_3(s) = \frac{1}{10}\ Fe_{10}O_{15} \cdot 9H_2O(s) + \frac{3}{5}\ H_2O(\ell)$$

$$\log K = -0.6$$

$$\log K = 0.6\ \log \mathbf{(H_2O)} \tag{3.25e}$$

$$3\ Fe(OH)_3(s) + H^+(aq) + e^-(aq) = Fe_3(OH)_8(s) + H_2O(\ell) \qquad \log K_R = 5.44$$

$$pE = 5.44 - pH \tag{3.25f}$$

$$\frac{3}{10}\ Fe_{10}O_{15} \cdot 9H_2O(s) + H^+(aq) + 0.8H_2O(\ell) + e^-(aq) = Fe_3(OH)_8(s)$$

$$\log K_R = 7.24$$

$$pE = 7.24 + 0.8\ \log \mathbf{(H_2O)} - pH \tag{3.25g}$$

where all solid-phase activities have been set equal to 1.0. For each chemical reaction, the pE–pH relation was obtained by writing down the equilibrium constant in terms of activities [with all solids assumed to be in their Standard States], taking the base 10 logarithm of both sides of the resulting equation and rearranging terms to solve for pE. For example, in order to derive Eq. 3.25d, one writes

$$K = \mathbf{(Fe^{2+})}^{3/2} / \mathbf{(H_2O)}^4 \mathbf{(H^+)}^4 \mathbf{(e^-)} = 10^{21.9}$$

$$\frac{3}{2}\ \log \mathbf{(Fe^{2+})} + 4\ \log \mathbf{(H_2O)} + 4\,pH + pE = \log K = 21.9 \tag{3.26}$$

and Eq. 3.25d follows after moving all terms but pE to the right side.

Reference to step 4, above as applied to Eq. 3.25a indicates that the boundary line separating $Fe^{3+}(aq)$ from Fe^{2+} (aq) in the pE–pH diagram will be a horizontal one at pE = 13, the maximum value observed in soil systems (see Section 2.2). Thus Fe^{3+} (aq) is not a likely redox species and can be excluded from the diagram. Equation 3.25e implies that equilibrium between ferric hydroxide and ferrihydrite is possible only at $\mathbf{(H_2O)} = 0.1$, regardless of pE or pH. If $\mathbf{(H_2O)} > 0.1$, ferrihydrite can also be excluded from the pE–pH diagram. This leaves Eqs. 3.25b, 3.25d, and 3.25f to define the stability fields of the species Fe^{2+} (aq), ferric hydroxide, and hydromagnetite. The resulting pE–pH diagram is shown in Fig. 3.8 for the case **(Fe^{2+})** = 10^{-5} [step 3 above requires a fixed Fe^{2+} (aq) activity in Eqs. 3.25d and 3.25f]. Under oxic and suboxic conditions, ferric hydroxide predominates, whereas under anoxic conditions, either Fe^{2+}(aq) or hydromagnetite is the stable species, depending on pH. Lowering (raising) the value of **(Fe^{2+})** would shift the two lower boundary lines in the diagram upward (downward), since the reductive dissolution of the two solid phases is thereby enhanced (inhibited).

3.3 Mixed Solid Phases

In soils, the solid phases typically are mixtures of chemical compounds. The generic term used to describe the processes by which these mixtures form is *coprecipitation*, the simultaneous precipitation of a compound in conjunction with other compounds by any mechanism at any rate. Three broad categories of coprecipitation phenomena have been identified in soils: *mixed solid formation, adsorption,* and *inclusion.*

Mixed solid (or *solid solution*) formation can occur, for example, when secondary minerals, such as carbonates, metal oxides, or clay minerals, precipitate from the soil solution during weathering. These solids often are characterized by a wide range of isomorphic substitutions, in which both cations

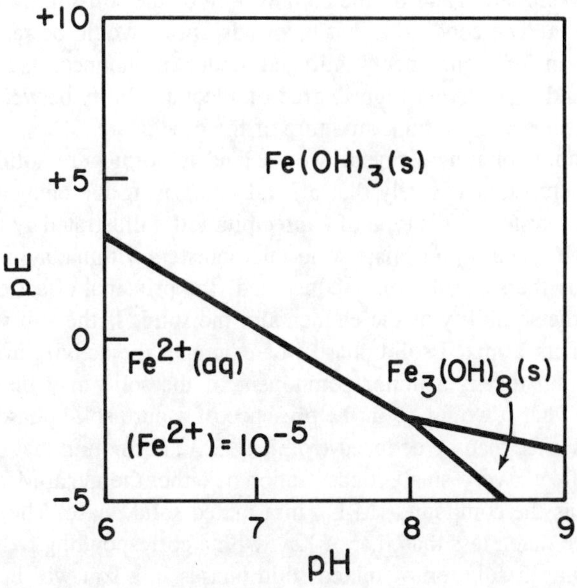

FIG. 3.8. A pE–pH diagram for Fe redox species based on Eq. 3.25. The activity of Fe^{2+}(aq) is fixed at 10^{-5}.

and anions in their structures may be replaced by ions of the same charge sign and comparable size.[20] Specific examples of this phenomenon occur during the formation of clay minerals, if metals replace Si in the tetrahedral sheet or Al in the octahedral sheet; during the formation of calcite, if Mg, Sr, Fe, Mn, or Na replaces Ca; and during the formation of hydroxyapatite [$Ca_5OH(PO_4)_3$], if Ca is replaced by Sr, or other metals, or if OH is replaced by F or other ligands. In coprecipitation through the formation of a mixed solid phase, the resulting solid is a homogeneous mass with its minor substituents distributed uniformly. Thus the basic requirements for this type of coprecipitation are the free diffusion and relatively high structural compatibility of the minor substituents with the precipitate as it is forming. These conditions often are met as well when minerals precipitate from a silicate melt to form the parent material of soils; feldspars and micas are well-known examples of primary-mineral mixed solid phases.[20]

If free diffusion within a precipitate on weathering time scales is not possible for a given cation or anion, that ion may still coprecipitate by an adsorption process. For example, if an Fe or Al oxide were precipitating in a soil solution to form colloidal material with a relatively large surface area and if the pH value were in the alkaline range, metal cations could adsorb on the surface of the precipitate. If these coprecipitated metals were not able to diffuse freely into the bulk solid, the development of a solid solution would be prevented. The same phenomenon could occur in the acidic pH range for an oxyanion in the soil solution, such as HPO_4^{2-} or SO_4^{2-}. Generally, it can be expected that coprecipitation through adsorption will be more dependent on the kinetics of precipitation and on the composition of the soil solution than mixed solid formation. The conditions that favor adsorption would be rapid precipitation of the major component, initiated under pronounced supersaturation conditions, and a relatively high degree of incompatibility between the minor adsorbing species and the bulk structure of the precipitate.

If the coprecipitating elements would tend to form pure solids with very different structures, it is likely that a solid comprising the elements will be a heterogeneous system. This type of coprecipitation is illustrated by the inclusion of separate TiO_2 and *o*-phosphate solid microclusters within clay minerals.

Insofar as the soil solution is concerned, the principal effect of coprecipitation is on the solubility of the elements in the solid. If the soil solution is in equilibrium with a mixed solid phase, for example, the activity in the aqueous phase of an ion that is a minor component of the solid may be significantly smaller than what it would be in the presence of a pure solid phase comprising the ion. This effect can be deduced from Eq. 3.3 after noting that $(\mathbf{M_aL_b}) << 1.0$ could reflect a very small concentration of either the metal M or the ligand L occurring as the compound, M_aL_b, in a mixed solid phase. The value of K_{so} would thus be much less than that of K_{dis}, with a corresponding reduction in the IAP. Often the dissolution of mixed solid phases in a soil will be dictated by complicated kinetic considerations, and the prediction of the composition of the soil solution as influenced by those solid phases will be quite difficult. On the other hand, if chemical equilibrium exists between solid and aqueous phases, or even if it is desired only to have a general understanding of reaction pathways, a thermodynamic description of a dissolving solid solution can be valuable.

Consider a solid solution formed because of the coprecipitation of two metal cations, M^{m+} and N^{n+}, with a ligand $L^{\ell-}$. The components ("end members") of the solid solution are the compounds M_aL_b and N_cL_d(s) where am = ℓb and cn = ℓd to ensure electroneutrality. For each component of the mixed solid, an expression like Eq. 3.2 can be developed:

$$(M^{m+})^a\,(L^{\ell-})^b = K_M(M_aL_b) \quad (3.27a)$$

$$(N^{n+})^c\,(L^{\ell-})^d = K_N(N_cL_d) \quad (3.27b)$$

where K_M and K_N are equilibrium constants for the dissolution of the two pure solid phases. If it is *assumed* that the mixed solid dissolves congruently while retaining a constant composition, its dissolution reaction can be expressed analogously to Eq. 3.1:

$$M_{a(1-x)}N_{cx}L_{b(1-x)+dx}\,(s) = a(1-x)M^{m+}(aq) + cxN^{n+}(aq) + [b(1-x) + dx]L^{\ell-}(aq) \quad (3.28)$$

where x is the mole fraction of N_cL_d (taken to be the minor component) in the mixed solid. This reaction describes a dissolution equilibrium state known as *stoichiometric saturation*.[21] This state is possible if the time scales for changes in the composition of the dissolving solid and for precipitation of any secondary solid phase (i.e., incongruent dissolution) are much longer than that for the congruent dissolution of the mixed solid. Its existence must be established experimentally.[22] If Eq. 3.28 is applicable, a corresponding dissolution equilibrium constant can be as follows:[23]

$$K_{dis}^{ss} \equiv \frac{(M^{m+})^{a(1-x)}\,(N^{n+})^{cx}\,(L^{\ell-})^{b(1-x)+dx}}{(M_{a(1-x)}N_{cx}L_{b(1-x)+dx})} \quad (3.29)$$

and a solubility product constant, K_{so}^{ss}, can be defined as the product of K_{dis}^{ss} with the activity of the mixed solid, analogous to Eq. 3.3. This approach treats the mixed solid as if it were a single phase whose activity under Standard-State T and P has unit value (cf. Table s1.1 in Chapter 1). So that this condition is defined precisely, the Standard-State chemical potential of the mixed solid can be expressed as follows:[23]

$$\mu_{ss}^0 \equiv (1-x)\mu_{ML} + x\mu_{NL} \quad (3.30)$$

where μ_{ML} and μ_{NL} are, respectively, the chemical potentials of the two component solids, M_aL_b and N_cL_d, in the mixture. Equation 3.30 permits the value of K_{dis}^{ss} to be calculated in the usual way with Standard-State chemical potentials (see Special Topic 1 in Chapter 1):

$$K_{dis}^{ss} \equiv \frac{1}{5.708}\,[\mu_{ss}^{0} - a(1-x)\mu^{0}(M^{m+}) - cx\,\mu^{0}(N^{n+})$$
$$- (b(1-x) + cx)\mu^{0}(L^{\ell-})] \qquad (T = 298.15\ K) \qquad (3.31)$$

Note that the value of K_{dis}^{ss} depends on the mole fraction x, and therefore *will vary with the composition of the mixed solid*. Unlike K_{dis} for a pure solid phase, K_{dis}^{ss} must be measured as a function of mixed solid composition in order to apply Eq. 3.29 to all possible states of stoichiometric saturation.

An important constraint on the composition dependence of K_{dis}^{ss} derives from the combination of Eqs. 3.27 and 3.29 applied to the mixed solid at fixed T and P:

$$K_{so}^{ss} = [K_M(\mathbf{M_aL_b})]^{1-x}[K_N(\mathbf{N_cL_d})]^{x} \qquad (3.32)$$

under the assumption that the Standard State for the mixed solid is its *actual* composition at T^0 and P^0 (i.e., unit mole fraction of the *mixture* and therefore unit activity of the mixed solid). Infinitesimal variations in ln K_{so}^{ss} produced by an infinitesimal change in the composition of the mixed solid are constrained by Eq. 3.32 to have the form

$$d\ln K_{so}^{ss} = (1-x)\,d\ln(\mathbf{M_aL_b}) - \ln(\mathbf{M_aL_b})\,dx$$
$$- \ln K_M dx + x\,d\ln(\mathbf{N_cL_d}) + \ln(\mathbf{N_cL_d})\,dx + \ln K_N\,dx \qquad (3.33a)$$

where the differentials of the solid-phase component activities are understood to result from an infinitesimal change in the mole fraction x. A second equation that constrains the variation of these two activities is the *Gibbs-Duhem equation* (at fixed T^0 and P^0):[24]

$$(1-x)\,d\ln(\mathbf{M_aL_b}) + x\,d\ln(\mathbf{N_cL_d}) = 0 \qquad (3.33b)$$

This thermodynamic constraint arises because the intensive properties that describe an equilibrium state of a single phase (e.g., T, P, and component activities) cannot all be varied independently. Equation 3.33 can be written in the compact form:

$$d\ln K_{so}^{ss} = [\ln(\mathbf{N_cL_d}) + \ln K_N - \ln(\mathbf{M_aL_b}) - \ln K_M]\,dx \qquad (3.34)$$

According to Eq. 3.32, the coefficient of dx on the right side of Eq. 3.34 is the same as $(1/x)[\ln K_{so}^{ss} - \ln K_M - \ln(\mathbf{M_aL_b})]$, so that[21,23]

$$\ln (M_aL_b) = -x\left(\frac{\partial \ln K^{ss}_{so}}{\partial x}\right)_{T,P} + \ln K^{ss}_{so} - \ln K_M \tag{3.35a}$$

upon expressing $d\ln K^{ss}_{so}$ directly in terms of dx. A similar equation can be derived for the second activity variable:[21,23]

$$\ln(N_cL_d) = (1 - x)\left(\frac{\partial \ln K^{ss}_{so}}{\partial x}\right)_{T,P} + \ln K^{ss}_{so} - \ln K_N \tag{3.35b}$$

Equation 3.35 shows that the activities of the solid-phase components can be calculated once the dependence of K^{ss}_{so} (as given by the numerator in Eq. 3.29) on the mole fraction, x, has been measured. The two activities then can be used to compute, for example, μ^0_{ss} from Eqs. s1.10 (Chapter 1) and 3.30.

An application of Eq. 3.35 to magnesian calcite is shown in Fig. 3.9.[22] The component solid phases in this case were chosen to be calcite (M = Ca, L = CO_3; a = b = 1) and dolomite (N = CaMg, L = CO_3; c = 0.5, d = 1). [It is equally possible to choose calcite and magnesite (N = Mg, L = CO_3, c = d = 1).] With the first-mentioned choice, Eq. 3.28 becomes

$$Ca_{1-x}(Ca_{0.5}Mg_{0.5})_xCO_3(s) = (1 - \tfrac{1}{2}x)Ca^{2+}(aq) + \tfrac{1}{2}x\,Mg^{2+}(aq) + CO_3^{2-}(aq) \tag{3.36}$$

for which

$$K^{ss}_{so} = (Ca^{2+})^{1-\frac{1}{2}x}(Mg^{2+})^{x/2}(CO_3^{2-}) \tag{3.37}$$

is the solubility product constant. The values of log K_{dis} for the two component solids are log K_{cal} = −8.48 and log K_{dol} = 8.545.[22] The solid-phase component, $N_cL_d \equiv Ca_{0.5}Mg_{0.5}CO_3$, represents half the unit-cell formula of dolomite but is perfectly acceptable in the macroscopic way that Eqs. 3.27–3.35 are applied to dissolution equilibria. Experimental measurements on the x-dependence of K^{ss}_{so} were fit by least-squares methods to the equation:[22]

$$\log K^{ss}_{so} = x(1 - x)\,[2.21 \pm 0.04 + 0.82 \pm 0.18\,(2x - 1)] + (1 - x)\log[K_{cal}(1 - x)] + x\log[K_{dol}\,x] \tag{3.38}$$

which then could be used to compute the first two terms on the right side of Eq. 3.35. [Note that base 10 logarithms (log) appear in Eq. 3.38, whereas base e logarithms (ln) appear in Eq. 3.35.] In this example, the component activities sometimes exceed the value 1.0, and the right side of Eq. 3.27 thus must be larger than K_{dis} for the pure solid component alone (except when the mole fraction of a component approaches zero in the mixed solid). This behavior leads to an IAP value for the solid component that generally exceeds the value of K_{so} for the corresponding pure solid phase.

Solid solutions of diaspore [α-AlOOH] and goethite [α-FeOOH] are commonly observed in highly weathered soils, with the value of x in the mixed solid, $Fe_{1-x}Al_xOOH$, ranging up to 0.33.[19,25] By analogy with Eq. 3.20, the dissolution reactions of the two solid-phase components can be expressed in the following form:

$$FeOOH(s) + 3H^+(aq) = Fe^{3+}(aq) + 2H_2O(\ell) \qquad \log K_{dis} = -1.69 \tag{3.39a}$$

$$AlOOH(s) + 3H^+(aq) = Al^{3+}(aq) + 2H_2O(\ell) \qquad \log K_{dis} = 7.58 \tag{3.39b}$$

These reactions do not quite fit into the format of Eq. 3.27, but they can be adapted to it *formally* by setting $\mathbf{(H_2O)}$ = 1.0 and $\mathbf{(L^{\ell-})}^b = \mathbf{(L^{\ell-})}^d \equiv \mathbf{(H^+)}^{-3}$:

$$\mathbf{(Fe^{3+})}\,\mathbf{(H^+)}^{-3} = 10^{-1.69}\,\mathbf{(FeOOH)} \tag{3.40a}$$

$$\mathbf{(Al^{3+})}\mathbf{(H^+)}^{-3} = 10^{7.58}\,\mathbf{(AlOOH)} \tag{3.40b}$$

Thus $\mathbf{(H^+)}^{-1}$ [which is equivalent to $\mathbf{(OH^-)}$ at unit water activity] plays the role

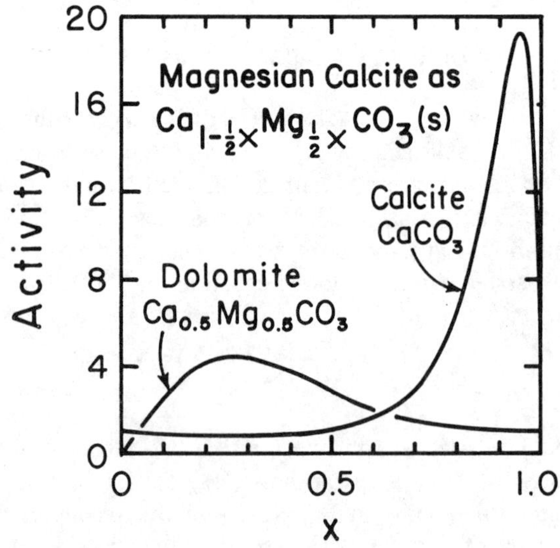

FIG. 3.9. Activities of the dolomite and calcite components of a magnesian calcite, expressed as functions of the mole fraction of the dolomite component.[22]

of an aqueous "ligand" activity. Equation 3.28 takes the following form:

$$Fe_{1-x}Al_xO(OH)(s) + 3H^+(aq) = (1 - x)Fe^{3+}(aq) + x\,Al^{3+}(aq) + 2H_2O(\ell) \quad (3.41)$$

with the mixed-solid solubility product constant

$$K_{so}^{ss} = (\mathbf{Fe^{3+}})^{1-x}(\mathbf{Al^{3+}})^x(\mathbf{H^+})^{-3} \quad (3.42)$$

As in the example of magnesian calcite, measurement of the IAP for Al-goethite as a function of the mole fraction of Al would permit the calculation of the activities of the "end-member" solid phases, goethite and diaspore, in the mixed solid (Eq. 3.35). Alternatively, simply by way of an illustration, it may be assumed that the two solid phases mix to form an *ideal solid solution,* which, by definition,[24] exhibits solid-phase activities always equal to the respective mole fractions of the components: **(FeOOH)** = 1 - x and **(AlOOH)** = x. Under this simplifying assumption, Eq. 3.34 can be integrated directly, with the following result:

$$\ln K_{so}^{idss} = (1 - x)\ln[K_M(1 - x)] + x\ln[K_N x] \quad (3.43)$$

This equation also can be derived from Eq. 3.32 after incorporating the definition of ideal mixing. For an ideal Al-goethite, Eqs. 3.39 and 3.43 yield the solubility product constant:

$$\log K_{so}^{idss} = (1 - x)\log[10^{-1.69}(1 - x)] = x\log[10^{7.58}x] \quad (3.44)$$

The effect of Al-goethite on Al solubility can be illustrated through a reconsideration of the activity-ratio diagram in Fig. 3.5, but with the system simplified to comprise only kaolinite and gibbsite in addition to *ideal* Al-goethite. For $(\mathbf{H_2O})$ = 1.0, Eqs. 3.21b and 3.21c yield the activity ratios for kaolinite and gibbsite:

$$\log[(\mathbf{kaolinite})/(\mathbf{Al^{3+}})] = -3.72 + 3\,pH - \log(\mathbf{H_4SiO_4^0}) \quad (3.45a)$$

$$\log[(\mathbf{gibbsite})/(\mathbf{Al^{3+}})] = -8.11 + 3\,pH \quad (3.45b)$$

Equation 3.40b can be developed in the same form as these two expressions after dividing both sides by **(AlOOH)***, where the asterisk refers to *pure* diaspore, and then setting **(AlOOH)/(AlOOH)*** equal to x:

$$\log[(\mathbf{diaspore})^*/(\mathbf{Al^{3+}})] = -7.58 + 3\,pH - \log x \quad (3.45c)$$

Inspection of Eq. 3.45 shows that the relative stabilities of the three solids will not depend on pH. Like gibbsite, the activity ratio for Al-goethite will plot as a horizontal line in a diagram similar to that in Fig. 3.5. Since $\log x < 0$, it satisfies the inequality $-7.58 + \log x > -8.11$, and ideal Al-goethite exhibits a larger activity ratio than gibbsite at any Al mole fraction. Thus ideal Al-goethite will be more stable than gibbsite, regardless of the activity of silicic acid or the pH value. The same conclusion would hold for ideal Al-goethite relative to kaolinite if $\log x < -3.86 - \log (\mathbf{H_4SiO_4^0})$, according to Eqs. 3.45a and 3.45c [with $(\mathbf{H_4SIO_4^0})$ assumed to be small enough to avoid competition from smectite]. At its maximum, $x = 0.33$. Then $(\mathbf{H_4SiO_4^0}) < 10^{-3.38}$ would be required to ensure Al-goethite stability against kaolinite. The largest reasonable value of $(\mathbf{H_4SiO_4^0})$ under soil weathering conditions is around $10^{-2.5}$, from which it follows that any $x < 0.04$ would result in Al-goethite stability over kaolinite. These conclusions, of course, refer to *ideal* Al-goethite and should be taken as illustrative, not predictive.

3.4 Reductive Dissolution Reactions

The dissolution reactions of Fe oxyhydroxides given in Eqs. 3.25b–d are more than special examples of the reaction in Eq. 3.1 because they involve *electron transfer*. As part of the process by which the solid phase dissolves, Fe (III) in the solid-phase reactant is transformed to Fe(II) in the aqueous solution phase. This change in oxidation state of the principal metal in the solid phase qualifies the processes described in Eqs. 3.25b–d as *reductive dissolution reactions*. Other examples of these reactions are given in Table 2.4, which lists reductive dissolution reactions for MnO_2 (row 7), goethite (row 10), magnetite (row 11), and hematite (row 12). In addition to these minerals, one could add oxyhydroxides of Cr or V, along with silicates or sulfides containing Fe(II), as candidates for *oxidative* dissolution reactions, in which aqueous Fe(III) species are the product.[26]

Equilibria involving reductive dissolution reactions add to the complexity of mineral solubility phenomena in just the way that pE–pH diagrams are more complicated than ordinary predominance diagrams, like that in Fig. 3.7. The electron activity or pE value becomes one of the "master variables" whose influence on dissolution reactions must be evaluated in tandem with other intensive "master variables," like pH or $p(\mathbf{H_4SiO_4^0})$. Moreover, the status of microbial catalysis under the suboxic conditions that facilitate changes in the oxidation states of transition metals has to be considered in formulating a thermodynamic description of reductive dissolution. This consideration is connected closely to the existence of labile organic matter and, in some cases, to the availability of photons.[26]

The kinetics of reductive dissolution reactions is made complicated (relative to the conceptual picture developed in Section 3.1) by electron transfer processes, similar to the way in which these processes bring complexity to reductive dissolution equilibria. In the absence of microbial catalysis, organic ligands (e.g., humic substances[27]) can serve as electron donors to reduce metals in oxyhydroxide solid phases.[26] Evidently, these metals would be bound into the mineral structure near its interface with a soil solution that contains the reductant ligand as an aqueous species. If the metal and ligand can form an inner-sphere complex, electron transfer would be facilitated, with a reduced metal and oxidized ligand as the resulting products. An *overall* reaction scheme to describe this process follows.[26,28]

$$\equiv M_{ox} - OH(s) + L_{Red}(aq) + H^{+}(aq) \underset{k_b}{\overset{k_f}{\rightleftharpoons}} \equiv M_{ox} - L_{Red}(s) + H_2O(\ell)$$

$$\equiv M_{ox} - L_{Red}(s) \overset{k_f'}{\rightarrow} \equiv M_{Red} - L_{ox}(s) \qquad (3.46)$$

This pair of coupled reactions depicts first a proton-promoted ligand exchange between a hydroxyl group that is bound to a metal in the solid structure (denoted $\equiv$ M) and is exposed at the mineral–soil solution interface, and a reductant organic ligand (denoted L) in the aqueous solution phase. Electron transfer then follows between the complexed ligand and the metal to yield a reduced species of the latter (denoted M_{Red}) and an oxidized ligand (denoted L_{ox}). The reaction sequence in Eq. 3.46 is analogous to that in Eq. 2.34, in that an inner-sphere complex formation between the metal and ligand is proposed to mediate electron transfer with negligible likelihood of dissociation back-reaction ($k_b' = 0$). At fixed proton and water concentrations, the coupled reactions in Eq. 3.45 can be regarded as a realization of the abstract reaction scheme in Eq. 1.52 ($A = \equiv M_{ox} - OH$; $B = L_{Red}$; $C = \equiv M_{ox} - L_{Red}$; $D = \equiv M_{Red} - L_{ox}$; E left undesignated). If the simplifying assumptions leading to the rate laws in Eq. 1.54 also are made, the kinetics of formation of the complex, $\equiv M_{ox} - L_{Red}$, can be described mathematically.[28]

Applied to Eq. 3.46 at fixed [H^+] and [H_2O], Eq. 1.54b takes the form

$$\frac{d[\equiv M_{ox} - L_{Red}]}{dt} = K_f[\equiv M_{ox} - OH][L_{Red}] - (K_b + k_f')[\equiv M_{ox} - L_{Red}] \qquad (3.47)$$

where $K_f = k_f[H^+]$ is a pseudo-second-order rate coefficient and $K_b = k_b\,[H_2O]$ is a pseudo-first-order rate coefficient. This rate law can be simplified further if the assumptions of a fixed [L_{Red}] and a very rapid (on the time scale of Eq. 3.47) transformation of L_{ox} in the complex, $\equiv M_{Red} - L_{ox}(s)$, to aqueous solution species are made.[26,28] Under these conditions, the pseudo-first-order rate constant $K_f^* = k_f[H^+][L_{Red}]$ can be introduced, and [$\equiv M_{ox}$ - OH] can be replaced by the simple mole-balance term $\{M_{oxT} - [\equiv M_{ox} - L_{Red}]\}$, where $M_{oxT} = [\equiv M_{ox} - OH] + [\equiv M_{ox} - L_{Red}]$ is the total (fixed) concentration of oxidized metal that can react with the reductant ligand. These substitutions put Eq. 3.47 into the same form as Eq. 1.55 ($c_A \Leftrightarrow [\equiv M_{ox} - L_{Red}]$; $k_b \Leftrightarrow K_f^*$; etc.):

$$\frac{d[\equiv M_{ox} - L_{Red}]}{dt} = -(K_f^* + K_b + k_f')[\equiv M_{ox} - L_{Red}] + K_f^* M_{oxT} \tag{3.48}$$

whose mathematical solution is then given by Eq. 1.56:

$$[\equiv M_{ox} - L_{Red}] = \frac{K_f^* M_{oxT}}{(K_f^* + K_b + k_f')} + \left\{[\equiv M_{ox} - L_{Red}]_0 - \frac{K_f^* M_{oxT}}{(K_f^* + K_b + k_f')}\right\} \exp[-(K_f^* + K_b + k_f')t] \tag{3.49}$$

where $[\equiv M_{ox} - L_{Red}]_0$ is the initial value of $[\equiv M_{ox} - L_{Red}]$. Equation 3.49 predicts the exponential change with time that is characteristic of pseudo-first-order reactions. Given the assumption of very rapid transformation of the complex, $\equiv M_{Red} - L_{ox}(s)$, in Eq. 3.46 to aqueous species, the right side of Eq. 3.49 also describes the time dependence of the rate of reductive dissolution of the solid phase containing M_{ox}.[28]

There are three time scales implicit in Eq. 3.48 because of the appearance of three first-order rate constants.[28] The time scale[29] for ligand complexation (i.e., adsorption) by M_{ox} (at fixed aqueous ligand concentration and pH) is $0.693/K_f^*$, whereas that for dissociation of the complex, $\equiv M_{ox} - L_{Red}(s)$, is $0.693/K_b$, and that for electron transfer is $0.693/k_f'$. Note that only the first of these three time scales depends explicitly on $[L_{Red}]$ and pH. The longest time scale (smallest rate constant) identifies the rate-limiting reaction of the scheme in Eq. 3.46. Under the condition, $t >> (K_f^* + K_b + k_f')^{-1}$, the concentration of $[\equiv M_{ox} - L_{Red}]$ achieves its equilibrium value irrespective of the rate-limiting reaction:

$$[\equiv M_{ox} - L_{Red}]_e = \frac{K_f^* M_{oxT}}{K_f^* + K_b + k_f'} \tag{3.50a}$$

as follows from Eq. 3.49 as $t \uparrow \infty$. This result can be written in the alternative form

$$[\equiv M_{ox} - L_{Red}]_e = M_{oxT} \frac{K_L[L_{Red}]_e}{1 + K_L[L_{Red}]_e} \tag{3.50b}$$

where

$$K_L = K_f/(K_b + k_f') \tag{3.51}$$

is a *Langmuir affinity parameter* and Eq. 3.50b is the *Langmuir adsorption isotherm equation*.[30] Equation 3.50b describes the way in which the concentration of adsorbed L_{Red} depends on the concentration of L_{Red} in aqueous solution (at fixed pH) when the rate of change of the former concentration is zero (cf. Eqs. 3.48 and 3.50b). Measured values of $[\equiv M_{ox} - L_{Red}]_e$ for selected values of $[L_{Red}]$ then can be used to determine the parameters M_{oxT} and K_L by fitting the data to a linear form of Eq. 3.50b:

$$\frac{[\equiv M_{ox} - L_{Red}]_e}{[L_{Red}]_e} = M_{oxT} K_L - K_L [\equiv M_{ox} - L_{Red}]_e \qquad (3.50c)$$

Equation 3.50c represents a line whose slope is $-K_L$ and whose x-intercept (left side equals zero) is M_{oxT}.

Given the assumption that the adsorbed species, $\equiv M_{Red} - L_{ox}(s)$, transforms very rapidly to aqueous species on the time scale of Eq. 3.48, it follows that $[\equiv M_{Red} - L_{ox}]$ will reach a constant value very rapidly and therefore that the rate of dissolution of the solid phase containing M_{ox} will be proportional to the rate at which $\equiv M_{Red} - L_{ox}(s)$ forms according to the second reaction in Eq. 3.46. This rate, in turn, is proportional to the concentration of the adsorbed species, $\equiv M_{ox} - L_{Red}(s)$ (cf. Eq. 1.54c for the case $k_b' = 0$). Therefore

$$\frac{d[M_{Red}]}{dt} = k[\equiv M_{ox} - L_{Red}] \qquad (3.52)$$

is the rate law for reductive dissolution under this condition, where k is a first-order rate coefficient and the concentration factor on the right side is given by Eq. 3.49. When $t >> (K_f^* + K_b + k_f')^{-1}$, the reductive dissolution rate becomes constant but dependent on pH and $[L_{Red}]$, as implied by Eq. 3.50b.

These concepts are illustrated in Fig. 3.10 for the reductive dissolution of hematite (α-Fe_2O_3) in the presence of ascorbic acid at pH 3.[26] In this example, M_{ox} = Fe(III), M_{Red} = Fe(II), and $L_{Red} = HA^-$, where A^{2-} is the ascorbate anion (log K = −4 for the dissociation of H_2A^0, but dissociation is invoked nonetheless to promote a ligand-exchange reaction). Equation 3.46 becomes

$$\equiv Fe^{III} - OH(s) + HA^-(aq) + H^+(aq) \underset{k_b}{\overset{k_f}{\rightleftharpoons}} \equiv Fe^{III} - HA(s) + H_2O(\ell)$$

$$\equiv Fe^{III} - HA(s) \xrightarrow{k_f'} \equiv Fe^{II} - HA^+(s) \qquad (3.53)$$

and Eq. 3.52 becomes

$$\frac{d[Fe(II)]}{dt} = k[\equiv Fe^{III} - HA] \qquad (3.54)$$

where

$$[\equiv Fe^{III} - HA]_e = Fe_T^{III} \frac{K_L[HA^-]_e}{1 + K_L[HA^-]_e} \tag{3.55}$$

is the equilibrium concentration of adsorbed ascorbic acid. The graph at the left in Fig. 3.10 shows measured values of the rate, d[Fe(II)]/dt (which was observed to be constant for $t > 1h$), divided by $a_s c_s$ and plotted against the total ascorbic acid concentration in aqueous solution. A concentration dependence qualitatively consistent with the right side of Eq. 3.55 is evident. The right side of Fig. 3.10 shows a plot of the dissolution rate against the concentration of adsorbed ascorbic acid that is consistent with Eq. 3.54. These data cannot be interpreted uniquely in terms of the illustrative reactions in Eq. 3.53 and the rate law in Eq. 3.54, however, unless measured ascorbic acid concentrations on the graphs in Fig. 3.10 refer directly to individual species of the organic ligand instead of its total aqueous or adsorbed concentrations. Suppose, as an example, it were assumed that the adsorbed species, $\equiv M_{Red} - L_{ox}(s)$, does not transform rapidly to aqueous solution species but is, in fact, detached very slowly on the time scale of Eq. 3.47. Then the concentration of this species would appear on the right side of Eq. 3.52 and, correspondingly, in Eq. 3.54. Quantitation of the total concentration of adsorbed ascorbic acid would not serve to distinguish this substitute rate law from that in Eq. 3.54. Alternative reactions and their rate laws must be tested by *species* quantitation.[31]

Reductive dissolution of minerals also can occur as a *photochemical* reaction in the presence of organic ligands.[32] An example of this light-mediated process is described briefly for hematite in the presence of oxalic acid in Eq. 1.9 and

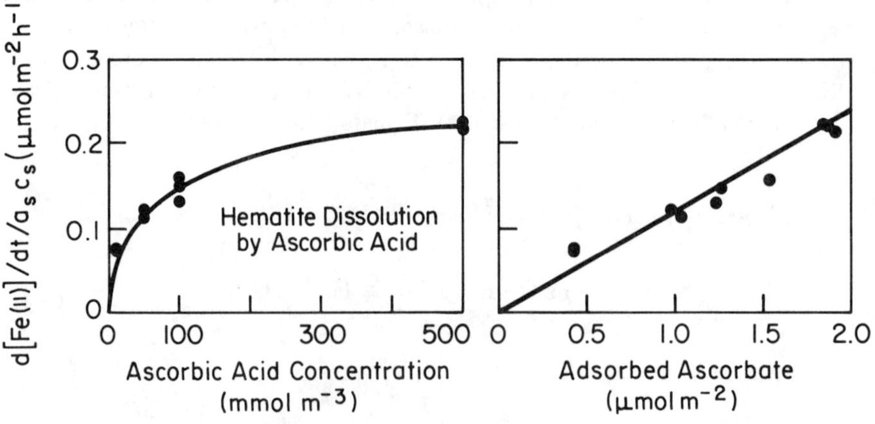

FIG. 3.10. Illustrations of Eqs. 3.54 and 3.55 for hematite dissolution [measured as the rate of soluble Fe(II) production] in the presence of ascorbic acid at pH 3.[26]

Fig. 1.2. An *overall* reaction sequence analogous to that in Eq. 3.46 can be developed from Fig. 1.2:[32]

$$Fe^{III}-OH(s) + C_2O_4^{2-}(aq) + H^+(aq) \underset{k_b}{\overset{k_f}{\rightleftharpoons}} \equiv Fe^{III}-C_2O_4^-(s) + H_2O($$

$$\equiv Fe^{III}-C_2O_4^-(s) \underset{h\upsilon}{\overset{k_f'}{\rightarrow}} \equiv Fe^{II}-\{C_2O_4^-\}_{ox} \qquad (3.56)$$

where $h\upsilon$ represents excitation of electron transfer by a photon of visible light to produce Fe(II) and the oxidized, adsorbed oxalate ligand, $\{C_2O_4^-\}_{ox}$. This illustrative reaction sequence is a special case of Eq. 1.52, if the concentrations of water and protons are constant, and Eq. 1.54 can be applied to derive rate laws for the Fe species ($c_A \Leftrightarrow [\equiv Fe^{III} - OH]$; $c_C \Leftrightarrow [\equiv Fe^{III} - C_2O_4^-]$; $c_D \Leftrightarrow [\equiv Fe^{II} - \{C_2O_4^-\}_{ox}]$ in Eq. 1.54). The rate of formation of the species, $\equiv Fe^{III} - C_2O_4^-$, is, by Eq. 1.54b, with $k_b' = 0$, as follows:

$$\frac{d[\equiv Fe^{III} - C_2O_4^-]}{dt} = K_f[\equiv Fe^{III} - OH][C_2O_4^{2-}] - (K_b + k_f')[\equiv Fe^{III} - C_2O_4^-] \qquad (3.57)$$

where K_f and K_b are pseudo-first-order rate coefficients pathways defined as in Eq. 3.47. Equations like Eqs. 3.48–3.51 also can be derived pertaining to the species, $\equiv Fe^{III} - C_2O_4^-$, if the conversion of $\equiv Fe^{II} - \{C_2O_4\}_{ox}^-$ to aqueous species is assumed to be very rapid on the time scale of Eq. 3.57. In general, however, the dissolution rate, as measured by d[Fe(II)]/dt in the aqueous solution phase, will depend on the reaction that is assumed specifically to be rate-limiting:[28,32]

$$\frac{d[Fe(II)]}{dt} = \begin{cases} k[\equiv Fe^{III} - C_2O_4^-] & (\text{formation of} \equiv Fe^{II} - \{C_2O_4^{2-}\}_{ox}) \\ k[\equiv Fe^{II} - \{C_2O_4\}_{ox}^-] & (\text{detachment of} \equiv Fe^{II} - \{C_2O_4\}_{ox}^-) \end{cases} \qquad (3.58)$$

Quantitation of surface oxalate species is required to test Eq. 3.58.

3.5 Dissolution Reaction Mechanisms

The chemical reaction for mineral dissolution in Eq. 3.1 (the forward reaction) represents the stoichiometric decomposition of a binary solid compound into aqueous ionic species. It is an *overall* reaction based on a chemical formula for the solid phase and the hypothesis that free ionic species in aqueous solution will be created in proportion to their stoichiometry in the solid for at least some time during the dissolution process (congruent dissolution). Implicit in this hypothesis are the notions that the composition of the dissolving solid is not affected by the creation of the aqueous species and that these species do not react to form new solid phases (incongruent dissolution). The status of Eq. 3.1 as an overall reaction is not affected, however, by the possibility of other aqueous species than free ions (solvation complexes) as products of decomposition, as long as species equilibration in the aqueous solution phase is completed on a time scale much smaller than that for the dissolution process. It is this condition, in fact, that enables the use of the IAP (Eq. 3.4) as a generic probe of dissolution kinetics.

The mechanistic significance of Eq. 3.1, despite its great utility in mineral solubility studies, is, on the other hand, almost nil. No implication of reaction pathways, other than the postulate of stoichiometric control of aqueous-phase products, can be made on the basis of an overall reaction alone, since it features only enough species to satisfy minimal equilibrium criteria. Any number of additional kinetic species can intervene to govern the reaction pathways, which may be parallel, sequential, or a combination of these two, and the detailed interactions of the solid phase with the aqueous phase are often unlikely to be represented accurately by a spontaneous decomposition reaction like Eq. 3.1. As a case in point, the dissolution reaction of calcite (the forward reaction in Eq. 3.14) may be considered. Given the existence of protons and carbonate species in the aqueous-solution phase, at least three overall reactions more specific than Eq. 3.14 can be postulated to epitomize the detailed interactions of calcite with aqueous species:[7,33,34]

$$CaCO_3(s) + H_2O(\ell) \overset{k_1}{\rightarrow} Ca^{2+}(aq) + HCO_3^-(aq) + OH^-(aq) \qquad (3.59a)$$

$$CaCO_3(s) + H_2CO_3^0(aq) \overset{k_2}{\rightarrow} Ca^{2+}(aq) + 2\,HCO_3^-(aq) \qquad (3.59b)$$

$$CaCO_3(s) + H^+(aq) \overset{k_3}{\rightarrow} Ca^{2+}(aq) + HCO_3^-(aq) \qquad (3.59c)$$

The first of these reactions is a hydrolysis process, the second is a carbonic acid–promoted dissolution, and the third is a proton-promoted dissolution. Equations 3.59b and 3.59c are the forward reactions in Eqs. 3.17 and 3.15, respectively. They provide a mechanistic underpinning for the dependence of k_d in Eq. 3.14 on pH or p_{CO_2}, as discussed in Section 3.1. Indeed, if Eq. 3.7 is applied to the forward reaction in Eq. 3.14 and rate laws for Eq. 3.59 are developed consistently with the hypothesis leading to Eq. 3.7, the result is[7,33,34]

$$\frac{d[Ca^{2+}]}{dt} = k_d = k_1[H_2O] + k_2[H_2CO_3^0] + k_3[H^+] \tag{3.60}$$

where the simple *additivity* of *parallel* reactions (cf. Eq. 2.6 in Section 2.1) has been invoked to obtain the second equality. Equation 3.60 can be evaluated by conducting dissolution experiments under varying pH, p_{CO_2}, and **(H_2O)** conditions.[7,33,34] The first-order rate coefficients on the right side will take on differing values in different experiments that reflect variable solid-phase structure and mechanistic details that are not explicit in Eq. 3.59.[7,34] For example, the proton-promoted dissolution of calcite could involve, as a first step, the adsorption of H^+ into the diffuse ion swarm[35] near the mineral surface and subsequent two-dimensional diffusion to a weakly acidic, exposed functional group; then the complexation of the proton by the functional group; then the polarization and weakening of ionic bonds in the mineral structure near the site of protonation (possibly during several additional proton adsorption steps); and, finally, detachment of ionic species from the solid surface and their (rapid) equilibration in the aqueous solution phase.[34,36] The time scales for these four steps could range from microseconds for adsorption to hours or days for detachment. The details of all of them, as well as the detailed morphology of the near-surface mineral structure, are lumped into the first-order rate coefficient k_3 in Eq. 3.60. Some clues as to their relative importance are provided by correlations with molecular data, like those in Fig. 3.4. It must not be forgotten, however, that a dissolution rate law that is first-order in $[H^+]$ can represent a transport-controlled process as well as a reaction-controlled process, and sensitive experimental testing methods are required to avoid misinterpreting data that have been fit to Eq. 3.60.[7,33,34,36,37]

Proton-promoted dissolution reactions are exemplified for carbonates, silicates, and metal oxyhydroxides by Eqs. 3.15, 3.18–3.20, 3.25, 3.39, 3.46, 3.53, 3.56, and 3.59c. The typical response of the rate of dissolution to varying pH is illustrated in Fig. 3.2, and this response is often hypothesized to be a result of the proton adsorption-bond-weakening structural detachment sequence described in connection with Eq. 3.60.[36] This sequence can be represented by the following generic reaction scheme:

$$\equiv C\text{-}A(s) + H^+(aq) \underset{k_{des}}{\overset{k_{ads}}{\rightleftarrows}} \equiv C\text{-}AH(s) \xrightarrow{k_{det}} C(aq) + HA(aq) \tag{3.61}$$

where $\equiv$ C-A(s) is a cation–anion unit at the mineral surface that is destabilized by proton adsorption and detaches to form the aqueous species C(aq) and HA(aq). Examples of this reaction scheme include the dissolution of Al oxyhydroxides,

$$\equiv Al - OH(s) + H^+(aq) \underset{k_{des}}{\overset{k_{ads}}{\rightleftarrows}} \equiv Al - OH_2^+(s) \xrightarrow{k_{det}} Al^{3+}(aq) + H_2O(\ell) \tag{3.62}$$

and that of olivine ($[Mg, Fe]_2 SiO_4$),

$$\equiv Mg - OH\,(s) + H^+(aq) \underset{k_{des}}{\overset{k_{ads}}{\rightleftarrows}} \equiv Mg - OH_2^+(s) \xrightarrow{k_{det}} Mg^{2+}(aq) + H_2O(\ell) \qquad (3.63)$$

Equation 3.61 is a special case of Eq. 1.52. If the rate of dissolution is equated to d[C(aq)]/dt, then a rate law analogous to Eq. 1.54c ($k_b' = 0$) can be postulated:

$$\frac{d[C(aq)]}{dt} = k_{det}\,[\equiv C - AH(s)]^{\alpha} \qquad (3.64)$$

where α is a partial reaction order *not* necessarily equal to 1.0, the stoichiometric coefficient of the species ≡ C-AH(s) in Eq. 3.61 (cf. Eq. 1.34 and the discussion following Eq. 3.11). Equation 3.64 predicts that a log–log plot of the rate of dissolution against [≡ C-AH(s)] will be a straight line whose scope equals the reaction order α. Data consistent statistically with this prediction are shown in Fig. 3.11 for the dissolution of metal oxyhydroxides and silicates. There is some evidence that, for simple binary solids, α is approximately the same as the oxidation number of the cationic species C in the molecular unit ≡ C-AH [e.g., $\alpha = 3$ for Al(III), or $\alpha = 4$ for Si(IV)].[36] This relationship, in turn, has been interpreted as depicting the number of prior protonation steps required to provoke the detachment step.

The dissolution reaction in Eq. 3.59b can be regarded as an example of a ligand-promoted process, in that adsorbed bicarbonate species are likely to play a role as intermediates in the kinetic analysis of the reaction.[5] Ligand-promoted dissolution reactions are a principal basis for the reductive dissolution processes described in Section 3.4 (see Eq. 3.46). The sequence of steps is analogous to that in proton-promoted dissolution:

$$\begin{aligned} \equiv C - A(s) + L(aq) &\underset{k_{des}}{\overset{k_{ads}}{\rightleftarrows}} \equiv C - L(s) + A(aq) \\ \equiv C - L(s) &\xrightarrow{k_{det}} CL(aq) \end{aligned} \qquad (3.65)$$

where L is a ligand that forms an inner-sphere complex with the cationic species C after exchange with the anionic species A. Examples of this sequence would be the reactions in Eqs. 3.53 and 3.56, if a detachment step were included, and the fluoride-promoted dissolution of Al oxyhydroxides:[38]

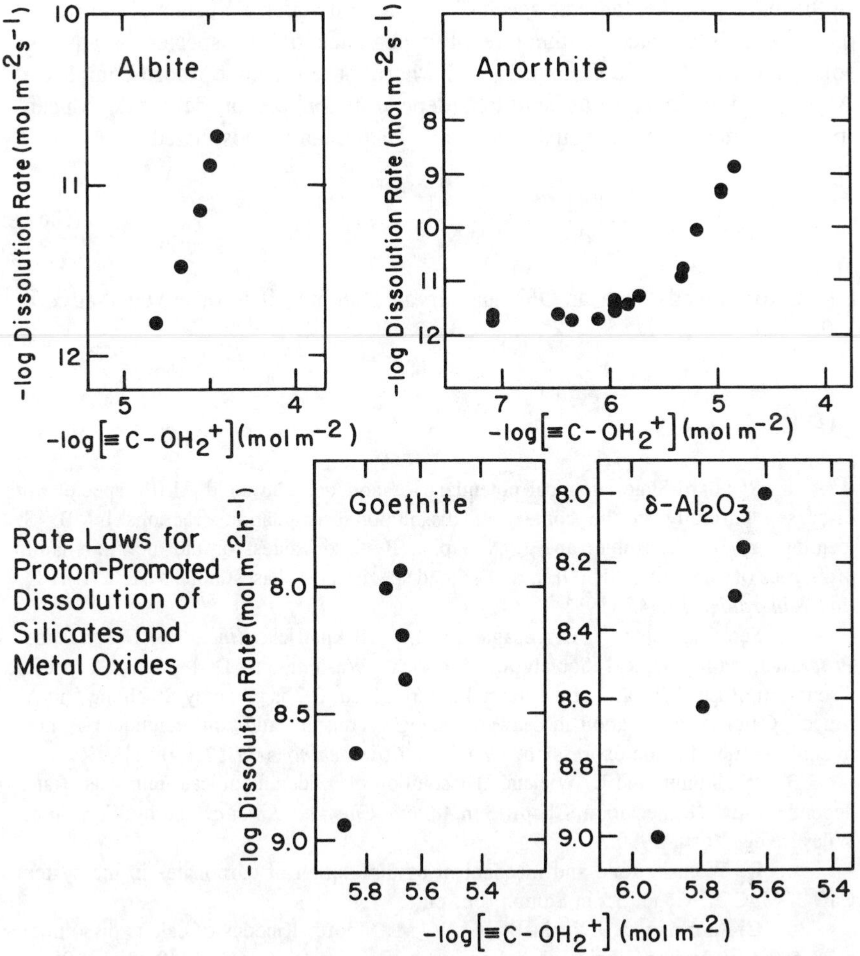

Fig. 3.11. Tests of the rate law in Eq. 3.64 for albite,[9] anorthite [C. Amrhein and D. L. Suarez, The use of a surface complexation model to describe the kinetics of ligand promoted anorthite dissolution, *Geochim. Cosmochim. Acta* **52**:2785 (1988)], goethite,[36] and δ-Al_2O_3.[36] A linear plot is consistent with the rate law.

$$\equiv Al-OH_2^+(s) + F^-(aq) \underset{k_{des}}{\overset{k_{ads}}{\rightleftarrows}} \equiv Al-F(s) + H_2O(\ell)$$

$$\equiv Al-F(s) \overset{k_{det}}{\rightarrow} AlF^{2+}(aq) \qquad (3.66)$$

In this example, the "anionic species" is a neutral water molecule. If the rate of dissolution is equated to the rate of appearance of CL species in aqueous solution (as in the case C = Al, L = F, where more than one fluoro-complex of Al is possible in aqueous solution, depending on pH and fluoride concentration[38]), then a rate law analogous to Eq. 3.64 can be postulated:

$$\frac{d[CL]}{dt} = k_{det}[\equiv C - L(s)]^{\alpha} \qquad (3.67)$$

Often, for ligands such as OH^- and organic anions, it is observed that $\alpha \approx 1.0$.[5,39]

NOTES

1. Standard-State chemical potentials for aqueous and solid Al(III) species are discussed carefully in the context of dissolution–precipitation reactions by B. S. Hemingway, R. A. Robie, and J. A. Apps, Revised values for the thermodynamic properties of boehmite, AlO(OH), and related species and phases in the system Al-H-O, *Am. Mineralog.* **76**:445 (1991).

2. See Chap. 1 in A. C. Lasaga and R. J. Kirkpatrick, *Kinetics of Geochemical Processes*, Mineralogical Society of America, Washington, DC, 1981. A useful discussion of Eq. 3.13 and of the overall nature of Eq. 3.7 is given by S. Zhong and A. Mucci, Calcite precipitation in seawater using a constant addition technique: A new overall reaction kinetic expression, *Geochim. Cosmochim. Acta* **57**:1409 (1993).

3. W. Stumm and E. Wieland, Dissolution of oxide and silicate minerals: Rates depend on surface speciation, Chap. 13 in *Aquatic Chemical Kinetics*, ed. by W. Stumm, Wiley, New York, 1990.

4. R. Wollast, Rate and mechanism of dissolution of carbonates in the system $CaCO_3$-$MgCO_3$, Chap. 15 in Stumm, op. cit.[3]

5. C. Amrhein, J. J. Jurinak, and W. M. Moore, Kinetics of calcite dissolution as affected by carbon dioxide partial pressure, *Soil Sci. Soc. Am. J.* **49**:1393 (1985).

6. W. P. Inskeep and P. R. Bloom, Kinetics of calcite precipitation in the presence of water-soluble organic ligands, *Soil Sci. Soc. Am. J.* **50**:1167 (1986).

7. L. Chou, R. M. Garrels, and R. Wollast, Comparative study of the kinetics and mechanisms of dissolution of carbonate minerals, *Chem. Geol.* **78**:269 (1989); and R. G. Compton and K. L. Pritchard, the dissolution of calcite at pH > 7: Kinetics and mechanism, *Phil. Trans. R. Soc. Lond.* **A330**:47 (1990).

8. W. P. Inskeep and P. R. Bloom, An evaluation of rate equations for calcite precipitation kinetics at p_{CO_2} less than 0.01 atm and pH greater than 8, *Geochim. Cosmochim. Acta* **49**:2165 (1985). See also S. Zhong and A. Mucci, op. cit.[2]

9. L. Chou and R. Wollast, Study of the weathering of albite at room temperature and pressure with a fluidized bed reactor, *Geochim. Cosmochim. Acta* **48**:2205 (1984).

10. W. H. Casey and H. R. Westrich, Control of dissolution rates of orthosilicate minerals by divalent metal–oxygen bonds, *Nature* **355**:157 (1992).

11. See Chap. 3 in A. C. Lasaga and R. J. Kirkpatrick, op. cit.,[2] and Chap. 7 in D. L. Sparks and D. L. Suarez, *Rates of Soil Chemical Processes*, Soil Science Society

of America, Madison, WI, 1991.

12. For an introduction to the two types of diagram, see, for example, Chaps. 5 and 6 in G. Sposito, *The Chemistry of Soils*, Oxford University Press, New York, 1989. Activity-ratio and predominance diagrams are discussed with many examples in Chaps. 5 and 7 of W. Stumm and J. J. Morgan, *Aquatic Chemistry*, Wiley, New York, 1981.

13. See for example, D. C. Melchior and R. L. Bassett, *Chemical Modeling of Aqueous Systems II*, American Chemical Society, Washington, DC, 1990, for a comprehensive discussion of speciation calculations related to mineral solubility.

14. See, for example, Chap. 1 in G. Sposito, op. cit.[12]

15. B. S. Hemingway, Gibbs free energies of formation for bayerite, nordstrandite, $Al(OH)^{2+}$, and $Al(OH)^{2+}$, aluminum mobility, and formation of bauxites and laterites, *Adv. Phys. Geochem.* **2**:285(1982). The mechanistic basis of the GLO Step Rule is discussed in Chap. 6 of W. Stumm, *Chemistry of the Solid–Water Interface*, Wiley, New York, 1992. See also Chap. 5 in W. Stumm and J. J. Morgan, op. cit.[12]

16. This feature of the GLO Step Rule can be turned to great advantage in experiments to study the precipitation of metastable solid phases. If the composition of the aqueous solution phase is controlled by titration while precipitation of the target solid is occurring ("constant composition method"), precise rate data as a function of $\Omega > 1$ can be obtained without significant transformation of the target solid to more stable phases. See, for example, J.-W. Zhang and G. Nancollas, Mechanisms of growth and dissolution of sparingly soluble salts, Chap. 9 in *Mineral–Water Interface Geochemistry*, ed. by M. F. Hochella and A. F. White, Mineralogical Society of America, Washington, DC, 1990.

17. H. M. May, D. G. Kinniburgh, P. A. Helmke, and M. L. Jackson, Aqueous dissolution, solubilities and thermodynamic stabilities of common aluminosilicate clay minerals: Kaolinite and smectites, *Geochim. Cosmochim. Acta* **50**:1667 (1986).

18. The concept of disorder and solubility "windows" is discussed in G. Sposito, Chemical models of weathering in soils, pp. 1–18 in *The Chemistry of Weathering*, ed. by J. I. Drever, D. Reidel, Dordrecht, The Netherlands, 1985.

19. See, for example, U. Schwertmann, Solubility and dissolution of iron oxides, *Plant and Soil* **130**:1 (1991), and A.P. Schwab and W. L. Lindsay, Effect of redox on the solubility and availability of iron, *Soil Sci. Soc. Am. J.* **47**:201 (1983).

20. See, for example, Chaps. 1 and 2 in G. Sposito, op. cit.,[12] for a discussion of isomorphic substitutions in primary and secondary minerals.

21. D. C. Thorstenson and L. N. Plummer, Equilibrium criteria for two-component solids reacting with fixed composition in an aqueous phase—Example: The magnesian calcites, *Am. J. Sci.* **277**:1203 (1977).

22. E. Busenberg and L. N. Plummer, Thermodynamics of magnesian calcite solid-solutions at 25 °C and 1 atm total pressure, *Geochim. Cosmochim. Acta* **53**:1189 (1989). This paper contains a prototypical approach to the experimental characterization of the dissolution reactions of solid solutions. See also E. Königsberger and H. Gamsjäger, Solid-solute phase equilibria in aqueous solution: VII. A re-interpretation of magnesian calcite stabilities, *Geochim. Cosmochim. Acta* **56**:4095 (1992).

23. P. D. Glynn and E. J. Reardon, Solid-solution aqueous-solution equilibria: Thermodynamic theory and representation, *Am. J. Sci.* **290**:164 (1990), **292**: 215 (1992); E. Königsberger and H. Gamsjäger, Solid-solution aqueous-solution equilibria: Thermodynamic theory and representation, *Am. J. Sci.* **292**:199 (1992). These papers contain comprehensive discussions of the thermodynamic description of solid solutions. See also L. N. Plummer, E. Busenberg, P. D. Glynn, and A. E. Blum, Dissolution of

aragonite–strontianite solid solutions in nonstoichiometric $Sr(HCO_3)_2$–$Ca(HCO_3)_2$–CO_2–H_2O solutions, *Geochim. Cosmochim. Acta* **56**:3045 (1992).

24. These basic thermodynamic concepts are discussed in Chaps. 1 and 2 of G. Sposito, *The Thermodynamics of Soil Solutions*, Clarendon Press, Oxford, U.K., 1981.

25. Y. Tardy and D. Nahon, Geochemistry of laterites, stability of Al-goethite, Al-hematite, and Fe^{3+}-kaolinite in bauxites and ferricretes: An approach to the mechanism of concretion formation, *Am. J. Sci.* **285**:865 (1985).

26. For a comprehensive review, see J. G. Hering and W. Stumm, Oxidative and reductive dissolution of minerals, Chap. 11 in M. F. Hochella and A. F. White, op. cit.[16]

27. See, for example, pp. 403–406 in E. M. Thurman, *Organic Geochemistry of Natural Waters*, Martinus Nijhoff, Dordrecht, The Netherlands, 1986, and T. D. Waite, Photoredox chemistry of colloidal metal oxides, Chap. 20 in *Geochemical Processes at Mineral Surfaces*, ed. by J. A. Davis and K. F. Hayes, American Chemical Society, Washington, DC, 1986.

28. A. T. Stone and J. J. Morgan, Reductive dissolution of metal oxides, Chap. 9 in *Aquatic Surface Chemistry*, ed. by W. Stumm, Wiley, New York, 1987. This chapter provides a comprehensive discussion of rate laws for reductive dissolution reactions, especially mathematical models.

29. See Note 7 at the end of Chapter 2 for a definition of the characteristic time scales of reactions.

30. See, for example, Chap. 8 in G. Sposito, op. cit.,[12] for an introductory discussion of adsorption isotherm equations.

31. The variety of mechanisms possible for reductive mineral dissolution processes that are mediated by organic ligands is discussed by J. G. Hering and W. Stumm, op. cit.,[26] A. T. Stone and J. J. Morgan, op. cit.,[28] and B. Sulzberger, D. Suter, C. Siffert, S. Banwart, and W. Stumm, Dissolution of Fe(III) (hydr)oxides in natural waters; Laboratory assessment on the kinetics controlled by surface coordination, *Mar. Chem.* **28**:127 (1989). The examples considered in this section are illustrative only.

32. See, for example, the reviews by B. Sulzberger, Photoredox reactions at hydrous metal oxide surfaces: A surface coordination chemistry approach, Chap. 14 in W. Stumm, op. cit.,[3] and T. D. Waite, op. cit.[27]

33. N. L. Plummer, T. M. L. Wigley, and D. L. Parkhurst, The kinetics of calcite dissolution in CO_2–water systems at 5–60°C and 0.0–1.0 atm CO_2, *Am. J. Sci.* **278**:179 (1978).

34. R. G. Compton and P. R. Unwin, The dissolution of calcite in aqueous solution at pH < 4: Kinetics and mechanism, *Phil. Trans. R. Soc. Lond.* **A330**:1 (1990).

35. The species of adsorbed ions (e.g., diffuse-layer species) are discussed in Chap. 7 of G. Sposito, op. cit.[12]

36. W. Stumm and R. Wollast. Coordination chemistry of weathering: Kinetics of the surface-controlled dissolution of oxide minerals, *Rev. Geophys.* **28**:53 (1990). See also A. E. Blum and A. C. Lasaga, The role of surface speciation in the dissolution of albite, *Geochim. Cosmochim. Acta* **55**:2193 (1990).

37. D. Rickard and E. L. Sjöberg, Mixed kinetic control of calcite dissolution rates, *Am. J. Sci.* **283**:815 (1983).

38. P. W. Schindler and G. Sposito, Surface complexation at (hydr)oxide interfaces, Chap. 4 in *Interactions at the Soil Colloid–Soil Solution Interface*, ed. by G. H. Bolt, M. F. De Boodt, M. H. B. Hayes, and M. B. McBride, Kluwer Academic Publ., Dordrecht, The Netherlands, 1991.

39. P. V. Brady and J. V. Walther, Controls on silicate dissolution rates in neutral

and basic pH solutions at 25°C, *Geochim. Cosmochim. Acta* **53**:2823 (1989).

FOR FURTHER READING

Brookins, D. G., *Eh–pH Diagrams for Geochemistry*, Springer-Verlag, Berlin, 1988. A comprehensive pictorial review of pE–pH diagrams for almost all of the Periodic Table, including compilations of the thermodynamic data used to construct the diagrams.

Glynn, P. D., and E. J. Reardon, Solid-solution aqueous-solution equilibria: Theory and representation, *Am. J. Sci.* **290**:164 (1990). An authoritative discussion of the chemical thermodynamics of homogeneous mixed solids.

Helgeson, H. C., The Robert M. Garrels Memorial Issue, *Geochim. Cosmochim. Acta* **56**, No. 8 (August 1992). A marvelous collection of research papers on mineral solubility issues, dedicated to one of the giants of aqueous geochemistry.

Hering, J. G., and W. Stumm, Oxidative and reductive dissolution of minerals, pp. 427–465 in *Mineral-Water Interface Geochemistry*, ed. by M. F. Hochella and A. F. White, Mineralogical Society of America, Washington, DC, 1990. A comprehensive survey of redox reactions as they influence mineral solubility.

Stumm, W., *Aquatic Chemical Kinetics*, Wiley, New York, 1990. Chapters 11–15 and 17 of this edited monograph provide excellent surveys of research on the kinetics and mechanisms of mineral dissolution processes investigated in both laboratory and field settings.

Stumm, W., and J. J. Morgan, *Aquatic Chemistry*, Wiley, New York, 1981. Chapter 5 of this standard textbook gives many examples of the concepts discussed in the present chapter. A broad conceptual picture of mineral dissolution kinetics and mechanisms is developed in the celebrated three-part paper "The Coordination Chemistry of Weathering":

I. Dissolution kinetics of δ-Al_2O_3 and BeO, *Geochim. Cosmochim. Acta* **50**:1847 (1986), by G. Furrer and W. Stumm; II. Dissolution of Fe(III) oxides, *Geochim. Cosmochim. Acta* **50**:1861 (1986), by B. Zinder, G. Furrer, and W. Stumm; III. A generalization on the dissolution rates of minerals, *Geochim. Cosmochim. Acta* **52**:1969 (1988), by E. Wieland, B. Wehrli, and W. Stumm.

Experimental features of mineral dissolution kinetics research are described in exemplary fashion in the following four papers:

Rickard, D., and E. L. Sjöberg, Mixed kinetic control of calcite dissolution rates, *Am. J. Sci.* **283**:815 (1983).

Chou, L., and R. Wollast, Steady-state kinetics and dissolution mechanisms of albite, *Am. J. Sci.* **285**:963 (1985).

Compton, R. G., and P. R. Unwin, The dissolution of calcite in aqueous solution at pH < 4: Kinetics and mechanism, *Phil. Trans. R. Soc. Lond.* **A330**:1 (1990).

Nagy, K. L., and A. C. Lasaga, Dissolution and precipitation kinetics of gibbsite at 80°C and pH 3: The dependence on solution saturation state, *Geochim. Cosmochim. Acta* **56**:3093 (1992).

PROBLEMS

1. Calculate the relative saturation with respect to calcite and gypsum ($CaSO_4 \cdot 2H_2O$; $K_{so} = 2.4 \times 10^{-5}$) for the soil solution whose speciation is described in Table 2.6. Take $I_{ef} = 10.5$ mol m^{-3}. (*Answer*: Log $\Omega_{cal} = -3.22 - 8.38 + 8.10 = -3.5$ and log $\Omega_{gyp} = -3.22 - 2.71 + 4.25 = -1.67$, indicating that undersaturation exists for both solids.)

2. The poorly crystalline Fe(III) mineral ferrihydrite ($Fe_{10}O_{15} \cdot 9H_2O$) forms with very small particle size, whereas geothite (α-FeOOH) and hematite (α-Fe_2O_3) form with particle sizes varying from about 5 nm to 1 μm in diameter. The effect of small particle size on these latter two Fe(III) minerals can be modeled mathematically with the empirical relationships:

$$\log \mathbf{(FeOOH)} = 27/d_{goe} \qquad \log \mathbf{(Fe_2O_3)} = 25/d_{hem}$$

where d is particle diameter in nm (d $\geq$ 4 nm). Use Eq. 3.25 and the appropriate data in Table 2.4 to develop chemical reactions that relate the three Fe(III) minerals, then derive corresponding equations relating $\mathbf{(H_2O)}$ and particle diameter at equilibrium. Determine whether a realizable water activity exists for ferrihydrite-goethite, ferrihydrite-hematite, and geothite-hematite equilibria with macrocrystalline (d $\uparrow \infty$) particles. Is there a particle size of goethite or hematite that is consistent with ferrihydrite equilibrium when $\mathbf{(H_2O)} = 1.0$? (*Hint*: The reactions are

$$Fe_{10}O_{15} \cdot 9H_2O(s) = 10\,FeOOH(s) + 4H_2O(\ell) \qquad \log K = 35$$

$$Fe_{10}O_{15} \cdot 9H_2O(s) = 5Fe_2O_3(s) + 9H_2O(\ell) \qquad \log K = 47$$

$$2FeOOH(s) = Fe_2O_3(s) + H_2O(\ell) \qquad \log K = 2.4$$

Expressions for log K in each case will relate solid-phase and water activities. Note that an infinitude of particle-size values is consistent with goethite–hematite equilibrium, but $d_{goe} > d_{hem}$ is required.)

3. Dissolution experiments for calcite in 300 mol m^{-3} KCl background

resulted in $(k_d/a_sc_s) = 2.1 \times 10^{-5}$ mol m^{-2} s^{-1} far from equilibrium and when the rate of dissolution achieved negligible values, the equilibrium concentrations, $[Ca^{2+}]_e = 0.432$ mol m^{-3}, $[CO_3^{2-}]_e = 0.0946$ mol m^{-3}. Compute a value of (k_p/a_sc_s) for these experiments. (*Answer*: 5.1×10^{-4} mol^{-1} m^4 s^{-1}.)

4. The carbonic acid-promoted dissolution of calcite (Eq. 3.59b) can be described in an open system by the sequence of heterogeneous reactions:

$$CO_2(g) + H_2O(\ell) \underset{k_b}{\overset{k_f}{\rightleftarrows}} H_2CO_3^0(aq)$$

$$CaCO_3(s) + H_2CO_3^0(aq) \rightarrow Ca^{2+}(aq) + 2HCO_3^-(aq)$$

where $k_f^* \equiv k_f[H_2O] \approx 4.54 \times 10^{-9}$ mol m^{-3} s^{-1} Pa^{-1}, $k_b \approx 5.2 \times 10^{-3}$ s^{-1}, and $k_2 \approx 0.027$ s^{-1}. Show that the rate of calcite dissolution under this scenario is given by

$$\frac{d[Ca^{2+}]}{dt} = 0.027\,[H_2CO_3^0]$$

where the concentration of carbonic acid changes according to the rate law

$$\frac{d[H_2CO_3^0]}{dt} = 46 \times 10^{-4}\,P_{CO_2} - 0.0322\,[H_2CO_3^0]$$

with p_{CO_2} in atm and all concentrations in mol m^{-3}. Show that the rate of calcite dissolution under constant p_{CO_2} is (t in s):

$$\frac{d[Ca^{2+}]}{dt} = \{0.027\ [H_2CO_3^0]_e + ([H_2CO_3^0]_0 - [H_2CO_3^0]_e)\exp(-0.0322t)\}$$

where $[H_2CO_3^0]_e = 1.43 \times 10^{-2}\,p_{CO_2}$ is the concentration of carbonic acid at equilibrium with $CO_2(g)$, and $[H_2CO_3^0]_0$ is its value at t = 0. (*Hint*: Apply Eqs. 1.54 and 1.55.)

5. Prepare activity-ratio and predominance diagrams for the Al(III) minerals whose dissolution reactions are described in Eqs. 3.18–3.20. Set pH = 6 for the activity-ratio diagram, but otherwise use fixed activity data as given in connection with Figs. 3.5 and 3.7. Repeat your calculations for **(H_2O)** = 0.5 instead of unit water activity. What is the effect on mineral stability? (*Answer*:

Equation 3.22 becomes the activity-ratio equations

$$\log\,[(\mathbf{montmorillonite})/(\mathbf{Al^{3+}})] = 21.29 + 2.6 \log\,(\mathbf{H_4SiO_4^0}) - 2.23 \log(\mathbf{H_2O})$$

$$\log\,[(\mathbf{kaolinite})/(\mathbf{Al^{3+}})] = 14.28 + \log\,(\mathbf{H_4SiO_4^0}) + \tfrac{1}{2} \log(\mathbf{H_2O})$$

$$\log\,[(\mathbf{gibbsite})/(\mathbf{Al^{3+}})] = 9.89 + 3 \log\,(\mathbf{H_2O})$$

and Eq. 3.24 becomes the boundary-line equations

$$\log\,(\mathbf{H_4SiO_4^0}) = -4.4 + 2.5 \log(\mathbf{H_2O})$$

$$\log\,(\mathbf{H_4SiO_4^0}) = -0.763 - 0.605\ \mathrm{pH} + 2.01 \log(\mathbf{H_2O})$$

$$\log\,(\mathbf{H_4SiO_4^0}) = 1.51 - 0.983 + 1.71 \log(\mathbf{H_2O})$$

Lowering the water activity stabilizes montmorillonite and destabilizes kaolinite and gibbsite, as can be inferred from Eqs. 3.18–3.20.)

6. Construct a pE–pH diagram for the species, $Fe^{3+}(aq)$, $Fe^{2+}(aq)$, α-$FeOOH(s)$, α-$Fe_2O_3(s)$. Use $(\mathbf{Fe^{2+}}) = 10^{-5}$, $3 \leq pH \leq 9$, and $-6 \leq pE \leq +13$. (*Hint*: Review Sections 2.2 and 3.2.)

7. Prepare a pE–pH diagram as in Problem 6, but with Al-substituted goethite instead of pure goethite. Include boundary lines corresponding to $x = 0.1$ and 0.3, with Al-goethite assumed to be an ideal solid solution. What is the effect of Al substitution on *ideal* Al-goethite stability? (*Hint*: The reduction half-reaction for ideal Al-goethite is

$$FeOOH(s) + 3H^+(aq) + e^-(aq) = Fe^{2+}(aq) + 2\ H_2O(\ell)$$

with $\log K_R = 11.3 + \log\,(1 - x)$. Thus, Al substitution tends to stabilize *ideal* Al-goethite against reduction.)

8. Prepare a graph of $\log K_{so}^{ss}$ against the mole fraction of dolomite for magnesian calcite. Use $0 \leq x \leq 0.2$ in the graph and find the value of x for which the solid phase has its smallest solubility product constant. Show that $\ln[(\mathbf{Mg^{2+}})/(\mathbf{Ca^{2+}})] = 2\,(\partial \ln K_{so}^{ss}/\partial x)_{T,\,P}$ and determine whether a value of x exists such that $(\mathbf{Mg^{2+}}) = (\mathbf{Ca^{2+}})$. (*Hint*: Combine Eqs. 3.27 and 3.35 aq applied to magnesian calcite in Eq. 3.36. Note that the slope of a $\log K_{so}^{ss}$ plot against x vanishes when the log solubility product constant has a minimum value.)

9. Lepidocrocite (γ-FeOOH), suspended in 10 mol m^{-3} NaCl at pH 4 and irradiated with UV light, dissolved at the constant rate of 8 μmole m^{-3} min^{-1} measured as d[Fe(II)]/dt. In the presence of oxalic acid (ethanedioic acid), the

rate of dissolution increased, and the data could be fit to the empirical equation

$$\frac{d\,[\mathrm{Fe(III)}]}{dt} = \frac{b\,[C_2O_4^{2-}]_e}{c + [C_2O_4^{2-}]_e} + \left[\frac{d\,[\mathrm{Fe(II)}]}{dt}\right]_0$$

where b = 28.6 μmol m^{-3} min^{-1}, c = 5 mmol m^{-3}, and the second term on the right side is the dissolution rate without oxalic acid present. Given that total concentration of Fe(III) that can react with oxalic acid is 840 μmol m^{-3}, calculate the first-order rate coefficient and the Langmuir affinity parameter for the reductive dissolution of lepidocrocite promoted by oxalic acid. (*Answer*: k = 5.67 $\times$ 10^{-4} s^{-1} and K_L = 200 m^3 mol^{-1}. Note that the dissolution rate "saturates" at oxalic acid concentrations >> 5 mmol m^{-3}.)

10. Experimental data on the pH dependence of proton adsorption by metal oxyhydroxides can often be represented mathematically by the logarithmic relationship:

$$\log\,[\equiv M - OH_2^+] = A + B(pH_0 - pH)$$

where A and B are adjustable parameters and pH_0 is the pH value at which $[\equiv M - OH_2^+]$ vanishes. [The equation applies for $(pH_0 - pH) \leq 1$.] Show that this relationship implies a power-law dependence of the rate of proton-promoted dissolution on the aqueous solution concentration of protons for pH < pH_0. (*Hint*: Combine the relationship with Eq. 3.64.)

4

SURFACE REACTIONS

4.1 Adsorption–Desorption Equilibria

Adsorption (or desorption) is the process by which a net accumulation (or loss) of a substance occurs at an interface between two phases. In a typical experiment, two phases are mixed intimately to provoke a chemical reaction leading to adsorption or desorption, and then a physical separation is made, with one of the separates being a single phase and the other, a mixture of the two reacted phases. For example, a solid-phase adsorbent and an aqueous solution could be mixed, and then separated by centrifugation into an aqueous phase (supernatant solution) and a slurry that contains both the solid adsorbent and some aqueous solution. If n_i is the moles of substance i in the reacted mixture and m_i is the molality of substance i in the separated aqueous phase, then the *relative surface excess*, $n_i^{(j)}$, of substance i, as compared to another substance j, is defined by[1]

$$n_i^{(j)} \equiv n_i - n_j(m_i/m_j) \qquad (4.1)$$

The conceptual meaning of Eq. 4.1 is that $n_i^{(j)}$ is the excess moles of substance i in the reacted mixture, relative to the content of a *reference substance* j in the mixture and to the composition of the separated aqueous solution, indicated by the molalities m_i and m_j. In the example of the slurry and supernatant solution, $n_i^{(j)}$ is the excess moles of i in the slurry, as compared to the content of the reference substance j and to an aqueous solution that has the mixed composition indicated by m_i and m_j. The right side of Eq. 4.1 can be positive (adsorption), zero (no surface excess), or negative (desorption), depending entirely on the *relative* behavior of the substances i and j when two phases containing them react. In applications of Eq. 4.1 to the reactions of soils with aqueous solutions, the reference substance j is invariably chosen to be *liquid water* (j = w):

$$n_i^{(w)} = n_i - n_w(m_i/m_w)$$
$$= n_i - M_w m_i$$
$$= \Delta m_i M_{wo} \tag{4.2}$$

where M_w is the mass of liquid water in the reacted slurry; Δm_i is the change in molality of substance i in the aqueous-solution phase because of its reaction with the soil, and M_{wo} is the total mass of water in the slurry plus the aqueous-solution phase. The second and third equalities in Eq. 4.2 connect the relative surface excess to convenient experimental variables. Implicit in Eq. 4.2 is the assumption that substance i does not form a mixed solid phase or a surface precipitate after reaction with the soil, an assumption that can be difficult to substantiate directly.[2] There is also the implicit assumption that liquid water adsorption is zero, since $n_w^{(w)} \equiv 0$ in Eq. 4.2. In more technical terms, this means that $n_i^{(w)}$ is the surface excess of a substance i at an interface on which no net accumulation or depletion of liquid water occurs.[1] Although the experimental picture is not entirely clear, for soil minerals consensus would place this interface at or very near (<1.0 nm away) the geometric boundary of a solid adsorbent.[3]

An *overall* reaction describing the adsorption or desorption of aqueous solution species by a solid adsorbent can be written as follows:

$$SR^{Z_{SR}}(s) + p\,M^{m+}(aq) + q\,L^{\ell-}(aq) + x\,H^+(aq)$$
$$+ y\,OH^-(aq) \underset{k_{des}}{\overset{k_{ads}}{\rightleftarrows}} SR'M_p(OH)_yH_xL_q^{Z_{SR'C}} + Q^{Z_Q}(aq) \tag{4.3}$$

where m is the valence of an adsorptive metal M; $-\ell$ is the valence of an adsorptive ligand L; Z_{SR} is the valence of a reactive surface moiety SR in the adsorbent, assumed to comprise a dissociable component Q of valence Z_Q, and an undissociable component SR′ of valence $Z_{SR} - Z_Q$; and $Z_{SR'C}$ is the valence of the adsorbent–adsorbate product, $SR'M_p(OH)_yH_xL_q(s)$. Electroneutrality in Eq. 4.3 requires the condition

$$Z_{SR} + pm + x - q\ell - y = Z_{SR'C} + Z_Q \tag{4.4}$$

Some special cases of the reaction in Eq. 4.3 are listed in Table 4.1. In each case, S represents the adsorbent structure not involved directly in an adsorption–desorption reaction. (It is equivalent to the symbol $\equiv M_{ox}$ in Eq. 3.46 or $\equiv C$ in Eq. 3.61, for example.) Other special cases are given in Eqs. 3.53, 3.56, 3.62, 3.63, and 3.66. Equation 4.3 can be generalized readily to permit more than 1 mol of the species $SR^{Z_{SR}}(s)$ to react, or to replace M^{m+} by a metal–hydroxy polymer (e.g., $Al_{13}(OH)_{32}^{7+}$), or to replace $L^{\ell-}$ by a polyanion (e.g., fulvic acid). Note that the adsorbent can be either inorganic or organic (cf. the third and fourth reactions in Table 4.1).

An equilibrium constant for the heterogeneous reaction in Eq. 4.3 can be defined analogously to K in Eq. 1.11:

$$K_{ads} \equiv (SR'C^{Z_{SR'C}})(Q^{Z_Q}) \,/\, (SR^{Z_{SR}})(M^{m+})^p(H^+)^x(L^{\ell-})^q(OH^-)^y \tag{4.5}$$

where $C \equiv M_p(OH)_yH_xL_q$ is the adsorbate and the reaction has been taken as an overall adsorption process. At equilibrium, the solid phase will be a mixture of the unreacted adsorbent, SR, and the product species, SR′C. For this reason, the conditional equilibrium constant corresponding to K_{ads} usually is expressed in terms of the mole fractions of these two species, x_{SR} and $x_{SR'C}$, as is conventional in the chemical thermodynamics of mixtures:[4]

$$K_{adsc} = x_{SR'C}\,(Q^{Z_Q}) \,/\, x_{SR}\,(M^{m+})^p(H^+)^x(L^{\ell-})^q(OH^-)^y \tag{4.6}$$

The relationship between mole fraction and activity, and therefore between K_{ads} and K_{adsc}, is made through rational activity coefficients (cf. Eq. 1.25):[4]

$$f_{SR'C} \equiv (SR'C^{Z_{SR'C}}) \,/\, x_{SR'C} \qquad f_{SR} \equiv (SR^{Z_R}) \,/\, x_{SR} \tag{4.7}$$

Thus

$$K_{ads} = f_{SR'C}\,K_{adsc} \,/\, f_{SR} \tag{4.8}$$

Equation 4.7 invites comparison with Eq. 1.12, as does Eq. 4.8 with Eq. 1.26. (Note, however, that the rational activity coefficient f is dimensionless.)

Table 4.1 Examples of the Reaction in Eq. 4.3

$SR^{Z_{SR}}$	M^{m+}	$L^{\ell-}$	Q^{Z_Q}	Reaction
SOH^0	–	–	H^+	$SOH^0(s) \leftrightarrows SO^-(s) + H^+(aq)$
SOH^0	–	–	–	$SOH^0(s) + H^+(aq) \leftrightarrows SOH_2^+(s)$
SO^-	Ca^{2+}	–	–	$SO^-(s) + Ca^{2+}(aq) \leftrightarrows SOCa^+(s)$
$SCOO^-$	K^+	–	–	$SCOO^-(s) + K^+(aq) \leftrightarrows SCOOK^0(s)$
SOH^0	Na^+	–	H^+	$SOH^0(s) + Na^+(aq) \leftrightarrows SONa^0(s) + H^+(aq)$
SO^-	Cu^{2+}	$HC_2O_4^-$	–	$SO^-(s) + Cu^{2+}(aq) + HC_2O_4^-(aq) \leftrightarrows SOCuHC_2O_4^0(s)$
SOH^0	Al^{3+}	–	H^+	$SOH^0(s) + Al^{3+}(aq) + OH^-(aq) \leftrightarrows SOAlOH^+(s) + H^+(aq)$
SOH^0	–	Cl^-	–	$SOH^0(s) + Cl^-(aq) + H^+(aq) \leftrightarrows SOH_2Cl^0(aq)$
SOH^0	–	$C_2O_4^{2-}$	OH^-	$SOH^0(s) + C_2O_4^{2-}(aq) + H^+(aq) \leftrightarrows SHC_2O_4^0(s) + OH^-(aq)$
SOH_2^+	–	$H_2PO_4^-$	H_2O	$SOH_2^+(s) + H_2PO_4^-(aq) \leftrightarrows SH_2PO_4^0(s) + H_2O(\ell)$

If the solid phase at equilibrium comprises only the two species, SR′C(s) and SR(s), then chemical thermodynamics can be applied to derive a general relationship between either $f_{SR'C}$ or f_{SR} and the conditional equilibrium constant, K_{adsc}.[5] To simplify notation, let 1 ≡ SR(s) and 2 ≡ SR′C(s), such that $f_{SR} = f_1$, $f_{SR'C} = f_2$, and $K_{ads} = K_{12}$. Because K_{12} depends only on temperature and pressure, infinitesimal changes in the mole fractions, x_1 (= x_{SR}) and x_2 (= $x_{SR'C}$), that are isothermal and isobaric are constrained by a condition, analogous to Eq. 3.33a, but applied to Eq. 4.8 after taking the base e (naperian) logarithm of both sides:

$$d\ln K_{12} = d\ln K_{12c} + d\ln f_2 - d\ln f_1 \equiv 0 \tag{4.9a}$$

This equation is supplemented by the constraint on activities imposed by the Gibbs-Duhem equation for isothermal, isobaric changes in composition (cf. Eq. 3.33b):

$$\begin{aligned}(1 - x_2)d\ln(f_1x_1) + x_2d\ln(f_2x_2) &= 0\\ = (1 - x_2)d\ln f_1 + x_2d\ln f_2 &\end{aligned} \tag{4.9b}$$

where the mole-balance expression, $x_1 + x_2 = 1$, has been used to obtain the second equality and Eq. 4.7 is introduced to represent activity in terms of the rational activity coefficient. The combination of Eqs. 4.9a and 4.9b yields parallel relationships between f_1 or f_2 and K_{12c}:

$$\begin{aligned}d\ln K_{12c} &= d\ln f_1 - d\ln f_2\\ &= d\ln f_1 + [(1 - x_2) / x_2]d\ln f_1\\ &= (1/x_2)d\ln f_1\\ &= -[1 / (1 - x_2)]d\ln f_2\end{aligned} \tag{4.10}$$

where the second and fourth steps result from an appeal to Eq. 4.9b. Expressions for ln f_1 and ln f_2 are derived from Eq. 4.10 by integration, with one limit chosen as the reference state; that is, $f_i = 1.0$ when $x_i = 1.0$, $i = 1, 2$. Therefore[5]

$$\begin{aligned}\ln f_1 &= \int_0^{x_2} x_2' d\ln K_{12c} = \int_0^{x_2} [d(x_2' \ln K_{12c}) - \ln K_{12c}\, dx_2']\\ &= x_2 \ln K_{12c} - \int_0^{x_2} \ln K_{12c}\, dx_2'\end{aligned} \tag{4.11a}$$

$$\ln f_2 = \int_{x_2}^{1} (1 - x_2)\, d\ln K_{12c} = \int_{x_2}^{1} \{d[(1 - x_2')\ln K_{12c}) + \ln K_{12c}] + \ln K_{12c}\, dx_2'\}$$

$$= -(1 - x_2)\ln K_{12c} + \int_{x_2}^{1} \ln K_{12c}\, dx_2' \tag{4.11b}$$

Equation 4.11 shows mathematically how measurements of $\ln K_{12c}$ as a function of x_2 at fixed T^o and P^o are used to calculate the rational activity coefficients. These measurements also can be used to compute the equilibrium constant:

$$\begin{aligned} \ln K_{12} &= \ln K_{12c} + \ln f_2 - \ln f_1 \\ &= \ln K_{12c} - (1 - x_2)\ln K_{12c} + \int_{x_2}^{1} \ln K_{12c}\, dx_2' \\ &\quad - x_2 \ln K_{12c} + \int_{0}^{x_2} \ln K_{12c}\, dx_2' \\ &= \int_{0}^{1} \ln K_{12c}\, dx_2 \end{aligned} \tag{4.12}$$

upon combining Eq. 4.8 in logarithmic form with Eq. 4.11. These relationships epitomize the thermodynamic description of the reaction in Eq. 4.3, given that the solid phase is a binary mixture of adsorbed species and that $\ln K_{12c}$ is known as a function of system composition.

To illustrate the application of Eqs. 4.11 and 4.12, the simple model expression:

$$\ln K_{12c} = a + bx_2 \tag{4.13}$$

can be introduced, where a, b are adjustable parameters. (Equation 4.13 is equivalent to the "mean field approach" in the molecular theory of adsorption.[6] It appears in a wide variety of mechanistic models of the reaction in Eq. 4.3.[7] In the present example, no mechanistic implications are involved in Eq. 4.13, which is taken solely as the result of a parametric analysis of data on K_{12c} as a function of composition.) The combination of Eqs. 4.11–4.13 produces the following model results:

$$\begin{aligned} \ln f_1 &= x_2(a + bx_2) - \int_{0}^{x_2} (a + bx_2')\, dx_2' \\ &= \tfrac{1}{2} b x_2^2 \end{aligned} \tag{4.14a}$$

$$\ln f_2 = -(1 - x_2)(a + bx_2) + \int_{x_2}^{1} (a + bx_2')\,dx_2'$$

$$= \tfrac{1}{2}b(1 - x_2)^2 = \tfrac{1}{2}bx_1^2 \tag{4.14b}$$

$$\ln K_{12} = \int_0^1 (a + bx_2)\,dx_2 = a + \tfrac{1}{2}b \tag{4.14c}$$

The ln f_i (i = 1, 2) turn out to be quadratic functions of the mole fractions and ln K_{12} is equal to the value of ln K_{12c} when $x_2 = x_1 = 0.5$. If b = 0, ln K_{12c} = ln K_{12} and $f_1 = f_2 = 1.0$, irrespective of the species composition. This corresponds to an ideal solid solution (cf. Section 3.3).

Equation 4.3 is formally similar to a complexation reaction between SR(s) and the aqueous solution species on the left side. Indeed, the solid-phase product on the right side can be interpreted on the molecular level as either an outer-sphere or an inner-sphere surface complex. The latter type of adsorbed species was invoked in connection with the generic adsorption–desorption reactions in Eqs. 3.46 and 3.61, which were applied to interpret mineral dissolution processes. In general, adsorbed species can be either diffuse-layer ions or surface complexes,[7] and both species are likely to be included in macroscopic composition measurements based on Eq. 4.2. Equation 4.3, being an overall reaction, does not imply any particular adsorbed species product, aside from its stoichiometry and the electroneutrality condition in Eq. 4.4.

The formal similarity between adsorption and complexation reactions can be exploited to incorporate adsorbed species into the equilibrium speciation calculations described in Sections 2.4 and 3.1. To do this, a choice of "adsorbent species" components ($SR^{Z_{SR}}$ in Eq. 4.3) must be made and equilibrium constants for reactions with aqueous ions must be available. A model for computing adsorbed species activity coefficients must also be selected.[8] Once these choices are made and the thermodynamic data are compiled, a speciation calculation proceeds by adding "adsorbent species" and adsorbed species ($SR'M_p(OH)_yH_xL_q$ in Eq. 4.3) to the mole-balance equations for metals and ligands, and then following the steps described in Section 2.4 for aqueous species. For compatibility of the units of concentration, $n_i^{(w)}$ in Eq. 4.2 is converted to an aqueous-phase concentration through division by the volume of aqueous solution.

An example of a speciation calculation involving metals and ligands that adsorb to form only inner-sphere surface complexes ("specific adsorption") is shown in Table 4.2 for a soil solution at pH 7.5. The generic adsorption reactions for these metals and ligands are exemplified by the third and tenth rows in Table 4.1:

$$SO^-(s) + M^{m+}(aq) = SOM^{(m-1)}(s) \tag{4.15a}$$

$$SOH_2^+(s) + L^{\ell-}(aq) = SL^{(1-\ell)}(s) + H_2O(\ell) \quad (4.15b)$$

Thus the "adsorbent species" selected as components are $SO^-(s)$ and $SOH_2^+(s)$, the first being classified formally as a "ligand" (because it is electron donating), whereas the second is classified formally as a "metal" (because it is electron accepting), insofar as a speciation calculation is concerned. Both of the adsorbed

Table 4.2 Results of a Chemical Speciation Calculation at pH 7.5 Involving "Specific Adsorption" (Inner-Sphere Surface Complexation)[a]

Total Concentrations

$Cu_T = 5.72$	$Cd_T = 5.85$	$Pb_T = 7.00$	$SOH_{2T} = 3.22$
$F_T = 5.69$	$PO_{4T} = 4.70$	$B(OH)_{4T} = 6.30$	$SO_T = 3.22$

Component Concentrations

$Cu^{2+} = 7.81$	$Cd^{2+} = 7.43$	$Pb^{2+} = 8.96$	$SOH_2^+ = 3.60$
$F^- = 5.72$	$PO_4^{3-} = 10.81$	$B(OH)_4^- = 9.59$	$SO^- = 4.62$

Species Distribution

Cu	
as a free metal	0.8%
complexed with CO_3	4.6
complexed with SO_4	0.2
complexed with SO	94.1[b]
complexed with Cl	0.3
Cd	
as a free metal	2.6
complexed with CO_3	0.2
complexed with SO_4	0.7
complexed with Cl	0.4
complexed with SO	96.1[b]
Pb	
as a free metal	1.1
complexed with CO_3	15.2
complexed with SO_4	0.4
complexed with SO	82.7[b]
complexed with OH	0.5
SOH_2	
as a free species	41.5
complexed with PO_4	3.1
complexed with OH	55.3[b]
F	
as a free ligand	93.1%
complexed with Ca	0.4
complexed with Mg	5.1
complexed with SOH_2	1.4[b]
PO_4	
complexed with Ca	0.5
complexed with Mg	1.8
complexed with Na	0.1
complexed with SOH_2	93.6[b]
complexed with H	4.0
$B(OH)_4$	
complexed with SOH_2	97.6[b]
complexed with H	2.3
SO	
as a free species	4.0
complexed with Cu	0.4
complexed with Cd	0.2
complexed with H	95.4[b]

[a]Concentrations are expressed as $-\log [\ ]_e$; for example, 5.72 means 1.91×10^{-6} mol dm^{-3}.
[b]*Inner-sphere* surface complex.

species on the right in Eq. 4.15 are interpreted molecularly as inner-sphere surface complexes involving surface hydroxyl groups. The calculation of activity coefficients for these species requires a model, and the one chosen in the present example was the Constant Capacitance model,[8,9] in which it is assumed that log activity coefficients are proportional to the total net surface charge density, σ_p[10]:

$$\sigma_p = (F/a_s c_s)\Big\{[SOH_2^+]_e + \sum_i (m_i - 1)\,[SOM_i^{(m_i-1)}]_e - [SO^-]_e - \sum_j (1 - \ell_j)\,[SL_j^{(1-\ell_j)}]_e\Big\} \tag{4.16}$$

where $[\ ]_e$ is an equilibrium concentration, F is the Faraday constant, a_s is specific surface area, and c_s is the concentration of adsorbent in suspension. The sums in Eq. 4.16 are over all adsorbed species.

In the present example, the metals Cu, Cd, and Pb, and the ligands, F, PO_4, and $B(OH)_4$ were permitted to undergo "specific adsorption" to form an inner-sphere surface complex with the species SO^- or SOH_2^+, respectively. All but F were found to be primarily adsorbed species at equilibrium.

4.2 Adsorption on Heterogeneous Surfaces

The adsorption–desorption reaction in Eq. 4.3 has been applied to soils in an average sense in a spirit very similar to that of the complexation reactions for humic substances, discussed in Section 2.3.[11] Although no assumption of uniformity is made, the use of Eq. 4.3 to describe adsorption or desorption processes in chemically heterogeneous porous media such as soils does entail the hypothesis that "effective" or average equilibrium (or rate) constants provide a useful representation of a system that in reality exhibits a broad spectrum of surface reactivity. This hypothesis will be an adequate approximation so long as this spectrum is unimodal and not too broad. If the spectrum of reactivity is instead multimodal, discrete sets of average equilibrium or rate constants—each connected with its own version of Eq. 4.3—must be invoked; and if the spectrum is very broad, the sets of these parameters will blend into a continuum (cf. the "affinity spectrum" in Eq. 2.38).

A simple model approach to the heterogeneity of surface reactivity in soils can be developed by assuming that Eq. 4.3 [without the dissociable species Q(aq)] applies to each member of a set of parallel adsorption–desorption reactions involving the same reactant aqueous species, but with differing adsorbent and adsorbate species. A rate law for each of the parallel reactions can be *postulated* in the following form (cf. Eq. 1.34):

$$\frac{d[SR_i]}{dt} = -k_{adsi}[C][SR_i] + k_{desi}(SR_{iT} - [SR_i]) \tag{4.17}$$

where i is an index for each reaction, $[C] \equiv [M^{m+}]^p[H^+]^x[L^{\ell-}]^q[OH^-]^y$, and $SR_{iT} \equiv [SR_i] + [SRC_i]$ is the total concentration of reactive adsorbent indexed by i (cf. the use of M_{oxT} in Eq. 3.48). Both [C] and SR_{iT} are assumed to be constant, whereas $[SR_i]$ decreases or increases with time from adsorption or desorption. The rate law for adsorption–desorption by the entire soil is then the sum of rate laws such as that in Eq. 4.17:

$$\frac{d[SR]}{dt} = -\sum_i k_{adsi}\,[C]\,[SR_i] + \sum_i k_{desi}\,(SR_{iT} - [SR_i]) \tag{4.18}$$

where $[SR] \equiv \Sigma_i\ [SR_i]$ is the total concentration of solid soil adsorbent. Equation 4.18 represents the kinetics of a set of *parallel* adsorption reactions. If the adsorption–desorption reactions instead are *coupled*, then Eq. 4.18 must be interpreted as the rate law for a *hypothetical* soil adsorbent, in the same spirit as were the rate laws in the quasiparticle model, described in Section 2.3 for humic substance complexation reactions.

The first term on the right side of Eq. 4.17 describes an adsorption reaction. Under experimental conditions such that [C] remains constant (e.g., flow-through reactions) and the second term can be ignored, the adsorption rate law has the exponential mathematical solution (cf. Eq. 1.56):

$$[SR_i] \propto \exp(-k_i t) \tag{4.19}$$

where $k_i \equiv k_{adsi}[C]$ is a pseudo-first-order rate coefficient. This exponential time dependence can be introduced into Eq. 4.18, again ignoring the time dependence of [C] and the desorption rate:

$$\frac{d[SR]}{dt} = -\sum_i k_i \exp(-k_i t) \tag{4.20}$$

By analogy with the "affinity spectrum" model in Eq. 2.38, the sum on the right side of Eq. 4.20 can be replaced with an integral over the *probability*, p(k) dk, that a rate coefficient k has a value between k and k + dk:[12]

$$\frac{d[SR]}{dt} = -\int_0^\infty kp(k)\exp(-kt)\,dk = \frac{d}{dt}\int_0^\infty p(k)\exp(-kt)\,dk \tag{4.21}$$

which implies that [SR] itself is proportional to the integral on the right side of the second equation (except possibly for a constant term). Once a choice of a mathematical model for the probability density function, p(k), has been made, the time dependence of [SR] as caused by adsorption processes is in principle determined fully. Note that after substitution into Eq. 4.21, the delta "function" probability density function (cf. Eq. 2.41),

$$p(k) = \sum_i \delta(k - k_i) \tag{4.22}$$

reestablishes the rate law in Eq. 4.20.

One possible model choice for p(k) that is of widespread use in statistical applications, because of its simplicity and flexibility, is the two-parameter gamma distribution:[13]

$$p(k) = [\beta^\alpha/\Gamma(\alpha)]\, k^{\alpha-1} \exp(-\beta k) \tag{4.23}$$

where α and β are positive constants and $\Gamma(\alpha)$ is the gamma function.[14] The definite integral[14]

$$\int_0^\infty x^a \exp(-bx)\, dx = \Gamma(a+1)/b^{a+1} = a\Gamma(a)/b^{a+1} \tag{4.24}$$

for $a > -1$ and $b > 0$, can be applied to show that

$$\int_0^\infty p(k)\, dk = 1 \qquad \text{(normalization)} \tag{4.25a}$$

$$\int_0^\infty k\, p(k)\, dk = \alpha/\beta \equiv \bar{k} \qquad \text{(mean value)} \tag{4.25b}$$

$$\int_0^\infty [k - (\alpha/\beta)]^2 p(k)\, dk = \alpha/\beta^2 \qquad \text{(variance)} \tag{4.25c}$$

Thus the mean and variance of a gamma distribution are sufficient to determine its two parameters, α and β. Note that the coefficient of variation (standard deviation divided by the mean) is equal to the square root of $1/\alpha$. The most probable value of k (the mode) occurs at $(\alpha - 1)/\beta$ if $\alpha > 1$, and as $\alpha \uparrow \infty$ the gamma distribution itself becomes the gaussian or "normal" distribution for the variable, βk.[13]

Reference again to Eq. 4.24 shows that d[SR]/dt is proportional to $\beta^\alpha\alpha/(\beta + t)^{\alpha+1} = \bar{k} / [1 + (t/\beta)]^{\alpha+1}$ and therefore that [SR] is proportional to $1/[1 + (t/\beta)]^\alpha$. (Make the associations $a \leftrightarrow \alpha$, $b \leftrightarrow \beta + t$, and $a \leftrightarrow \alpha - 1$, $b \leftrightarrow \beta + t$ in Eq. 4.24.) These results, in turn, imply a power-law relationship between d[SR]/dt and [SR], that is,[15]

$$d[SR]/dt \propto \bar{k}[SR]^{(\alpha+1)/\alpha} = \bar{k}_{ads}[C]\,[SR]^{(\alpha+1)/\alpha}$$

The same line of reasoning can be applied to the desorption term in Eq. 4.18 after replacing its left side with $-d(SR_T - [SR])/dt$ (to which it is equal because of mole balance) and letting p(k) become a probability density function for the rate coefficient k_{desi}. If Eq. 4.23 is chosen again to represent the distribution of desorption rate coefficients, and if the coefficient of variation (although *not* necessarily the mean or the variance) is assumed to be the same as for the distribution of adsorption rate coefficients, then d[SR]/dt for the desorption process will also be proportional to

$$\bar{k}_{des}(SR_T - [SR])^{(\alpha+1)/\alpha}$$

where $SR_T \equiv \Sigma_i\, SR_{i_T}$.[16]

The rate law in Eq. 4.18 now can be expressed as a power-law expression:[16]

$$\frac{d[SR]}{dt} = -[C]\bar{k}_{ads}[SR]^{(\alpha+1)/\alpha} + \bar{k}_{des}(SR_T - [SR])^{(\alpha+1)/\alpha} \tag{4.26a}$$

or, equivalently,

$$\frac{d[SRC]}{dt} = \bar{k}_{ads}[C](SR_T - [SRC])^{(\alpha+1)/\alpha} - \bar{k}_{des}[SRC]^{(\alpha+1)/\alpha} \tag{4.26b}$$

by mole balance, where $[SRC] \equiv \Sigma_i\, [SRC_i] = SR_T - [SR]$. Equation 4.26 applies to a set of parallel adsorption–desorption reactions for which both the adsorption and desorption rate coefficients are distributed according to the gamma probability density function with the same coefficient of variation, $\alpha^{-1/2}$, but with different mean and variance.

At equilibrium, the left side of Eq. 4.26 vanishes and the rate law defines an *adsorption isotherm equation*:

$$[SRC]_e = \frac{SR_T A [C]_e^b}{1 + A[C]_e^b} \tag{4.27}$$

where $[\ ]_e$ is an equilibrium concentration and

$$A \equiv (\bar{k}_{ads}/\bar{k}_{des})^b \qquad b \equiv \alpha/(\alpha + 1) \quad (0 \le b \le 1) \tag{4.28}$$

Equation 4.27 has the form of the *generalized Freundlich isotherm equation.*[17] If the product $A[C]_e^b << 1$, it reduces to the *van Bemmelen–Freundlich isotherm equation* and, if b = 1, it becomes the *Langmuir isotherm equation* (cf. Eq. 3.50b).[18] Thus Eq. 4.18 and a gamma distribution of the two rate coefficients it contains are sufficient to generate three very common adsorption isotherm equations, with the effect of kinetics appearing in the "affinity parameter" A and the breadth of the distribution of rate coefficients reflected in the exponent b. Note that this latter parameter has unit value only in the limiting case, $\alpha \uparrow \infty$, which corresponds roughly to an extremely narrow, gaussian "spike" distribution for the two rate coefficients. Thus the smaller is the exponent b, the broader is the distribution of rate coefficients among the set of parallel adsorption–desorption reactions represented by Eq. 4.26.

The chemical significance of Eqs. 4.18, 4.26, or 4.27 depends entirely on whether the several mechanistic assumptions (uncoupled reactions, gamma distribution of rate coefficients with the same coefficient of variation, etc.) that were used to derive them can be verified *independently* by molecular-scale experiments. Unless this kind of substantiation is possible, the adherence of data on adsorption–desorption reactions to these equations has no unique mechanistic implication.[2]

4.3 Adsorption Relaxation Kinetics

In analogy with Eqs. 1.50 and 2.5, the overall surface ligand-exchange reactions in Eqs. 3.46, 3.53, 3.56, 3.65, 3.66, and 4.15b can be dissected into steps by applying the concept of the Eigen-Wilkins-Werner mechanism, discussed in Section 2.1. Following this perspective, one would decompose the overall surface complexation reaction in Eq. 4.15b into a set of coupled reactions (cf. Eq. 1.50):

$$SOH_2^+(s) + L^{\ell -}(aq) \underset{k_b}{\overset{k_f}{\rightleftarrows}} SOH_2^+ - - L^{\ell -}(s) \underset{k_b'}{\overset{k_f'}{\rightleftarrows}} SL^{(1-\ell)}(s) + H_2O(\ell) \tag{4.29}$$

where the intermediate species, $SOH_2^+ - - L^{\ell -}$, refers to an outer-sphere complex between the generic surface site, S^+, and the ligand, $L^{\ell -}$. This outer-sphere surface complex is subsequently transformed into the inner-sphere surface complex, $SL^{1-\ell}$. The sequence of reactions in Eq. 4.29 is a special case of the abstract reaction scenario in Eq. 1.52 ($A = SOH_2^+$, $B = L^{\ell -}$, $C = SOH_2^+ - - L^{\ell -}$, $D = SL^{(1-\ell)}$, and $E = H_2O$). A set of rate laws for this scenario appears in Eq. 1.54, and a pair of nonredundant rate laws is given in Eqs. 1.54a and 1.54c.

A full mathematical description of the time dependence of the concentrations of the species in Eq. 4.29 based on the rate laws in Eqs. 1.54a and 1.54c can be accomplished by computer calculation, but often a good approximation can be obtained without extensive numerical analysis by *linearization* of the rate laws. This approach assumes that a suitable experimental method exists to detect very small perturbations from equilibrium among species linked by a set of coupled reactions. The rate laws then are expressed as functions of the small concentration deviations that occur in response to a perturbation, and products of these deviations are neglected, relative to a single deviation, in order to derive a set of coupled, *linearized* rate laws.

The linearization of Eqs. 1.54a and 1.54c proceeds as follows. Let $c_A = c_A^{eq} + \Delta c_A$, $c_B = c_B^{eq} + \Delta c_B$, etc., where c^{eq} is an equilibrium species concentration and Δc is a small deviation from an equilibrium value. This decomposition of each species concentration in Eqs. 1.54a and 1.54c produces the following set of equations:

$$\frac{d\Delta c_A}{dt} = -k_f c_A^{eq} c_B^{eq} - k_f(c_A^{eq}\Delta c_B + c_B^{eq}\Delta c_A) - k_f \Delta c_A \Delta c_B + k_b c_C^{eq} + k_b \Delta c_C \tag{4.30a}$$

$$\frac{d\Delta c_D}{dt} = -k_f' c_C^{eq} + k_f' \Delta c_C - k_b' c_D^{eq} c_E^{eq} - k_b'(c_D^{eq}\Delta c_E + c_E^{eq}\Delta c_D) - k_b' \Delta c_D \Delta c_E \tag{4.30b}$$

where the fact that $dc_A^{eq}/dt = dc_D^{eq}/dt \equiv 0$ has been used to simplify the left side. Without approximation, all terms containing only equilibrium concentrations can be deleted from the right side because they must sum algebraically to zero:

$$c_C^{eq}/c_A^{eq}c_B^{eq} = k_f/k_b \equiv K_c \tag{4.31a}$$

$$c_D^{eq}c_E^{eq}/c_C^{eq} = k_f'/k_b' \equiv K_c' \tag{4.31b}$$

where K_c and K_c' are the respective conditional equilibrium constants for the first and second reactions in the sequence. The linearization approximation involves dropping the term in $\Delta c_A \Delta c_B$ from Eq. 4.30a and that in $\Delta c_D \Delta c_E$ from Eq. 4.30b:

$$\frac{d\Delta c_A}{dt} \approx -k_f(c_A^{eq}\Delta c_B + c_B^{eq}\Delta c_A) + k_b \Delta c_C \tag{4.32a}$$

$$\frac{d\Delta c_D}{dt} \approx k_f' \Delta c_C - k_b'(c_D^{eq}\Delta c_D + c_E^{eq}\Delta c_D) \tag{4.32b}$$

A final simplification of the rate laws can be made by incorporating mole-balance conditions that follow from the stoichiometry of the reactions in Eq. 1.52 (cf. Eq. 1.29):

$$\Delta c_A = \Delta c_B \qquad \Delta c_C = -\Delta c_A - \Delta c_D \qquad \Delta c_D = \Delta c_E \tag{4.33}$$

where the second equality reflects the same mole-balance constraint as that used to eliminate Eq. 1.54b from the set of rate laws in Eq. 1.54. The combination of Eqs. 4.32 and 4.33 yields the coupled, *linearized* rate laws:

$$\frac{d\Delta c_A}{dt} = -\left[k_f(c_A^{eq} + c_B^{eq}) + k_b\right]\Delta c_A - k_b\,\Delta c_D \tag{4.34a}$$

$$\frac{d\Delta c_D}{dt} = -k_f'\,\Delta c_A - \left[k_b'(c_D^{eq} + c_E^{eq}) + k_f'\right]\Delta c_D \tag{4.34b}$$

Were it not for the coupling terms, $k_b\Delta c_D$ and $k'_f\Delta c_A$, Eq. 4.34 would have the same form as Eq. 1.55 (neglecting its constant term on the right side), with an exponential-decay solution typical of first-order reactions (Eqs. 1.56 and 4.19). Because the coupling terms are linear in the Δc, however, it is always possible to find a solution to Eq. 4.34 by postulating that a pair of "time constants," τ_1 and τ_2, exists such that the Δc still show an exponential time dependence ("relaxation"):[19]

$$\Delta c_A = \alpha_1 \exp(-t/\tau_1) + \alpha_2 \exp(-t/\tau_2) \tag{4.35a}$$

$$\Delta c_D = \beta_1 \exp(-t/\tau_1) + \beta_2 \exp(-t/\tau_2) \tag{4.35b}$$

where α_i, β_i ($i = 1, 2$) are adjustable parameters whose values are chosen according to the initial conditions imposed on the Δc. The substitution of Eq. 4.35 into Eq. 4.34 produces the algebraic equations:

$$\sum_{i=1}^{2}\left[\alpha_i\left(a_{11} - \frac{1}{t_i}\right) + \beta_i a_{12}\right]\exp\left(\frac{-t}{\tau_i}\right) = 0 \tag{4.36a}$$

$$\sum_{i=1}^{2}\left[\beta_i\left(a_{22} - \frac{1}{t_i}\right) + \alpha_i a_{21}\right]\exp\left(\frac{-t}{\tau_i}\right) = 0 \tag{4.36b}$$

where

$$a_{11} = k_f(c_A^{eq} + c_B^{eq}) + k_b \tag{4.36c}$$

$$a_{12} = k_b \qquad a_{21} = k_f' \tag{4.36d}$$

$$a_{22} = k_b'(c_D^{eq} + c_E^{eq}) + k_f' \tag{4.36e}$$

The terms in square brackets in Eqs. 4.36a and 4.36b must vanish *independently* for each value of τ in order that these two equations be valid for all values of t. Therefore the τ-values must be chosen such that (i = 1, 2):

$$\alpha_i \left(a_{11} - \frac{1}{\tau_i} \right) + \beta_i a_{12} = 0 \tag{4.37a}$$

$$\alpha_i a_{21} + \beta_i \left(a_{22} - \frac{1}{\tau_i} \right) = 0 \tag{4.37b}$$

The theory of simultaneous algebraic equations[19] then provides the derivation of the necessary and sufficient condition for the existence of α_i, β_i, and τ_i satisfying Eq. 4.37:

$$\frac{1}{\tau_1} = \frac{1}{2} \left\{ (a_{11} + a_{22}) + [(a_{11} + a_{22})^2 - 4(a_{11}a_{22} - a_{12}a_{21})]^{1/2} \right\} \tag{4.38a}$$

$$\frac{1}{\tau_2} = \frac{1}{2} \left\{ (a_{11} + a_{22}) - [(a_{11} + a_{22})^2 - 4(a_{11}a_{22} - a_{12}a_{21})]^{1/2} \right\} \tag{4.38b}$$

with the a_{ij} (i, j = 1, 2) given by Eqs. 4.36c–4.36e. Equation 4.38 prescribes the values of the "time constants" that appear in the model expressions in Eq. 4.35. Thus, because of the linearity of Eq. 4.34 (which makes Eq. 4.37 possible), the Δc can be expressed *mathematically* as if they resulted from uncoupled, parallel reactions (cf. Eqs. 4.20 and 4.35), provided that a pair of "time constants" (equivalent to "renormalized" first-order rate coefficients) is defined by Eq. 4.38. If there were no coupling, $a_{12} = a_{21} = 0$ and the "time constants" would be simply a_{11} and a_{22}, as is evident from both Eqs. 4.34 and 4.38.

As in the applications of Eqs. 1.53 and 1.54 discussed in Section 1.5, it often occurs that the time scales of the component reactions in a coupled sequence, like that in Eq. 4.29, are widely separated, with one reaction equilibrating long before the other. When this is true, Eq. 4.38 can be simplified. For example, if the first reaction in the sequence in Eq. 1.52 or 4.29 comes to equilibrium much more rapidly than the second one, the condition that $k_f(c_A^{eq} + c_B^{eq}) + k_b >> k_f'$ or $k_b'(c_D^{eq} + c_E^{eq})$ is met, and Eq. 4.38 reduces to the approximations:[20]

$$\frac{1}{\tau_1} \approx a_{11} \qquad \frac{1}{\tau_2} \approx a_{22} - \left(\frac{a_{12}a_{21}}{a_{11}} \right) \tag{4.39}$$

Table 4.3 lists the result in Eq. 4.39 along with two other common examples of coupled reaction sequences and the approximate equations for their "time constants" based on linearization analysis and the assumption of "fast" and "slow" components.[20] In each example, τ_1 is the smaller "time constant."

Equation 4.35 shows that the concentration deviations based on a linearization analysis of the rate laws in Eqs. 1.54a and 1.54c will decay to zero exponentially ("relax") as governed by the two "time constants," τ_1 and τ_2. These two parameters, in turn, are related to the rate coefficients for the coupled reactions whose kinetics the rate laws describe (Eqs. 4.36c–4.36e and 4.38). If the rate coefficients are known to fall into widely different time scales for each of the coupled reactions, their relation to the "time constants" can be simplified mathematically (Eq. 4.39 and Table 4.3). Thus an experimental determination of the "time constants" leads to a calculation of the rate coefficients.[20] In the example of the metal complexation reaction in Eq. 1.50, with the assumptions that the outer-sphere complexation step is much faster than the inner-sphere complexation step and that dissociation of the inner-sphere complex is negligible ($k'_b = 0$ in Eq. 1.54c), the results for τ_1 and τ_2 in the first row of Table 4.3 can be applied. The expression for $1/\tau_1$ indicates that measurements of this parameter as a function of differing equilibrium concentrations of the complexing metal and ligand will produce a straight line whose slope is k_f and whose y-intercept is k_b. The measured values of $1/\tau_2$ at these same two equilibrium concentrations then lead to a calculation of k'_f.

Experimental methodologies for perturbing a chemical reaction at equilibrium are well developed and descriptions of them are widely available.[20,21] The choice of method depends on the time scale of the reaction kinetics and the kinds of chemical species whose concentration deviations are to be measured. Techniques as simple as the dilution of one or more chemical species or as complicated as electromagnetic field pulsing can be involved (Fig. 4.1). The basic principles, regardless of methodology, are that an external perturbation (e.g., a change in applied pressure) occurs over a time interval that is very much smaller than the time scales of the reaction kinetics; that the mechanism

Table 4.3 Approximate "Time Constants" in Two-Step, Coupled-Reaction Sequences[20]

Sequence	Conditions[a]	$1/\tau_1$	$1/\tau_2$
$A + B \leftrightarrows C \leftrightarrows D + E$	$k_f(c_A^{eq} + c_B^{eq}) + k_b >> k'_f,\ k'_b(c_D^{eq} + c_E^{eq})$	$k_f(c_A^{eq} + c_B^{eq}) + k_b$	$k'_b(c_D^{eq} + c_E^{eq}) + [k'_f k_f(c_A^{eq} + c_B^{eq}) /(k_f(c_A^{eq} + c_B^{eq}) + k_b)]$
$A + B \leftrightarrows C \leftrightarrows D$	$k_f(c_A^{eq} + c_B^{eq}),\ k_b << k'_f + k'_b$	$k'_f + k'_b$	$k_f(c_A^{eq} + c_B^{eq}) + [k_b k'_b/(k'_f + k'_b)]$
$A + B \leftrightarrows C$ $C + D \leftrightarrows E$	$k_f(c_A^{eq} + c_B^{eq}) + k_b >> k'_f(c_C^{eq} + c_D^{eq}),\ k'_b$	$k_f(c_A^{eq} + c_B^{eq}) + k_b$	$k'_f c_C^{eq} + k'_b + [k'_f k_f(c_A^{eq} + c_B^{eq})c_D^{eq} /(k_f(c_A^{eq}+c_B^{eq}) + k_b)]$

[a]In each example, k_f, k_b apply to the first reaction and k'_f, k'_b apply to the second reaction in the sequence.

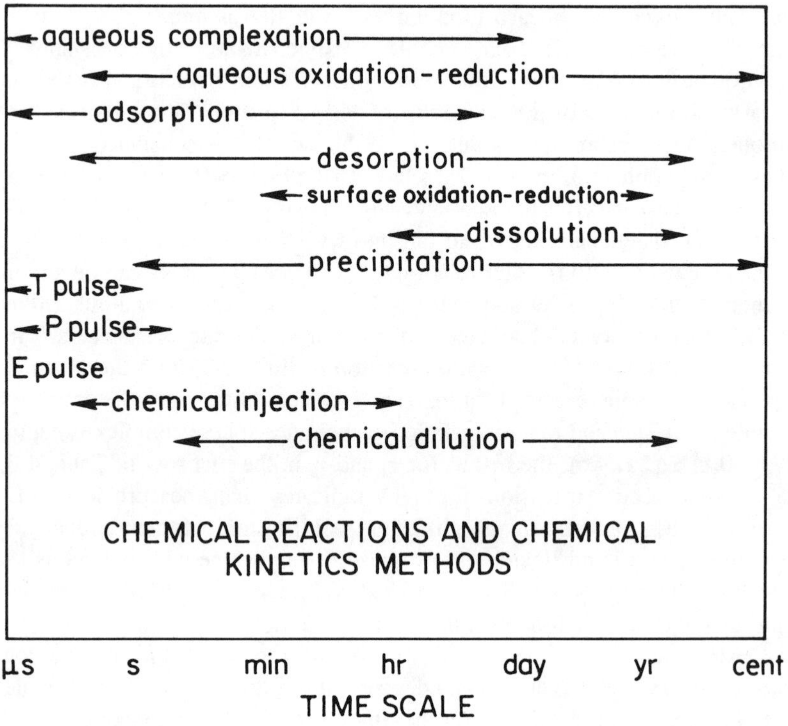

FIG. 4.1. Characteristic time scales for soil chemical reactions and for some typical experimental kinetics techniques.

of response to this external perturbation is identical to that which operates when the same kind of perturbation occurs internally as a spontaneous fluctuation (e.g., a local fluctuation in the molal volume); and that the response is not so intense as to require the full nonlinearity of the rate laws in order to describe it (i.e., linearization of the rate laws is a good approximation).

A prototypical example of Eq. 4.29 occurs with the "specific adsorption" of oxyanions by metal oxyhydroxides.[9] Taking MoO_4^{2-} as the oxyanion and goethite (α-FeOOH) as the metal oxyhydroxide, one can write down the reaction sequence (cf. Eqs. 3.53 and 3.56):

$$\equiv Fe - OH_2^+(s) + MoO_4^{2-}(aq) \underset{k_b}{\overset{k_f}{\rightleftarrows}} \equiv Fe - OH_2^+ - - MoO_4^{2-}(s)$$

$$\underset{k_b'}{\overset{k_f'}{\rightleftarrows}} \equiv Fe - MoO_4^-(s) + H_2O(\ell) \qquad (4.40)$$

This sequence is a realization of the abstract scenario in the first row of Table 4.3 (A = $\equiv Fe - OH_2^+(s)$; C = $\equiv Fe - OH_2^+ - - MoO_4^-$; D = $\equiv Fe - MoO_4^-$, etc.). If the assumption is made that the first step is very much faster than the second, the expressions for the "time constants" in the first row of the table can be applied:

$$1/\tau_1 = k_f([\equiv Fe - OH_2^+]_e + [MoO_4^{2-}]_e) + k_b \tag{4.41a}$$

$$\begin{aligned} 1/\tau_2 = \Big\{ k_f^{\prime} k_f([\equiv Fe - OH_2^+]_e + [MoO_4^{2-}]_e)/(k_f([\equiv Fe - OH_2^+]_e \\ + [MoO_4^{2-}]_e) + k_b) \Big\} + k_b^{\prime}([\equiv Fe - MoO_4^-]_e + [H_2O]_e) \end{aligned} \tag{4.41b}$$

Given that the concentration deviation for "species E," H_2O, will be negligible, the second term on the right side of Eq. 4.32b will reduce to $-k'_b \Delta c_D$, making the corresponding term in Eq. 4.34b equal to $-(k'_b + k'_f)\Delta c_D$, and a_{22} equal to $(k'_b + k'_f)$ in Eq. 4.36e. From Eq. 4.39 it then follows that the second term on the right in Eq. 4.41b will be simply k'_b, without the equilibrium concentrations of "species D and E," in this case. Thus, if outer-sphere surface complexation is much faster than inner-sphere surface complexation and if the effect of any perturbation of the reactions in Eq. 4.40 on the concentration of water is negligible, the linear relationships

$$1/\tau_1 = k_f S + k_b \qquad 1/\tau_2 = k_f^{\prime} k_f S \tau_1 + k_b^{\prime} \tag{4.42}$$

should apply, where S = $[\equiv Fe - OH_2^+]_e + [MoO_4^{2-}]_e$. These relationships have been found to describe experimental graphs of $1/\tau_1$ or $1/\tau_2$ versus S or $k_f S \tau_1$, respectively.[22] Because the reactions in Eq. 4.40 occur on time scales in the range of 1–200 ms, they can be perturbed effectively by applying a pressure pulse (Fig. 4.1) on a microsecond time scale and then observing the response of the aqueous species in the system via a conductivity measurement.[20-22] The exponential decay of the conductivity over the millisecond time scale is resolved into components according to Eq. 4.35 in order to determine $1/\tau_1$ and $1/\tau_2$.

A reaction sequence analogous to that in Eq. 4.40 can also be developed for the "specific adsorption" of bivalent metal cations (e.g., Cu^{2+}, Mn^{2+}, or Pb^{2+}) by metal oxyhydroxides.[21] In this application the abstract scenario in the first row of Table 4.3 is realized with A = $\equiv$Al-OH, B = M^{2+}, C = $\equiv$Al-OH - - M^{2+}, D = $\equiv$Al-OM$^+$, and E = H^+, where M is the metal complexed by an OH group on the surface of an aluminum oxyhydroxide. Analysis of pressure-pulse relaxation kinetics data leads to a calculation of the second-order rate coefficient k_f, under the assumption that the first step in the sequence in Eq. 4.40 is rate determining. Like k_d, the rate coefficient for the dissolution of a metal-containing solid (Section 3.1; cf. Fig. 3.4), measured values of k_f correlate positively in a log–log plot with k_{wex}, the rate coefficient for water exchange on the metal cation M (Eq. 2.3). This correlation implies that desolvation of the complexed metal cation figures importantly in the adsorption mechanism.

The left side of Eq. 4.29 can itself be resolved into additional steps involving protonation and outer-sphere surface complexation:

$$SOH(s) + H^+(aq) \underset{k_d}{\overset{k_a}{\rightleftarrows}} SOH_2^+(s)$$

$$SOH_2^+(s) + L^{\ell-}(aq) \underset{k_b}{\overset{k_f}{\rightleftarrows}} SOH_2^+ \text{ -- } L^{\ell-}(s) \tag{4.43}$$

This subsequence is useful to consider if the time scale for proton adsorption–desorption reactions is comparable to or longer than that for outer-sphere surface complexation. It is a special case of the abstract scenario listed third in Table 4.3. Under the conditions given there, the protonation–proton dissociation reaction ($A = SOH$, $B = H^+$, $C = SOH_2^+$) is assumed to be much faster than outer-sphere surface complexation–dissociation, such that ($k_f \leftrightarrow k_a$, $k_b \leftrightarrow k_d$, $k'_f \leftrightarrow k_f$, $k'_b \leftrightarrow k_b$ here)

$$1/\tau_1 = k_a([SOH]_e + [H^+]_e) + k_d \tag{4.44a}$$

$$1/\tau_2 = k_f[SOH_2^+]_e + k_b + k_f k_a([SOH]_e + [H^+]_e)\,[L^{\ell-}]_e \tau_1 \tag{4.44b}$$

If proton dissociation is, in addition, a slow process relative to outer-sphere surface complexation (i.e., if $K_c = [SOH_2^+]_e/[SOH]_e[H^+]_e = k_a/k_d$ is very large), then k_d can be neglected in Eq. 4.44a, and Eq. 4.44b simplifies to the expression

$$1/\tau_2 = k_f([SOH_2^+]_e + [L^{\ell-}]_e) + k_b \tag{4.44c}$$

which is the same as the expression for $1/\tau_1$ in Eqs. 4.41a and 4.42. The latter equation was obtained under the assumption of "instantaneous" and permanent (on the time scale for outer-sphere surface complexation) protonation that is implicit in Eqs. 4.29 and 4.40. Equation 4.44c has been applied to the reaction of sulfate anions with goethite:[23]

$$\equiv Fe - OH(s) + H^+(aq) \underset{k_d}{\overset{k_a}{\rightleftarrows}} \equiv Fe - OH_2^+(s) \tag{4.45a}$$

$$\equiv Fe - OH_2^+(s) + SO_4^{2-}(aq) \underset{k_b}{\overset{k_f}{\rightleftarrows}} \equiv Fe - OH_2^+ - - SO_4^{2-}(s) \qquad (4.45b)$$

Because of the millisecond time scale for these reactions, pressure-pulse perturbation (Fig. 4.1) with conductivity detection of the response can be used, as in the molybdate adsorption example. Evidently, the inner-sphere surface complexion step for sulfate occurs on time scales very much longer than those for its outer-sphere surface complexation, and therefore it was not observed experimentally with the method used.

For anions that complex weakly with surface hydroxyl groups (e.g., perchlorate), it is possible that the outer-sphere surface complexation step in Eq. 4.43 will be faster than both protonation and proton dissociation. This condition is just the opposite of that given for the third reaction sequence in Table 4.3. Its effect on the "time constants" for the sequence can be derived by applying the approach in Eqs. 4.30–4.38.[20] In place of Eqs. 4.36c–e, one finds the factors in multiplying Δc_A (= $\Delta[SOH]$ = $\Delta[H^+]$) and Δc_D (= $\Delta[L^{\ell-}]$) in Eq. 4.34 to be

$$a_{11} = k_a([SOH])_e + [H^+]_e) + k_d \qquad (4.36f)$$

$$a_{12} = -k_d \qquad a_{21} = -k_f[L^{\ell-}]_e \qquad (4.36g)$$

$$a_{22} = k_f([SOH_2{}^+]_e + [L^{\ell-}]_e) + k_b \qquad (4.36h)$$

These factors are introduced into Eq. 4.38 to calculate the "time constants." If the condition in Table 4.3 holds, then Eq. 4.39 applies, and Eq. 4.44 results as the approximate expression for the "time constants." If, on the other hand, the opposite condition holds, then $a_{22} >> a_{11}$ and Eq. 4.38 reduces to the approximations (cf. Eq. 4.39):

$$1/\tau_1 \approx a_{11} - (a_{12}a_{21}/a_{22}) \qquad 1/\tau_2 \approx a_{22} \qquad (4.46)$$

and Eq. 4.44 becomes

$$1/\tau_1 = k_a([SOH]_e + [H^+]_e) + k_d(k_f[SOH_2^+]_e + k_b)\tau_2 \qquad (4.47a)$$

$$1/\tau_2 = k_f([SOH_2^+]_e + [L^{\ell-}]_e) + k_b \qquad (4.47b)$$

Equation 4.47a is illustrated in Fig. 4.2 for the protonation of hematite (α-Fe_2O_3) suspended in acidic perchlorate solution.[24] The variable plotted along the x-axis is obtained by factoring k_a on the right side of Eq. 4.47a and making use of Eq. 4.31:

$$1/\tau_1 = k_a \left\{ [SOH]_e + [H^+]_e + K_c^{-1} \frac{([SOH_2^+]_e + K_{c1}^{-1})}{([SOH_2^+]_e + [L^{\ell-}]_e + K_{c1}^{-1})} \right\} \qquad (4.48)$$

where

$$K_c = [SOH_2^+]_e \,/\, [SOH]_e[H^+]_e = k_a/k_d \qquad (4.49a)$$

$$K_{c1} = [SOH_2^+ - - L^{\ell-}]_e \,/\, [SOH_2^+]_e[L^{\ell-}]_e = k_f/k_b \qquad (4.49b)$$

and $L^{\ell-} = ClO_4^-$ in this example. The conditional equilibrium constants can be determined in the experiments used to determine the equilibrium concentrations, thus making it possible to use the factor in curly brackets in Eq. 4.48 as the x-axis variable in Fig. 4.2. The slope of the resulting line is equal to the protonation rate coefficient k_a, and k_d can then be computed by using Eq. 4.49a.

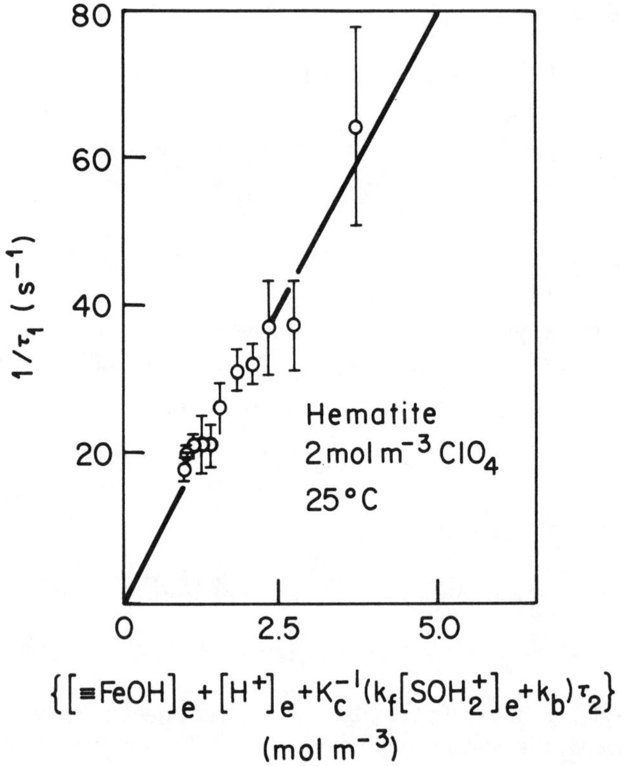

FIG. 4.2. Experimental test of Eq. 4.47a for the reaction in Eq. 4.45a as applied to hematite (α-Fe_2O_3) suspended in 2 mol m^{-3} perchlorate solution at pH 2–5 (data from R. D. Astumian et al.[24]). A linear plot is consistent with Eq. 4.47a.

A similar methodology can be applied to Eq. 4.47b to calculate k_b using Eq. 4.49b after k_f has been determined from measurements of $1/\tau_1$.[25]

4.4 Surface Oxidation–Reduction Reactions

Surface oxidation–reduction reactions involve electron-transfer processes between surface species. Examples of these reactions are considered in Section 3.4 as part of the mechanistic basis for the reductive dissolution of metal oxyhydroxide solid phases. The two-step, overall coupled-reaction sequence in Eq. 3.46 describes the reduction of a metal at the solid surface as mediated by the adsorption and subsequent oxidation of a ligand drawn from the aqueous solution phase. The adsorption of the ligand by the mechanism of inner-sphere surface complexation with the metal facilitates electron transfer and therefore is an essential step in the overall dissolution process. Other adsorption mechanisms (e.g., outer-sphere surface complexation or incorporation of the ligand into the diffuse-ion swarm) can also serve to facilitate electron transfer if the metal and ligand can be brought together in sufficiently close proximity and immobilized long enough.[26]

The surface reactions in reductive dissolution processes are illustrated for a Fe(III) oxyhydroxide solid phase in Eqs. 3.53–3.55, with data shown in Fig. 3.10 for the organic ligand ascorbic acid. Other examples include Mn(IV) or Mn(III) oxyhydroxide solids and ligands, such as quinones, phenols, and inorganic oxyanions. Taking birnessite (δ-MnO_2) and selenite (SeO_3^{2-}) as a case in point, one can adapt Eq. 3.53 in the form[27]

$$\equiv Mn^{IV} - OH(s) + HSeO_3^-(aq) + H^+(aq) \underset{k_b}{\overset{k_f}{\rightleftarrows}} \equiv Mn^{IV} - OSeOOH(s) + H_2O(\ell)$$

$$\overset{k_f'}{\rightarrow} \equiv Mn^{II} - OSeO_2OH(s) + 2H^+(aq) \tag{4.50}$$

This overall sequence differs a little from Eq. 3.53 (and from the reaction in Eq. 3.46) because two electrons are transferred when Mn(IV) is reduced to Mn(II) while Se(IV) is oxidized to Se(VI) to form selenate (SeO_4^{2-}). A similar situation occurs in the oxidation of As(III) to As(V) or Cr(III) to Cr(VI) on Mn oxyhydroxide solids.[27,28] At fixed proton and water concentrations, the rate law for the formation of the surface species, $\equiv Mn^{IV} - OSeOOH$, can still be described by Eq. 3.47, but with the *three* pseudo-rate coefficients, $K_f = k_f[H^+]$, $K_b = k_b[H_2O]$, $K_f' = k_f'[H_2O]$, the last of which replaces k_f' in Eq. 3.47. Equations 3.48–3.51 then can be applied with this minor modification.

As discussed in connection with the dissolution rate law in Eq. 3.58, the relationship of Eq. 4.50 to the rate at which $Mn^{2+}(aq)$ appears as a result of Mn reduction depends on a more detailed picture of the surface reactions involved. A generalization of the coupled reactions in Eq. 4.50 to permit the backward reaction in the second step and to allow the detachment of Mn^{2+} and SeO_4^{2-} to proceed sequentially can be expressed by following the scenario:[27]

$$\equiv Mn^{IV} - OH(s) + HSeO_3^-(aq) + H^+(aq) \underset{k_b}{\overset{k_f}{\rightleftarrows}} \equiv Mn^{IV} - OSeOOH(s) + H_2O(\ell)$$

$$\underset{k_b^{/}}{\overset{k_f^{/}}{\rightleftarrows}} \equiv Mn^{II} - OSeO_2OH(s) + 2H^+(aq)$$

$$\equiv Mn^{II} - OSeO_2OH(s) + H_2O(\ell) \overset{K_{det}}{\rightarrow} \equiv Mn^{II} - OH(s) + SeO_4^{2-}(aq) + 2H^+(aq)$$

$$\equiv Mn^{II} - OH(s) + H^+(aq) \underset{k_{des}}{\overset{k_{ads}}{\rightleftarrows}} \equiv Mn^{II} - OH_2^+(s)$$

$$\overset{k_{det}^{/}}{\rightarrow} Mn^{2+}(aq) + H_2O(\ell) \tag{4.51}$$

where the first detachment step is analogous to the backward reaction in the ligand-exchange process involving selenite and the second detachment step is an example of the proton-promoted dissolution reaction scheme in Eq. 3.61. Five principal (nonredundant) rate laws can be written to describe the kinetics of the scenario in Eq. 4.51 at constant proton and water concentrations:

$$\frac{d[HSeO_3^-]}{dt} = -K_f[\equiv Mn^{IV} - OH][HSeO_3^-] + K_b[\equiv Mn^{IV} - OSeOOH] \tag{4.52a}$$

$$\frac{d[\equiv Mn^{IV} - OSeOOH]}{dt} = -\frac{d[HSeO_3^-]}{dt} - K_f^{/}[\equiv Mn^{IV} - OSeOOH] + K_b^{/}[\equiv Mn^{II} - OSeO_2OH] \tag{4.52b}$$

$$\frac{d[SeO_4^{2-}]}{dt} = K_{det}[\equiv Mn^{II} - OSeO_2OH] \tag{4.52c}$$

$$\frac{d[\equiv Mn^{II} - OH]}{dt} = \frac{d[SeO_4^{2-}]}{dt} - K_{ads}[\equiv Mn^{II} - OH] + k_{des}[\equiv Mn^{II} - OH_2^+] \quad (4.52d)$$

$$\frac{d[Mn^{2+}]}{dt} = k'_{det}[\equiv Mn^{II} - OH_2^+] \quad (4.52e)$$

where $K_f = k_f[H^+]$, $K_b = k_b[H_2O]$, $K'_f = k'_f[H_2O]$, $K'_b = k'_b[H^+]^2$, $K_{det} = k_{det}[H_2O]$, and $K_{ads} = k_{ads}[H^+]$ are pseudo-rate coefficients. Equation 4.52 comprises five coupled ordinary differential equations in five independent concentration variables, and therefore a mathematical solution for the kinetics of the reaction scenario in Eq. 4.51 is possible, either approximately in closed form or "exactly" by numerical analysis. (Note that $[\equiv Mn^{IV} - OH]$ can be related to the total concentration of reducible Mn and to $[\equiv Mn^{IV} - OSeOOH]$, and that $[\equiv Mn^{II} - OH]$ can be related to the total concentration of reduced Mn, $[\equiv Mn^{II} - OSeO_2OH]$, and $[\equiv Mn^{II} - OH_2^+]$, by the use of mole balance, as done in Eq. 3.47.)

If a clear separation of time scales exists for Eqs. 4.52a and 4.52b, as compared to Eqs. 4.52c–4.52e, then the kinetics of surface oxidation–reduction can be decoupled from those of surface species detachment. For example, if Mn(II) oxidation is negligibly slow, $K'_b = 0$ and Eqs. 4.52a and 4.52b can be solved approximately, as is described in either Section 3.4 in connection with Eqs. 3.48 and 3.49, or Section 4.3 in connection with Eq. 4.30. The first approach applies to constant $[HSeO_3^-]$, whereas the second one requires small deviations of the concentrations of the four chemical species in Eqs. 4.52a and 4.52b from their equilibrium values.

On the (assumed) much longer time scale over which SeO_4^- and Mn^{2+} begin to appear in the aqueous-solution phase from the decomposition of $\equiv Mn^{II} - OSeO_2OH$, Eqs. 4.52c–4.52e can be solved under an appropriate imposed condition regarding the time variation of $[\equiv Mn^{II} - OSeO_2OH]$ based on the surface oxidation–reduction kinetics. For example, under steady-state conditions that yield constant concentrations of the adsorbed and dissolved selenite species, Eqs. 4.52a and 4.52b lead to a constant concentration of adsorbed selenate and therefore a constant rate of selenate detachment from the mineral surface (Eq. 4.52c). If the reasonable assumption is also made that the proton reaction with $\equiv Mn^{II} - OH$ equilibrates rapidly, then

$$\frac{d[\equiv Mn^{II} - OH_2^+]}{dt} = K_{ads}[\equiv Mn^{II} - OH] - (k_{des} + k'_{det})[\equiv Mn^{II} - OH_2^+] \approx -k'_{det}[Mn^{II} - OH_2^+] \quad (4.52f)$$

which is the (redundant) rate law for the protonated surface species, implies an exponential decay with time for $[\equiv Mn^{II} - OH_2^+]$ and therefore the rate of mineral dissolution (Eq. 4.52e).[29]

The overall surface oxidation–reduction reactions in Eqs. 3.46 and 4.50 exhibit the premise that inner-sphere complexes can mediate electron-transfer processes effectively. This premise originates in the elementary oxidation–reduction reactions for aqueous-solution species that appear in Eq. 2.34: associative metal exchange takes place on a ligand offering lone-pair electrons with which to complex a metal ion that will, in turn, reduce the metal ion already bound to the ligand (see Fig. 2.5).[30] The role of the "bridging ligand" in mediating electron transfer is critical, as is illustrated also by Problem 6 in Chapter 2, which shows that Fe(II) can be oxidized by O_2 much more rapidly as a hydrolytic species than as a solvation complex. Indeed, a mechanistic basis for the rate law in Eq. 2.36 can be provided by the following parallel scheme (cf. Problems 5 and 6 in Chapter 2):[31]

$$Fe^{2+}(aq) + O_2(g) \xrightarrow{k_f} Fe^{3+}(aq) + O_2^-(aq) \tag{4.53a}$$

$$Fe^{2+}(aq) + H_2O(\ell) \overset{^*K_1'}{=} FeOH^+(aq) + H^+(aq) \tag{4.53b}$$

$$FeOH^+(aq) + O_2(g) \xrightarrow{k_f'} FeOH^{2+}(aq) + O_2^-(aq) \tag{4.53c}$$

$$Fe^{2+}(aq) + 2H_2O(\ell) \overset{^*\beta_2}{=} FeOH_2^0(aq) + 2H^+(aq) \tag{4.53d}$$

$$Fe(OH)_2^0(aq) + O_2(g) \xrightarrow{k_f''} Fe(OH)_2^+(aq) + O_2^-(aq) \tag{4.53e}$$

where $O_2^-(aq)$ is the superoxide anion and the hydrolysis reactions are assumed to equilibrate much more quickly than the oxidation–reduction reactions. The equilibrium constants for the reactions in Eqs. 4.53b and 4.53d are

$$^*K'_1 = \mathbf{(FeOH^+)(H^+)\,/(Fe^{2+})(H_2O)} \approx 10^{-9.3} \tag{4.54a}$$

$$^*\beta_2 = \mathbf{(Fe(OH)_2^0)(H^+)^2\,/(Fe^{2+})(H_2O)^2} \approx 10^{-20.6} \tag{4.54b}$$

whereas the forward rate constants for the reactions in Eqs. 4.53a, 4.53c, and 4.53e have the values[31] $k_f \approx 10^{-8}$ atm^{-1} s^{-1}, $k_f' \approx 0.032$ atm^{-1} s^{-1}, and $k_f'' \approx 10^4$ atm^{-1} s^{-1}.

The overall rate law for the oxidation of Fe(II) based on Eq. 4.53 then takes the form

$$\frac{-\,d[\mathrm{Fe(II)}]}{dt} = \left\{ k_f[\mathrm{Fe}^{\,+}] + k_f'[\mathrm{FeOH}^{+}] + k_f''[\mathrm{Fe(OH)}_2^0] \right\} p_{O_2}$$

$$= \left\{ k_f + k_f'\,{}^*K_1'[\mathrm{H}^+]^{-1} + k_f''\,{}^*\beta_2[\mathrm{H}^+]^{-2} \right\} p_{O_2}\,[\mathrm{Fe}^{2+}] \tag{4.55}$$

upon neglecting ionic strength effects on the hydrolysis equilibrium constants. At pH < 9, $[Fe^{2+}]$ on the right side of Eq. 4.55 can be equated accurately to [Fe(II)]. If pH ≈ 3, only the reaction in Eq. 4.53a is important, and Fe(II) oxidation occurs on a time scale equal to $0.693/k_f p_{O_2} \approx 3.3 \times 10^8$ s, or *years* (Fig. 4.3). If pH ≈ 6, on the other hand, the reaction in Eq. 4.53e dominates, and Fe(II) oxidation occurs on the time scale $0.693/k_f''{}^*\beta_2[H^+]^{-2}p_{O_2} \approx 1.3 \times 10^5$ s, or *days*, as mentioned in connection with Eq. 2.37. Thus complex formation increases the rate of Fe(II) oxidation by several orders of magnitude. This fact is also reflected by a strong positive correlation between the forward rate coefficients and the equilibrium constants for the reactions in Eqs. 4.53a, 4.53c, and 4.53e.[31]

This same line of reasoning can be extended to surface oxidation reactions

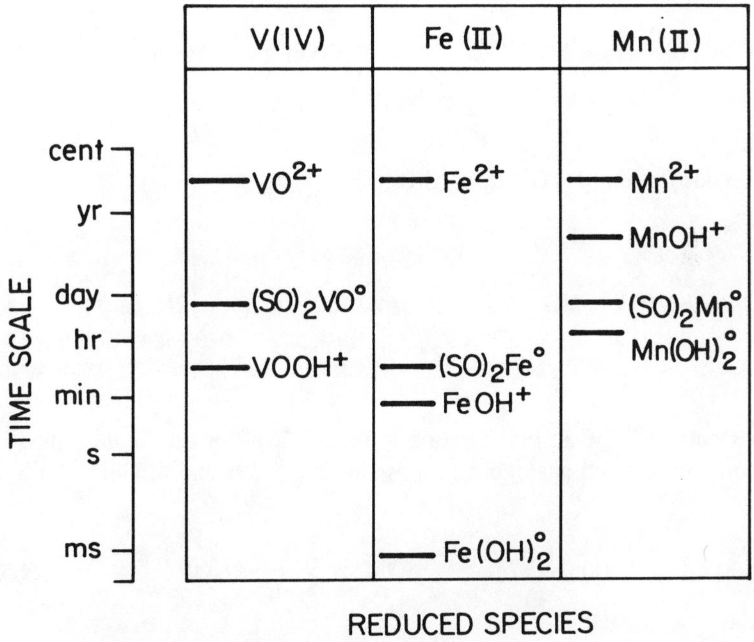

FIG. 4.3. Characteristic time scales [= 0.693/(rate coefficient × p_{O_2}), where rate coefficient = k_f, k_f', or k_f'', as in Eq. 4.53] for the oxidation of aqueous or surface species of V(IV), Fe(II), and Mn(II) at p_{O_2} = 1 atm (data from W. Stumm, B. Sulzberger, and J. Sinniger, The coordination chemistry of the oxide-electrolyte interface; The dependence of surface reactivity (dissolution, redox reactions) on surface structure, *Croatica Chimica Acta* **63**:277 (1990)).

by following the same logical path used to adapt the Eigen-Wilkins-Werner mechanism in Eqs. 1.6 and 1.7 to the surface complexation reaction in Eq. 4.29.[31,32] The analogs of Eqs. 4.53b and 4.53d for the inner-sphere surface complexation of a reduced cationic species are examples of Eq. 4.3 applied to an hydroxylated surface:

$$\mathrm{SOH^o(s) + M^{m+}(aq) \underset{k_{des}}{\overset{k_{ads}}{\rightleftarrows}} SOM^{(m-1)+}(s) + H^+(aq)} \tag{4.56a}$$

$$\mathrm{2SOH^o(s) + M^{m+}(aq) \underset{k'_{des}}{\overset{k'_{ads}}{\rightleftarrows}} (SO)_2M^{(m-2)+}(s) + 2H^+(aq)} \tag{4.56b}$$

Special cases of these reactions include the adsorption of Mn^{2+} by goethite (α-FeOOH):[32]

$$\mathrm{\equiv FeOH(s) + Mn^{2+}(aq) \underset{k_{des}}{\overset{k_{ads}}{\rightleftarrows}} \equiv FeOMn^+(s) + H^+(aq)} \tag{4.57}$$

and the adsorption of VO^{2+} by δ-Al_2O_3:[33]

$$\mathrm{2 \equiv AlOH(s) + VO^{2+}(aq) \underset{k'_{des}}{\overset{k'_{ads}}{\rightleftarrows}} (\equiv AlO)_2VO^o(s) + 2H^+(aq)} \tag{4.58}$$

The oxidation of the adsorbed cation in Eqs. 4.56–4.58 can be described by a reaction with oxygen analogous to those in Eqs. 4.53c and 4.53e:[32]

$$\mathrm{SOM^{(m-1)+}(s) + O_2(g) \overset{k'_f}{\rightarrow} SOM^{(m-2)+}{-}{-}{-}O_2^-(s)} \tag{4.59a}$$

$$\mathrm{(SO)_2M^{(m-2)+}(s) + O_2(g) \overset{k''_f}{\rightarrow} (SO)_2M^{(m-3)+}{-}{-}{-}O_2^-(s)} \tag{4.59b}$$

where the dashed line on the right side denotes a complex between the oxidized, adsorbed cationic species and the superoxide anion. The rate law for these two parallel reactions is analogous to that in Eq. 4.55:

$$-\frac{d[M_{red}]}{dt} = \left\{ k_f' [SOM^{(m-1)+}] + k_f'' [(SO)_2M^{(m-2)+}] \right\} p_{O_2} \qquad (4.60)$$

where $[M_{red}]$ is the sum of the concentrations of the two surface species of the reduced cation M^{m+}. Figure 4.4 shows experimental data consistent with this rate law. On the left side of the figure is a graph of $\log[\equiv(AlO)_2VO^o]$ against time as obtained from experiments on the oxidation of VO^{2+} adsorbed by δ-Al_2O_3 at pH 6.5 and $p_{O_2} = 1$ atm.[33] The linearity of the plot indicates conformity to the pseudo-first-order rate law (cf. Note 7 in Chapter 2):

$$\frac{d[(\equiv AlO)_2VO^o]}{dt} = -k_f'' p_{O_2} [(\equiv AlO)_2VO^o] \qquad (4.61a)$$

with $k_f'' \approx 3.5 \times 10^{-5}$ atm^{-1} s^{-1}. This value of the rate coefficient may be compared with $k_f' \approx 1.3 \times 10^{-3}$ atm^{-1} s^{-1} for the oxidation of $VOOH^+$(aq) and $k_f \approx 10^{-8}$ atm^{-1} s^{-1} for the oxidation of VO^{2+}(aq), as described by a rate law like that in Eq. 4.55 for Fe(II).[33] The right side of Fig. 4.4 shows a graph of the

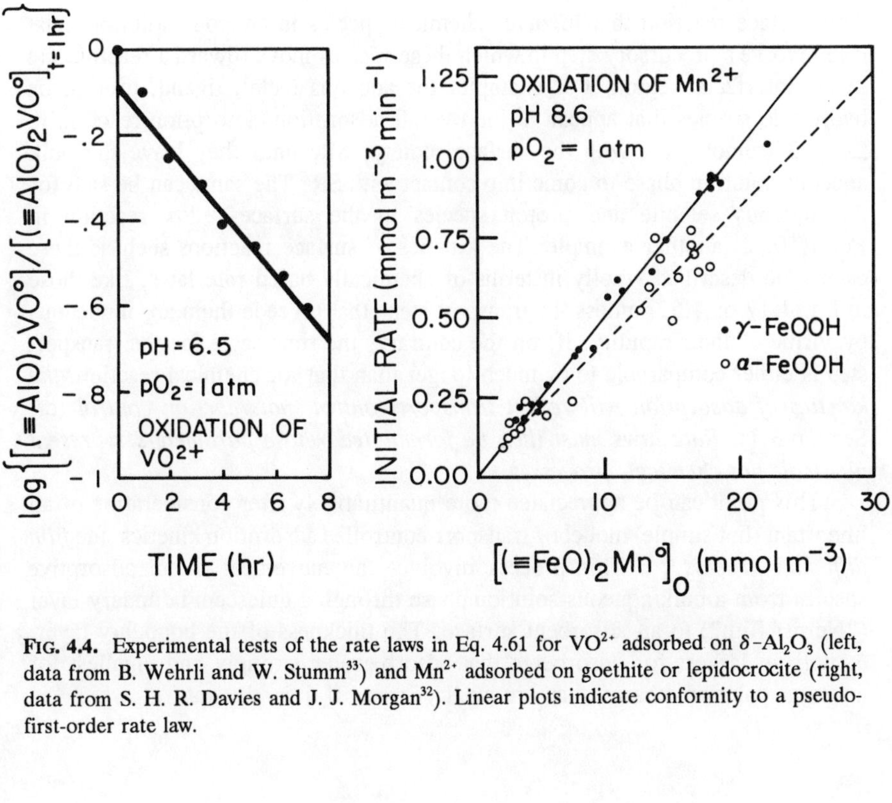

Fig. 4.4. Experimental tests of the rate laws in Eq. 4.61 for VO^{2+} adsorbed on δ-Al_2O_3 (left, data from B. Wehrli and W. Stumm[33]) and Mn^{2+} adsorbed on goethite or lepidocrocite (right, data from S. H. R. Davies and J. J. Morgan[32]). Linear plots indicate conformity to a pseudo-first-order rate law.

initial rate of oxidation of Mn^{2+} adsorbed on goethite (α-FeOOH) or lepidocrocite (γ-FeOOH) as a function of $[(\equiv FeO)_2Mn^o]$.[32] These data, collected at pH 8.6 and p_{O_2} = 1 atm, are consistent with the pseudo-first-order rate law:

$$\frac{d[(\equiv FeO)_2Mn^o]}{dt} = -k_f'' p_{O_2} [(\equiv FeO)_2Mn^o] \tag{4.61b}$$

with $k_f'' \approx 1.3 \times 10^{-5}$ atm^{-1} s^{-1}. This value is considerably smaller than k_f' in Eq. 4.55, but is about one thousand times larger than k_f in that equation.

For these two examples the rate coefficient for the oxidation of a cationic species complexed with two surface hydroxyls is much larger than that for the oxidation of the corresponding solvation complex, but much smaller than that for the oxidation of the corresponding aqueous hydrolytic species (Fig. 4.3). The enhancement of the oxidation rate by surface complexation is likely to be related to the effect of the "bridging ligand," whereas the failure of the rate to achieve the high value for the corresponding aqueous hydrolytic species may be related to a geometric disadvantage that adsorption brings by lowering the effective cross section for collisions with oxygen molecules.[31]

4.5 Transport-Controlled Adsorption Kinetics

Any surface reaction that involves chemical species in aqueous solution must also involve a precursory step in which these species move toward a reactive site in the interfacial region. For example, the aqueous metal, ligand, proton, or hydroxide species that appear in the overall adsorption–desorption reaction in Eq. 4.3 cannot react with the surface moiety, SR, until they leave the bulk aqueous solution phase to come into contact with SR. The same can be said for the aqueous selenite and proton species in the surface redox reaction in Eq. 4.50, as another example. The kinetics of surface reactions such as these cannot be described wholly in terms of chemically based rate laws, like those in Eq. 4.17 or 4.52, unless the transport steps that precede them are innocuous by virtue of their rapidity. If, on the contrary, the time scale for the transport step is either comparable to or much longer than that for chemical reaction, *the kinetics of adsorption will reflect transport control, not reaction control* (cf. Section 3.1). *Rate laws must then be formulated whose parameters represent physical, not chemical, processes.*

This point can be appreciated more quantitatively after consideration of an important (but simple) model of transport-controlled adsorption kinetics, the *film diffusion process*.[34,35] This process involves the movement of an adsorptive species from a bulk aqueous-solution phase through a quiescent boundary layer ("Nernst film") to an adsorbent surface. The thickness of the boundary layer, δ, will be largest for adsorbents that adsorb water strongly and smallest for aqueous solution phases that are well stirred. If j is the rate at which an adsorptive arrives at the adsorbent surface, per unit area of the latter (the adsorptive *flux* to the adsorbent, in units of mol m^{-2} s^{-1}), and if diffusion is the mechanism by which the adsorptive makes its way through the boundary layer, the *Fick rate law*[36] can be invoked to describe adsorptive transport:

$$j = \frac{D}{\delta}([i]_{bulk} - [i]_{surf}) \quad (4.62)$$

where $[i]_{bulk}$ is the concentration of adsorptive species i in the bulk aqueous-solution phase, $[i]_{surf}$ is its concentration at the boundary layer-adsorbent surface interface, and D is its *diffusion coefficient* (units of m^2 s^{-1}) in the boundary layer. The Fick rate law is based on the premise that a difference in adsorptive concentration across the boundary layer "drives" the adsorptive to move through the layer. The transport parameter D is a quantitative measure of the effectiveness of the molecular and hydrodynamic processes that respond (instantaneously) to the concentration difference in order to bring the adsorptive to the adsorbent surface. The rate of adsorption based on Eq. 4.62 is equal to the product of the adsorptive flux and the adsorbent surface area per unit volume of aqueous-solution phase:

$$\text{rate of adsorption} = (D a_s c_s/\delta)([i]_{bulk} - [i]_{surf}) \quad (4.63a)$$

where a_s is the specific surface area of the adsorbent and c_s is its concentration in the aqueous phase.

The film diffusion process provides a supply of adsorptive molecules at the adsorbent surface to engage in a chemical reaction leading to adsorption (Eq. 4.3). The rate law for this reaction is developed, for example, in conjunction with the adsorption step in the sequential reaction schemes that appear in Eqs. 3.46, 3.56, and 4.51. Prototypical expressions are in Eqs. 3.47, 3.57, and 4.52a; a generic rate law for reaction-controlled adsorption is in Eq. 4.17. For the present example the rate of adsorption can be described by the equation

$$\text{rate of adsorption} = k_{ads}[i]_{surf}[SR] - k_{des}[SRi] \quad (4.63b)$$

where SR is a reactive surface moiety and SRi is the adsorbed form of species i (cf. Eq. 4.3). The film diffusion process supplies this species at a rate that is matched by the subsequent chemical reaction through adjustment of the value of $[i]_{surf}$ to a steady-state value determined by the mass-balance condition

$$(D a_s c_s/\delta)([i]_{bulk} - [i]_{surf}) = k_{ads}[i]_{surf}[SR] - k_{des}[SRi] \quad (4.64)$$

The effects of transport and reaction controls on adsorption can be evaluated readily after solving Eq. 4.64 for $[i]_{surf}$:

$$[i]_{surf} = \frac{k_{diff}[i]_{bulk} + k_{des}[SRi]}{k_{diff} + k_{ads}[SR]} \tag{4.65}$$

where

$$k_{diff} \equiv D a_s c_s / \delta \tag{4.66}$$

is a first-order *film diffusion rate coefficient*. Equation 4.65 can be substituted into either Eq. 4.63a or 4.63b to calculate the rate of adsorption. If the former equation is selected, the result is

$$\text{rate of adsorption} = k_{diff}\left[\frac{k_{ads}[i]_{bulk}[SR] - k_{des}[SRi]}{k_{diff} + k_{ads}[SR]}\right] \tag{4.67}$$

A comparison between the effects of film diffusion and the adsorption reaction can be made by examining the denominator in Eq. 4.67. Under the condition $k_{diff} >> k_{ads}[SR]$, transport through the boundary layer is much more rapid than the adsorption reaction, and Eq. 4.67 takes the approximate form

$$\text{rate of adsorption} \underset{k_{diff}\uparrow\infty}{\sim} k_{ads}[i]_{bulk}[SR] - k_{des}[SRi] \tag{4.68a}$$

which is the same rate law as appears in Eqs. 3.47 and 3.57 (with $k_f' \equiv 0$), in terms of the *bulk* concentration of the adsorptive species. In this limiting case the adsorption kinetics are fully *reaction-controlled*. Under the opposite condition, $k_{diff} << k_{ads}[SR]$, transport through the boundary layer is very slow as compared to the adsorption reaction and Eq. 4.67 takes the approximate limiting form

$$\text{rate of adsorption} \underset{k_{diff}\downarrow 0}{\sim} k_{diff}\left([i]_{bulk} - \frac{k_{des}[SRi]}{k_{ads}[SR]}\right) \tag{4.69}$$

The significance of the second term on the right side of Eq. 4.69 is seen after setting the left side of Eq. 4.63b equal to zero and solving for $[i]_{surf}$:

$$[i]_{surf}^{eq} = k_{des}[SRi]/k_{ads}[SR] \tag{4.70}$$

which is the concentration of adsorptive i produced at the adsorbent surface when the adsorption–desorption reaction has come to equilibrium. Thus Eq. 4.69 can be expressed in the alternate form

$$\text{rate of adsorption} \underset{k_{diff} \downarrow 0}{\sim} k_{diff}([i]_{bulk} - [i]_{surf}^{eq}) \quad (4.68b)$$

In this limiting case the adsorption reaction produces the steady-state value of $[i]_{surf}$ and the adsorption kinetics are wholly *transport-controlled*. Measurement of the rate of adsorption accordingly provides little or no *chemical* information about the adsorption process.[35,37]

The film diffusion process assumes that reactive surface groups are exposed directly to the aqueous-solution phase and that the transport barrier to adsorption involves only the healing of a uniform concentration gradient across a quiescent adsorbent surface boundary layer. If instead the adsorbent exhibits significant microporosity at its periphery, such that aqueous solution can effectively enter and adsorptives must therefore traverse sinuous microgrottos in order to reach reactive adsorbent surface sites, then the transport control of adsorption involves *intraparticle diffusion*.[35,38] A simple mathematical description of this process based on the Fick rate law can be developed by generalizing Eq. 4.62 to the partial differential expression[36]

$$j = D\partial[i]/\partial x \quad (4.71)$$

where the position variable x is measured *inward* along a perpendicular axis to the adsorbent surface. (Equation 4.62 is a finite-difference form of Eq. 4.71.) Equation 4.71 is applied to the interior of the adsorbent, such that D is the diffusion coefficient of the adsorptive i within the adsorbent, whose solution-filled porosity is n_s. No boundary layer exists in this case, and the adsorptive flux at the adsorbent–aqueous solution interface is simply equal to the rate of adsorptive mass loss from aqueous solution per unit adsorbent surface area:[39]

$$(\partial[i]/\partial t)_{surf} = D a_s c_s (\partial[i]/\partial x)_{surf} \quad (4.72)$$

Mathematical analysis of the diffusion problem in this case for a rectangular parallelepiped adsorbent leads to an equation for the total moles of adsorptive that have entered the adsorbent pores by the elapsed time t (the time integral of the right side of Eq. 4.72):[39]

$$\frac{M(t)}{M_\infty} = 1 - \sum_{n=1}^{\infty} \frac{2\alpha(1+\alpha)}{1 + \alpha + \alpha^2 q_n^2} \exp(-Dq_n^2 t/\ell^2) \quad (4.73a)$$

where $\alpha = \rho_s/n_s c_s$, ρ_s being the mass density of the adsorbent; 2ℓ is the thickness of the adsorbent along the x-axis; $M_\infty \equiv M_0/(1+\alpha)$, M_0 being the initial total moles of adsorptive in the aqueous solution phase; and q_n is the nth positive root of the equation (for fixed α):

$$\tan q_n = -\alpha q_n \quad (4.74a)$$

For very dilute suspensions, $\alpha \uparrow \infty$ and $q_n = (n + 1/2)\pi$. A graph of $M(t)/M_\infty$ plotted against the dimensionless time parameter $\sqrt{Dt/\ell^2}$ is shown in Fig. 4.5[39] for several values of α. Note that the uptake of adsorptive, for a given initial adsorptive concentration, is more rapid when the volume ratio of bulk aqueous solution to adsorbent pore space (i.e., α) is small.

The intraparticle diffusion problem for a spherical adsorbent can also be solved to develop an expression similar to Eq. 4.73:[39]

$$\frac{M(t)}{M_\infty} = 1 - \sum_{n=1}^{\infty} \frac{6\alpha(1+\alpha)}{9 + 9\alpha + \alpha^2 q_n^2} \exp(-Dq_n^2 t/R^2) \tag{4.73b}$$

where R is the sphere radius and q_n is a nonzero root of the equation:

$$\tan q_n = \frac{3\, q_n}{3 + \alpha q_n^2} \tag{4.74b}$$

This model approach, which produces curves much like those in Fig. 4.5, is applicable to the intraparticle diffusion of organic adsorptives in natural colloids that are approximately spherical.[40] It can be coupled to an appropriate adsorption

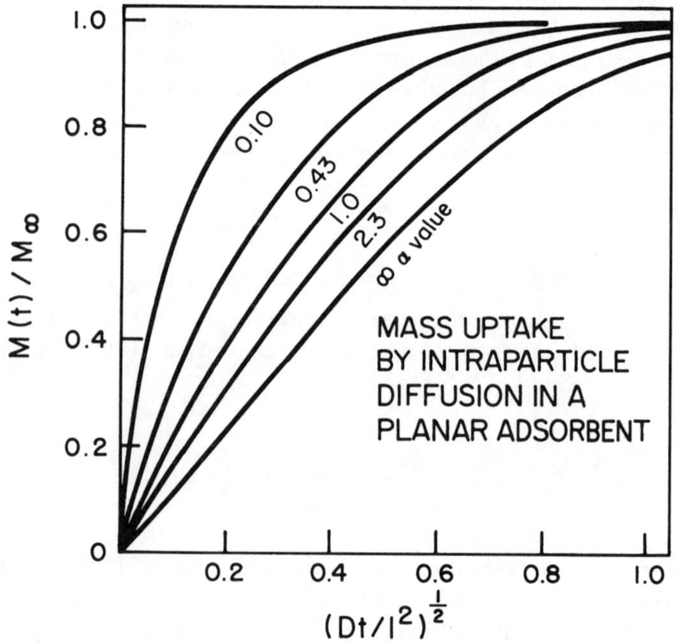

FIG. 4.5. Graphs of Eq. 4.73a with fixed values of the volume ratio of bulk aqueous solution to adsorbent pore space (α). (Adapted by permission of Oxford University Press from J. Crank[39].)

reaction in a way similar to what was done to derive Eq. 4.64.

NOTES

1. Concepts and terminology of adsorption processes on solid adsorbents are discussed by D. H. Everett, Reporting data on adsorption from solution at the solid/solution interface, *Pure Appl. Chem.* **58**:967 (1986). See also D. H. Everett, *Manual of Symbols and Terminology for Physicochemical Quantities and Units. Appendix II: Definition, Terminology and Symbols in Colloid and Surface Chemistry*, Butterworths, London, 1972 [published in *Pure Appl. Chem.* **31**:577 (1972)]. The need for a *relative* definition of the amount of adsorbed substance stems from the fact that the actual position of an interface cannot be specified with absolute precision, even conceptually.

2. G. Sposito, Distinguishing adsorption from surface precipitation, Chap. 11 in *Geochemical Processes at Mineral Surfaces*, ed. by J. A. Davis and K. F. Hayes, American Chemical Society, Washington, DC, 1986.

3. See, for example, Chap. 2 in G. Sposito, *The Surface Chemistry of Soils*, Oxford University Press, New York, 1984. The location of an interface is a molecular-scale concept that macroscopic definitions like Eq. 4.1 cannot make precise. That the interface is likely to be located within three molecular diameters of the periphery of an adsorbent solid is sufficient detail for the application of the concepts in the present section. See D. H. Everett, op. cit.,[1] for additional discussion of the interface to which Eq. 4.1 applies (known technically as a *Gibbs dividing surface*).

4. See, for example, Chap. 9 in K. Denbigh, *The Principles of Chemical Equilibrium*, Cambridge University Press, Cambridge, 1981. The IUPAC recommendation for the symbol to represent rational activity coefficients is γ_x, which is not used in this book in order to make the distinction between solid solutions and aqueous solutions more evident. In strict chemical thermodynamics, however, all activity coefficients are based on the mole fraction scale, with the definition for aqueous species (Eq. 1.12) actually being a variant that reflects better the ionic nature of electrolyte solutions and the dominant contribution of liquid water to these mixtures. (See, for example, Chap. 2 in R. A. Robinson and R. H. Stokes, *Electrolyte Solutions*, Butterworths, London, 1970.)

5. This relationship is derived for the more general case of arbitrary stoichiometric coefficients for SR and SR′C in Section 3.7 of G. Sposito, *The Thermodynamics of Soil Solutions*, Clarendon Press, Oxford, 1981. In that case, Eq. 4.12 remains formally unchanged, whereas the left sides of Eqs. 4.11a and 4.11b are multiplied by the stoichiometric coefficients of SR and SR′C, respectively, and x_2 is replaced by the quantity: (stoichiometric coefficient of SR)x_2 / [(stoichiometric coefficient of SR)x_2 + (stoichiometric coefficient of SR′C)x_1].

6. See, for example, Section 5.2 in G. Sposito, op. cit.,[3] for a discussion of the mean field approach in the molecular theory of adsorption.

7. G. Sposito, Molecular models of ion adsorption on mineral surfaces, Chap. 6 in *Mineral–Water Interface Geochemistry*, ed. by M. F. Hochella and A. F. White, Mineralogical Society of America, Washington, DC, 1990.

8. A variety of these models is reviewed by S. Goldberg, Use of surface complexation models in soil chemical systems, *Adv. Agron.* **47**:233 (1992).

9. P. W. Schindler and W. Stumm, The surface chemistry of oxides, hydroxides,

and oxide minerals, Chap. 4 in *Aquatic Surface Chemistry*, ed. by W. Stumm, Wiley, New York, 1987.

10. The total net surface charge density is proportional to net valence times concentration for each reactant adsorbent species and product adsorbent–adsorbate species in Eq. 4.15. Its presence in the equation for the activity coefficient reflects a model concept, that these charged surface species create an average electric field that influences ion adsorption. See, for example, G. Sposito, op. cit.[7]

11. These applications are described comprehensively for several different mechanistic models of the adsorption–desorption process at equilibrium by S. Goldberg, op. cit.[8] Kinetics applications are discussed by D. L. Sparks and D. L. Suarez, *Rates of Soil Chemical Processes*, Soil Science Society of America, Madison, WI, 1991.

12. The concept of an "affinity spectrum" model for the rate constants that describe the complexation reactions of humic substances is discussed by M. S. Shuman, B. J. Collins, P. J. Fitzgerald, and D. L. Olson, Distribution of stability constants and dissociation rate constants among binding sites on estuarine copper–organic complexes: Rotated disk electrode studies and an affinity spectrum analysis of ion-selective electrode and photometric data, Chap. 17 in *Aquatic and Terrestrial Humic Materials*, ed. by R. F. Christman and E. T. Gjessing, Ann Arbor Science, Ann Arbor, MI, 1983. See also Section 6.4 in J. Buffle, *Complexation Reactions in Aquatic Systems*, Wiley, New York, 1988. The integral on the right side of Eq. 4.21 is a *Laplace transform*; thus the rate law is specified fully once the Laplace transform of the probability density function p(k) is calculated.

13. For a lucid discussion of the gamma distribution, see Chap. 17 in N. L. Johnson and S. Kotz, *Continuous Univariate Distributions*, Vol. 1, Wiley, New York, 1970.

14. See, for example, Chap. 6 in M. Abramowitz and I. A. Stegun, *Handbook of Mathematical Functions*, Dover, New York, 1972.

15. The power-law relationship between d[SR]/dt and [SR] based on Eq. 4.20 is derived in detail by R. R. D. Kemp and B. W. Wojciechowski, The kinetics of mixed feed reactions, *Ind. Eng. Chem. (Fundamentals)* **13**:332 (1974).

16. P. J. Crickmore and B. W. Wojciechowski, Kinetics of adsorption on energetically heterogeneous surfaces, *J. C. S. Faraday I* **73**:1216 (1977). If the coefficients of variation for the two gamma distributions are not the same, Eq. 4.26b takes the more general form

$$\frac{d[SRC]}{dt} = \bar{k}_{ads}[C](SR_T - [SRC])^{(\alpha+1)/\alpha} - \bar{k}_{des}[SRC]^{(\delta+1)/\delta}$$

where δ is the coefficient of variation for the distribution of k_{desi} values. The equilibrium condition then becomes

$$[SRC]_e A[C]_e^b + [SRC]_e^{b'} = A[C]_e^b$$

instead of Eq. 4.27, where $b' = \delta(\alpha + 1)/\alpha(\delta + 1)$.

17. Equation 4.27 and its special cases are discussed, with applications, by D. G. Kinniburgh, General purpose adsorption isotherms, *Environ. Sci. Technol.* **20**:895 (1986).

18. See, for example, Section 4.1 in G. Sposito, op. cit.,[3] where an alternate

derivation of Eq. 4.27 is given. It proceeds directly from the *equilibrium* condition as applied to Eq. 4.18 to generate a Langmuir isotherm equation for each adsorbate species, $SR'C_i$. See also Chap. 4 in W. Rudzinski and D. H. Everett, *Adsorption of Gases on Heterogeneous Surfaces,* Academic Press, San Diego, CA, 1992.

19. See, for example, Chap. 2 in G. Goertzel and N. Tralli, *Some Mathematical Methods of Physics*, McGraw-Hill, New York, 1960. Because Eq. 4.34 is a set of *linear* rate laws, although coupled, it is possible to express their solutions as the superposition of solutions of uncoupled (i.e., parallel-reaction) rate laws, as in Eq. 4.35. The number of terms in the superposition will be the same as the number of rate laws (two in the present case). The parameters in Eq. 4.35 are then chosen to make the solutions meet all mathematical conditions imposed by the problem to be solved.

20. C. F. Bernasconi, *Relaxation Kinetics*, Academic Press, New York, 1976. This standard textbook discusses comprehensively the linearization of rate laws for a wide variety of reaction sequences as well as experimental methods of measuring "time constants."

21. See especially Chaps. 2 and 3 in D. L. Sparks and D. L. Suarez, op. cit.[10] A summary review of chemical relaxation methods is given by T. Yasunaga and T. Ikeda, Adsorption-desorption kinetics at the metal-oxide-solution interface studied by relaxation methods, Chap. 12 in J. A. Davis and K. F. Hays, op. cit.[2]

22. P.-C. Zhang and D. L. Sparks, Kinetics and mechanisms of molybdate adsorption/desorption at the goethite/water interface using pressure-jump relaxation, *Soil Sci. Soc. Am. J.* **53**:1028 (1989). The rate coefficients in Eq. 4.42 need not be modified by the model surface-species activity coefficients adopted in this paper.

23. P.-C. Zhang and D. L. Sparks, Kinetics and mechanisms of sulfate adsorption/desorption on goethite using pressure-jump relaxation, *Soil Sci. Soc. Am. J.* **54**:1266 (1990). The rate coefficients in Eq. 4.44c need not be modified by the model surface-species activity coefficients adopted in this paper.

24. R. D. Astumian, M. Sasaki, T. Yasunaga, and Z. A. Schelly, Proton adsorption–desorption kinetics on iron oxides in aqueous suspensions, using the pressure-jump method, *J. Phys. Chem.* **85**:3832 (1981).

25. M. Sasaki, M. Morlya, T. Yasunaga, and R. D. Astumian, A kinetic study of ion-pair formation on the surface of α-FeOOH in aqueous suspensions using the electric field pulse technique, *J. Phys. Chem.* **87**:1449 (1983).

26. A. T. Stone and J. J. Morgan, Reductive dissolution of metal oxides, Chap. 9 in W. Stumm, op. cit.[9]

27. M. J. Scott, Kinetics of adsorption and redox processes on iron and manganese oxides: Reactions of As(III) and Se(IV) at goethite and birnessite surfaces, Ph.D. dissertation, California Institute of Technology, Pasadena, CA, 1991. Environmental Quality Laboratory Report No. 33. Strictly, the two protons appearing on the right side of the last step in Eq. 4.50 and the first step in Eq. 4.51 are consumed in the protonation of the oxygen bridges in the birnessite structure that link $\equiv Mn^{II}$ to $\equiv Mn^{IV}$.

28. C. A. Johnson and A. G. Xyla, The oxidation of chromium(III) to chromium(VI) on the surface of manganite (γ-MnOOH), *Geochim. Cosmochim. Acta* **55**:2861 (1991). See also A. Manceau and L. Charlet, X-ray absorption spectroscopic study of the sorption of Cr(III) at the oxide–water interface. I. Molecular mechanism of Cr(III) oxidation on Mn oxides, *J. Colloid Interface Sci.* **148**:425 (1992).

29. The set of differential equations provided by Eqs. 4.52d–f, without the constraint of rapid proton reaction equilibrium, have been solved analytically for the case in which the rate of selenate detachment is zero. The result is presented in Chap. 1 of

C. H. Bamford and C. F. H. Tipper, *Comprehensive Chemical Kinetics*, *Vol. 2, The Theory of Kinetics*, Elsevier, Amsterdam, 1969 (cf. Eqs. 4.34 and 4.35):

$$[\equiv Mn^{II} - OH] = \alpha\{[(\lambda - K') / \lambda d] \exp(-\lambda\tau) + [(K' - \lambda') / \lambda' d] \exp(-\lambda'\tau)\}$$

$$[\equiv Mn^{II} - OH_2^+] = (\alpha/d) [\exp(-\lambda'\tau) - \exp(-\lambda\tau)]$$

where α is an arbitrary scale factor, $d = \lambda - \lambda'$, $\tau = K_{ads}t$,

$$2\lambda = 1 + K + K' + [(1 + K + K')^2 - 4K']^{1/2}$$

$$2\lambda' = 1 + K + K' - [(1 + K + K')^2 - 4K']^{1/2}$$

and $K = k_{des}/K_{ads}$, $K' = k'_{det}/K_{ads}$. The time dependence of $[Mn^{2+}]$ is found by introducing the result for $[\equiv Mn^{II} - OH_2^+]$ into Eq. 4.52e and integrating:

$$[Mn^{2+}] = (\alpha K'/\lambda' d) [1 - \exp(-\lambda'\tau)] + (\alpha K'/\lambda d) [1 - \exp(-\lambda\tau)]$$

These three concentration–time equations describe the proton-promoted dissolution reaction in Eq. 4.51 under the assumption that the selenate detachment reaction is already at equilibrium. The case $K' << K$ corresponds to the approximate equality in Eq. 4.52f. Note that the rate of mineral dissolution, $d[Mn^{2+}]/dt$, will be constant for observation times much smaller than $1/k'_{det}$.

30. See, for example, Chap. 12 in J. Burgess, *Ions in Solution*, Wiley, New York, 1988, for a discussion of this mechanism.

31. B. Wehrli, Redox reactions of metal ions at mineral surfaces, Chap. 11 in *Aquatic Chemical Kinetics*, ed. by W. Stumm, Wiley, New York, 1990. See also W. Stumm and B. Sulzberger, The cycling of iron in natural environments, *Geochim. Cosmochim. Acta* **56**:3233 (1992).

32. S. H. R. Davies and J. J. Morgan, Manganese(II) oxidation kinetics on metal oxide surfaces, *J. Colloid Interface Sci.* **129**:63 (1989).

33. B. Wehrli and W. Stumm, Oxygenation of vanadyl(IV). Effect of coordinated surface hydroxyl groups and OH^-, *Langmuir* **4**:753 (1988).

34. W. Nernst, Theorie der Reaktiongeschwindigkeit in heterogenen Systemen, Zeitschr. Phys. Chem. **47**:52 (1904).

35. Film and intraparticle diffusion processes are described in detail in Chap. 6 of F. Helfferich, *Ion Exchange*, McGraw-Hill, New York, 1962.

36. See, for example, Chap. 1 in J. Crank, *The Mathematics of Diffusion*, Oxford University Press, New York, 1975. The partial differential form of the Fick rate law is

$$j = - D\partial[i]/\partial x$$

for a diffusing species i, where x is a position coordinate measured *outward* along an axis perpendicular to the adsorbent surface. Thus, properly, a positive concentration gradient leads to an inward (i.e., *negative*) diffusive flux toward the adsorbent surface.

Equation 4.62 is a finite-difference form of the expression above with the convenient definition of a *positive* diffusive flux toward an adsorbent surface.

37. The issue of "mixed" film diffusion and reaction control of adsorption is discussed in the context of special cases of Eq. 4.63b by D. Rickard and E. L. Sjöberg, Mixed kinetic control of calcite dissolution rates, *Am. J. Sci.* **283**:815 (1983), and by R. A. Ogwada and D. L. Sparks, Kinetics of ion exchange on clay minerals and soil: II. Elucidation of rate-limiting steps, *Soil Sci. Soc. Am. J.* **50**:1162 (1986). A generalization of the case of a nonquiescent boundary layer with a nonuniform concentration gradient is described by R. G. Compton and P. R. Unwin, The dissolution of calcite in aqueous solution at pH < 4: Kinetics and mechanism, *Phil. Trans. R. Soc. Lond.* A**330**:1 (1990).

38. See, for example, M. L. Brusseau and P.S.C. Rao. Sorption nonideality during organic contaminant transport in porous media, *Crit. Rev. Environ. Control* **19**:33 (1989) and W. W. Wood, T. F. Kraemer, and P. P. Hearn, Jr., Intragranular diffusion: An important mechanism influencing solute transport in clastic aquifers? *Science* **247**:1569 (1990).

39. See, for example, Chaps. 4 and 6 in J. Crank, op. cit.[36] The "diffusion problem" is formulated mathematically in the partial differential equation

$$\partial[i]/\partial t = D\partial^2[i]/\partial x^2$$

under the initial condition $[i]_0 = 0$, and the boundary condition in Eq. 4.72. The exponential decay in Eq. 4.73a comes about because each side of the expression above is proportional to [i], if it is assumed to factor into a product of some function of t and some function of x ("separation of variables assumption"). Thus the left side of the expression is proportional to (the time-dependent part of) [i] and a first-order rate law results, leading to the usual exponential decay with elapsed time. The "natural" units of measure of this elapsed time are given by ℓ^2/D, the time required to diffuse a distance ℓ.

40. See, for example, S.-C. Wu and P. M. Gschwend, Sorption kinetics of hydrophobic organic compounds to natural sediments and soils, *Environ. Sci. Technol.* **20**:717 (1986); W. P. Ball and P. V. Roberts, Long-term sorption of halogenated organic chemicals by aquifer material. 2. Intraparticle diffusion, *Environ. Sci. Technol.* **25**:1237 (1991); and M. L. Brusseau, R. E. Jessup, and P. S. C. Rao, Nonequilibrium sorption of organic chemicals: Elucidation of rate-limiting processes, *Environ. Sci. Technol.* **25**:124 (1991). See also Chap. 11 in R. P. Schwarzenbach, P. M. Gschwend, and D. M. Imboden, *Environmental Organic Chemistry*, Wiley, New York, 1993.

FOR FURTHER READING

Bolt, G. H., M. F. DeBoodt, M. H. B. Hayes, and M. B. McBride, *Interactions at the Soil Colloid–Soil Solution Interface*, Kluwer, Dordrecht, The Netherlands, 1991. The first 13 chapters of this edited treatise provide advanced reviews of the thermodynamics, kinetics, and molecular mechanisms of adsorption by soil colloids.

Goldberg, S., Use of surface complexation models in soil chemical systems, *Adv. Agron.* **47**:233 (1992). A comprehensive review of the methods of chemical modeling for ion adsorption equilibria in soils.

Rudzinski, W., and D. H. Everett, *Absorption of Gases on Heterogeneous*

Surfaces, Academic Press, San Diego, CA, 1992. An exhaustive review of the "site probability" approach to describe adsorption isotherms for heterogeneous adsorbents, analogous to the "rate coefficient probability" approach discussed in Section 4.2.

Sparks, D. L., and D. L. Suarez, *Rates of Soil Chemical Processes*, Soil Science Society of America, Madison, WI, 1991. The 11 chapters of this edited volume provide recent reviews of all the topics discussed in the present chapter, including experimental methodologies.

Schwarzenbach, R. P., P. M. Gschwend, and D. M. Imboden, *Environmental Organic Chemistry*, Wiley, New York, 1993. A fine textbook on the chemistry of synthetic organic compounds in aqueous systems, with Chapter 11 serving as a useful introduction to transport versus reaction control of adsorption reactions.

Stumm, W., *Aquatic Surface Chemistry*, Wiley, New York, 1987. This edited volume offers advanced reviews of adsorption, surface oxidation–reduction, and mineral dissolution reactions with emphasis on metal oxyhydroxide adsorbents.

Stumm, W., *Chemistry of the Solid–Water Interface*, Wiley, New York, 1992. A gem of a textbook on surface chemistry applied to natural particles by one of the modern-day masters: a "must read" for all serious students.

Stumm, W., and B. Sulzberger, The cycling of iron in natural environments: Considerations based on laboratory studies of heterogeneous redox processes, *Geochim. Cosmochim. Acta* **56**:3233 (1992). A comprehensive review of surface-oxidation–reduction processes on iron oxyhydroxide solids.

PROBLEMS

1. Equilibrium data on the adsorption of Al^{3+} by goethite (α-FeOOH) can be described by the reaction

$$\equiv FeOH(s) + Al^{3+}(aq) + OH^{-}(aq) = \equiv FeOAlOH^{+}(s) + H^{+}(aq)$$

which is a special case of Eq. 4.3. The composition dependence of the conditional equilibrium constant for this reaction is given by

$$\log K_{adsc} = 12.29 - 3.49 x_{Al}$$

where x_{Al} is the mole fraction of the species $\equiv FeOAlOH^{+}(s)$. Calculate the rational activity coefficient for this latter species and the equilibrium constant for the adsorption reaction. (*Answer*: $f_{Al} = \exp(-4.02\, x_{FeOH}^{2})$ and $K_{ads} = 6.3 \times 10^{10}$, where x_{FeOH} is the mole fraction of the species, $\equiv FeOH(s)$.)

2. Develop the implications of Eq. 4.27 for the equilibrium constant

describing the reaction in Eq. 4.3 [without the dissociable species $Q^{Z_Q}(aq)$]. Show that Eq. 4.27 leads to the following results:

$$\text{(a)}\quad K_{12c} = A^{1/b}\,(x_2/x_1)^{(b-1)/b}$$

$$\text{(b)}\quad K_{12} = \bar{k}_{ads}\,/\,\bar{k}_{des}$$

$$\text{(c)}\quad \mathbf{(SRC)} = x_2^{1/b} \text{ and } \mathbf{(SR)} = x_1^{1/b}$$

where the notation developed in connection with Eqs. 4.11 and 4.12 has been used and A and b are the parameters in Eq. 4.28. The expressions for the activities in (c) are known as the *Rothmund-Kornfeld model*. (*Hint*: Show that Eq. 4.27 is equivalent to the relationship $x_2/x_1 = A[C]^b$ and then substitute this result into Eq. 4.6, noting that $\mathbf{(Q^{Z_Q})}$ is to be deleted. Next, apply Eq. 4.12 using the relations, $\int \ln x\, dx = x(\ln x - 1)$ and $x_1 = 1 - x_2$. Finally, compare the results in (a) and (b), noting Eq. 4.7.)

3. The data in Table 4.4 (page 178) apply to the adsorption of Al^{3+} by the clay mineral kaolinite. Develop a rate law for the decrease in $Al^{3+}(aq)$ concentration based on these data and calculate the Arrhenius activation energy for the adsorption process. (*Hint*: Assume that $d[Al^{3+}]/dt = d[SR]/dt$ and consider Eq. 4.26. The kinetics data lead to $k_{ads} = 3 \times 10^{-6}$ m s^{-1} at 25°C and 2.5×10^{-6} m s^{-1} at 10°C, with $E_a = 8.62$ kJ mol^{-1}. The rate law is first-order with respect to $[Al^{3+}]$ and [SR], zero-order with respect to $[H^+]$.)

4. Generalize the derivation of Eq. 4.27 to include the dissociable species $Q^{Z_Q}(aq)$ in Eq. 4.3. Show that the result is the generalized *competitive Freundlich isotherm equation*:

$$[SR^{\prime}C]_e = \frac{SR_T A([C]_e/[Q]_e)^b}{1 + A([C]_e/[Q]_e)^b}$$

with parameters A and b defined as in Eq. 4.28. The adsorption of Ca^{2+} by ionizable surface hydroxyls on ferrihydrite ($Fe_{10}O_{15} \cdot 9H_2O$) at pH 8 can be described by the isotherm equation

$$n_{Ca}^{(w)} = \frac{0.054\,(10^{-4.8}[Ca^{2+}]_e/[H^+]_e)^{0.8}}{1 + (10^{-4.8}[Ca^{2+}]_e/[H^+]_e)^{0.8}}$$

where the relative surface excess of Ca, $n_{Ca}^{(w)}$, is in units of mol Ca (mol Fe)$^{-1}$. Calculate $\bar{k}_{ads}/\bar{k}_{des}$ and the coefficient of variation for the gamma distribution of the two rate coefficients. (*Hint*: Generalize Eq. 4.17 to include [Q] as a factor in the desorption reaction and note the role of [C] in the derivation of Eq. 4.26.)

Table 4.4 Kinetics Data for Al^{3+} Adsorption by Kaolinite

$-d[Al^{3+}]/dt$ (mol m^{-3} s^{-1})	$[Al^{3+}]$ (mol m^{-3})	$[H^+]$ (mol m^{-3})	[SR][a] (m^{-1})
	T = 283.15 K		
$10^{-3.22}$	$10^{-1.64}$	$10^{-0.03}$	1.02×10^4
$10^{-3.22}$	$10^{-1.64}$	$10^{-0.45}$	1.02×10^4
$10^{-3.22}$	$10^{-1.64}$	$10^{-0.79}$	1.02×10^4
$10^{-3.22}$	$10^{-1.64}$	$10^{-1.14}$	1.02×10^4
$10^{-3.52}$	$10^{-1.64}$	$10^{-0.76}$	5.30×10^3
$10^{-3.21}$	$10^{-1.64}$	$10^{-0.76}$	1.02×10^4
$10^{-2.92}$	$10^{-1.64}$	$10^{-0.76}$	2.12×10^4
$10^{-3.22}$	$10^{-1.64}$	$10^{-0.76}$	1.02×10^4
$10^{-2.92}$	$10^{-1.34}$	$10^{-0.76}$	1.02×10^4
$10^{-3.52}$	$10^{-1.95}$	$10^{-0.76}$	1.02×10^4
	T = 298.15 K		
$10^{-3.15}$	$10^{-1.64}$	$10^{-0.03}$	1.02×10^4
$10^{-3.15}$	$10^{-1.64}$	$10^{-0.45}$	1.02×10^4
$10^{-3.14}$	$10^{-1.64}$	$10^{-0.79}$	1.02×10^4
$10^{-3.13}$	$10^{-1.64}$	$10^{-1.14}$	1.02×10^4
$10^{-3.22}$	$10^{-1.64}$	$10^{-0.76}$	5.30×10^3
$10^{-3.14}$	$10^{-1.64}$	$10^{-0.76}$	1.02×10^4
$10^{-2.83}$	$10^{-1.64}$	$10^{-0.76}$	2.12×10^4
$10^{-3.14}$	$10^{-1.64}$	$10^{-0.76}$	1.02×10^4
$10^{-2.82}$	$10^{-1.34}$	$10^{-0.76}$	1.02×10^4
$10^{-3.45}$	$10^{-1.95}$	$10^{-0.76}$	1.02×10^4

[a][SR] is taken to be equal to $a_s c_s$ (a_s = specific surface area, c_s = adsorbent concentration), the adsorbent surface area per unit volume of aqueous phase, because the value of $-d[Al^{3+}]/dt$ is an *initial* rate of adsorption. Data from W. J. Walker, C. S. Cronan, and H. H. Patterson, A kinetic study of aluminum adsorption by aluminosilicate clay minerals, *Geochim. Cosmochim. Acta* **52**: 55(1988).

5. The adsorption of selenate (SeO_4^{2-}) by goethite (α-FeOOH) is thought to result primarily in outer-sphere surface complexes, whereas the adsorption of selenite (SeO_3^{2-}) is thought to result primarily in inner-sphere surface complexes. Develop equations for the "time constants" τ_1 and τ_2 in the adsorption kinetics of these two species and compare the resulting equations for τ_2. (*Hint*: Consider Eqs. 4.40–4.45.)

6. Measurements of the "time constant" τ_1 for the adsorption of protons by hematite (α-Fe_2O_3) at pH < 3.5 in perchlorate background solution indicate that the second term on the right side of Eq. 4.47a is negligible and that the adsorption rate coefficient $k_a \approx 15.1$ mol^{-1} m^3 s^{-1}. Given $[\equiv FeOH]_e \approx 0.3$ mol m^{-3}, calculate τ_1 over the pH range between 2.0 and 3.0. (*Answer*: The value of τ_1 increases from about 6 ms to about 50 ms over this pH range.)

7. The oxidation of VO^{2+} in the presence of anatase (TiO_2) can be described by the reaction scheme in Eqs. 4.56 and 4.59, with $(TiO)_2VO^0(s)$ being the principal adsorbed species. The Arrhenius activation energy for the oxygen reaction in Eq. 4.59b is $E_a = 56.5$ kJ mol^{-1}. Given that $[(TiO)_2VO^0] = 0.05$ mol m^{-3} and $p_{O_2} = 1$ atm, calculate the rate of decrease in the concentration of V(IV) in solution at 20°C and 30°C if $k''_f = 6.4 \times 10^{-5}$ atm^{-1} s^{-1} at 25°C. (*Answer*: $-d[V(IV)]/dt = 2.2$ μmol m^{-3} s^{-1} at 20°C and 4.7 μmol m^{-3} s^{-1} at 30°C. The temperature dependence of k''_f is given by $\ln k''_f = -9.6566 + 6.795\,[3.354 - (1000/T)]$ at absolute temperature T.)

8. Solve Eqs. 4.52a and 4.52b approximately under the assumption that they can be decoupled completely from Eqs. 4.52c–d. Consider the following conditions: $Mn(IV)_T = 0.5$ mol m^{-3}, pH 4, $k_f = 0.025$ mol^{-2} m^6 s^{-1}, $K_b = 0.002$ s^{-1}, $K'_f = 1.3 \times 10^{-6}$ s^{-1}, and $K'_b \approx 0$. Calculate the time dependence of $[HSeO_3^-]$ and $[\equiv Mn^{IV} - OSeOOH]$, given $[HSeO_3^-]_0 = 0.1$ mol m^{-3} and $[\equiv Mn^{IV} - OSeOOH]_0 = 0$. (*Hint*: Equations 4.52a and 4.52b have the same *mathematical* form as Eqs. 4.52d and 4.52f under the condition that $[\equiv Mn^{IV} - OH] \approx Mn(IV)_T$ in Eq. 4.52a and $d[SeO_4^{2-}]/dt = 0$ in Eq. 4.52d. Thus, the results given in Note 29 in the present chapter can be applied. The time dependence of the two species concentrations is then found by the substitutions: $[\equiv Mn^{II} - OH] \leftrightarrow [HSeO_3^-]$; $[\equiv Mn^{II} - OH_2^+] \leftrightarrow [\equiv Mn^{IV} - OSeOOH]$; $K = K_b / K_f Mn(IV)_T = 1.60$; $K' = K'_f / K_f Mn(IV)_T = 1.04 \times 10^{-3}$; and $\tau = K_f Mn(IV)_T t = t/800$ (t in s). In the present example, $K' << K$ and therefore $\lambda \approx 1 + K = 2.60$; $\lambda' \approx K'/(1 + K) = 4.00 \times 10^{-4}$; $d \approx \lambda$; $(\lambda - K')/\lambda \approx 1$; and $(K' - \lambda')/\lambda' \approx K$. The time dependence of the two species concentrations is accordingly

$$[HSeO_3^-] \approx \alpha\,[0.39 \exp(-t/308) + 0.61 \exp(-t/2.0 \times 10^6)]$$

$$[\equiv Mn^{IV} - SeOOH] \approx 0.39\alpha\,[\exp(-t/2.0 \times 10^6) - \exp(-t/308)]$$

with $\alpha = [HSeO_3^-]_0 = 0.1$ mol m^{-3}. The exponential term with the smaller "time constant" represents the adsorption–desorption kinetics, whereas that with the larger "time constant" represents the electron-transfer kinetics.)

9. Estimate the minimum boundary-layer thickness for significant influence on the rate of adsorption of Al^{3+} by kaolinite as described in Problem 3. Take $D = 10^{-9}$ m^2 s^{-1}. (*Answer*: $k_{diff}/\bar{k}_{ads}[SR] \approx 1$ implies $\delta \approx D/\bar{k}_{ads} \approx 3 \times 10^{-4}$ m $= 300$ μm. Typical particle size is 1.0 μm.)

10. Derive an equation for a "mixed-kinetics" rate of biselenite adsorption by goethite, as described in Problem 8. Take $D = 10^{-9}$ m^2 s^{-1}, $a_s c_s = 1.44 \times 10^4$ m^{-1}, and $\delta = 10$ nm. (*Answer*: Applying the methodology in Section 4.5, one derives the rate law:

$$\text{rate of adsorption} = \left[\frac{3.6[\mathrm{HSeO_3^-}]\,[\equiv\mathrm{Mn^{IV}-OH}] - 2.88[\equiv\mathrm{Mn^{IV}-OSeOOH}]}{1440 + 0.0025\,[\equiv\mathrm{Mn^{IV}-OH}]} \right]$$

where the rate is in units of mol m^{-3} s^{-1} and all concentrations are expressed in units of mol m^{-3}. Note that $k_{diff} / k_{ads}[\equiv Mn^{IV} - OH]_0 \approx 10^6$, so transport control is nil for the initial rate of adsorption.)

5

ION EXCHANGE REACTIONS

5.1 Ion Exchange as an Adsorption Reaction

Prototypical ion exchange reactions can be expressed by the chemical equations[1]

$$bAX_a(s) + aB^{b+}(aq) \rightleftarrows aBX_b(s) + bA^{a+}(aq) \qquad (5.1)$$

$$dCY_c(s) + cD^{d-}(aq) \rightleftarrows cDY_d(s) + dC^{c-}(aq) \qquad (5.2)$$

where a, b, c, d are stoichiometric coefficients related to the valences of the cations A^{a+}, B^{b+}, or the anions, C^{c-}, D^{d-}, and X or Y represents 1 mol of negative or positive charge carried by a solid exchanger. Thus Eq. 5.1 describes cation exchange on the exchanger X^-, whereas Eq. 5.2 describes anion exchange on the exchanger Y^+. For example, the exchange of Na^+ with Mg^{2+} on the particular montmorillonite whose dissolution reaction is described in Eq. 3.18 can be represented by the reaction

$$MgX_2(s) + 2Na^+(aq) \rightleftarrows 2NaX(s) + Mg^{2+}(aq) \qquad (5.3)$$

where $X^- \equiv 2.4\,[Si_{3.82}Al_{0.18}](Al_{1.29}Fe(III)_{0.335}Mg_{0.335})O_{10}(OH)_2^{0.415-}$. The solid exchanger is the negatively charged montmorillonite layer, and Mg^{2+} or Na^+ is the interlayer exchangeable cation. In this example, 1 mol of exchanger charge, X, is carried by 1.2 mol of montmorillonite unit cells, which are based conventionally on 24 oxygen ions.

The reactions in Eqs. 5.1 and 5.2, taken in the broadest geochemical sense, refer only to the replacement of one ion by another existing in a solid structure. Thus, for example, Eq. 5.1 could be applied to provide a description of solid solution composition alternate to that in Section 3.3. Instead of basing the composition of Al-goethite on the solid components diaspore and goethite, as in Eq. 3.41, one could describe a continuum of possible compositions by combining Eqs. 3.39a and 3.39b into the cation exchange reaction

$$FeY_3(s) + Al^{3+}(aq) \rightleftarrows AlY_3\,(s) + Fe^{3+}(aq) \qquad (5.4)$$

where $Y^- \equiv \frac{1}{3}OOH^{3-}$, a formal one-third of the anionic unit OOH^{3-}. In soil chemistry, however, applications of Eqs. 5.1 and 5.2 are restricted to *adsorption reactions*, with the exchanging ions identified as adsorptives and the solid exchanger compound identified as the adsorbent. Therefore *ion exchange reactions can be derived as examples of the adsorption–desorption reaction in Eq. 4.3*. Formal congruence between Eq. 5.1 or 5.2 and Eq. 4.3 is achieved by designating $SR \equiv SR'Q$, setting $Z_{SR} = Z_{SR'C} \equiv 0$, and permitting more than 1 mol of SR' to combine with Q or with $C \equiv M_p(OH)_yH_xL_q$, if the valences of these latter two adsorptive species happens to be different (*heterovalent ion exchange*). With these conditions imposed and with the formation reaction for the complex ion C added, that is,

$$pM^{m+}(aq) + qL^{\ell-}(aq) + xH^+(aq) + yOH^-(aq)$$
$$\rightarrow M_p(OH)_yH_xL_q{}^{(pm+x-q\ell-y)}(aq) \equiv C^{Z_C}(aq) \qquad (5.5)$$

Eq. 4.3 takes the form

$$|Z_C|\;(SR')_{|Z_Q|}Q(s) + |Z_Q|\;C^{Z_C}(aq) \underset{k_{des}}{\overset{k_{ads}}{\rightleftarrows}} |Z_Q|\;(SR')_{|Z_C|}C(s) + |Z_C|\;Q^{Z_Q}(aq) \qquad (5.6)$$

The correspondence between Eq. 5.6 and Eqs. 5.1 and 5.2 is apparent. Special cases of Eq. 5.6 appear in the fifth and tenth rows of Table 4.1.

In soil chemistry, at least historically, a further restriction of Eqs. 5.1 and 5.2 to include only fully solvated adsorbed ions (i.e., ions in outer-sphere surface complexes or in the diffuse ion swarm) often has been imposed.[1] Under this restriction, the ions in Eqs. 5.1 and 5.2 are termed *readily exchangeable*. Ion exchange reactions that involve inner-sphere surface complexation (e.g., those in Eqs. 3.56, 3.66, and 4.50) are thereby excluded and given a separate designation as "specific adsorption" processes. "Specific adsorption," which might better be termed "highly selective adsorption," actually refers to the selectivity, for a given cation or anion, of complexing functional groups at the surface of an inorganic or organic solid adsorbent. If the surface functional groups form very stable complexes with a given ion, the functional groups are said to be selective for that ion, and the interaction between the ion and the solid that bears the functional groups is called "specific adsorption." Usually, specific adsorption phenomena are reflected by their very small sensitivity to changes in the ionic strength of the aqueous solution contacting the exchanger, since these changes affect primarily the diffuse ion swarm. Examples of cations that are typically considered to be adsorbed selectively by soil minerals and organic matter are the bivalent transition metal and Group IB and IIB metal cations.

Examples of selectively adsorbed anions include F^- and the oxyanions SeO_3^{2-}, PO_4^{3-}, $C_2O_4^{2-}$, BO_3^{3-}, and MoO_4^{2-}, along with their protonated forms.

Because "specific adsorption" phenomena, as described above are expected to produce highly selective ion exchange behavior, there has been an unfortunate tendency in the soil chemistry literature to regard this behavior as unique evidence for highly selective surface complexation. This point of view reflects a misunderstanding of the relationship between molecular mechanisms and macroscopic data. The mechanism of surface complexation may be employed to interpret ion exchange reactions, but the mechanism itself must be established by separate experiments. The exchange reaction is a strictly macroscopic concept that is consistent with several different kinds of exchange mechanism at the molecular level (cf. Section 4.1).

Experimental composition data on ion exchange reactions are obtained as described in Section 4.1. Often these data are summarized in exchange isotherms. An *exchange isotherm* is a graph, at a fixed temperature and applied pressure, of the charge fraction of an ion on an exchanger solid versus its charge fraction in the contiguous aqueous solution. In terms of the variables in Eq. 4.2, these charge fractions are defined by the equations

$$E_A \equiv |Z_A| n_A^{(w)} \qquad \tilde{E} \equiv |Z_A| m_A / \tilde{Q} \tag{5.7a}$$

where $n_A^{(w)}$ is the relative surface excess, m_A is the molality, Z_A is the valence of an ion A (*not* necessarily a cation), and

$$Q \equiv \Sigma_i |Z_i| n_i^{(w)} \qquad \tilde{Q} \equiv \Sigma_i |Z_i| m_i \tag{5.7b}$$

are, respectively, the *total moles of adsorbed charge* and the *total moles of adsorptive charge* contributed by all ions competing in simultaneous adsorption reactions according to Eq. 5.1 or 5.2. Examples of exchange isotherms for the cation exchange reactions

$$2NaX(s) + Mg^{2+}(aq) = MgX_2(s) + 2Na^+(aq) \tag{5.8a}$$

$$CaX_2(s) + Mg^{2+}(aq) = MgX_2(s) + Ca^{2+}(aq) \tag{5.8b}$$

on a Vertisol are shown in Fig. 5.1. For these examples, $T = 298$ K, $P = 1$ atm, and $\tilde{Q} = 50 \text{ mol}_c \text{ m}^{-3}$.

Exchange isotherms can be used to calculate an *exchange selectivity coefficient*,[1] the conventional terminology for the conditional equilibrium constant associated with Eq. 5.1 or 5.2. For example, in the case of the cation exchange reaction in Eq. 5.1,

$$K_{exc} \equiv x_B^a (\mathbf{A}^{a+})^b / x_A^b (\mathbf{B}^{b+})^a = K_{ex} f_A^b / f_B^a \tag{5.9}$$

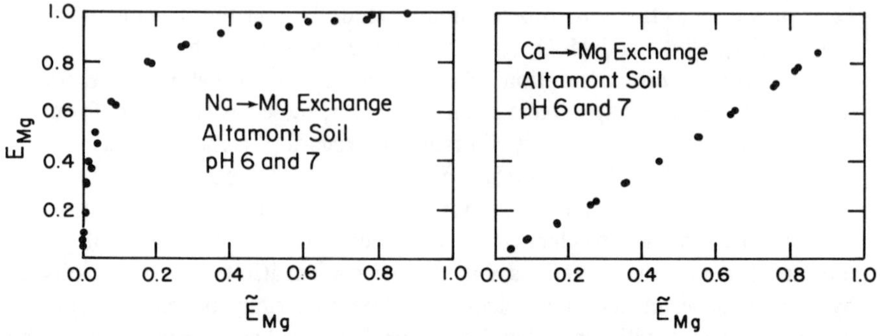

FIG. 5.1. Exchange isotherms corresponding to the reactions in Eq. 5.8 for the (silt + clay) fraction of a Vertisol, with data at pH 6 and 7 combined [data from P. F. Fletcher, G. Sposito, and C. S. LeVesque, Sodium-calcium-magnesium exchange reactions on a montmorillonitic soil: I. Binary exchange reactions, *Soil Sci. Soc. Am. J.* **48**:1016 (1984)].

where, by analogy with Eq. 4.7,

$$f_A = (\mathbf{AX}_a)/x_A \qquad f_B = (\mathbf{BX}_b)/x_B \tag{5.10}$$

Thus the dependence of K_{exc} on exchanger composition is the same as that of the ratio of activity coefficients $f_A{}^b/f_B{}^a$. In Section 4.1, the Reference State in which the activity coefficient of an adsorbate component has unit value was implied to be that of the pure component ("homoionic exchanger") at T = 298.15 K and P = 1 atm (see also Note 4 in Chapter 4). This specification of the Reference State is not wholly adequate, however, because ion exchange reactions invariably are carried out in aqueous media. The question arises therefore as to precisely what relationship the Reference State of a homoionic exchanger should have to the aqueous solution that must be brought into contact with it in order to initiate an exchange reaction. If the prescription in Section 4.1 is followed literally, the homoionic exchanger should be free of all water except structural water when it is in the Reference State. This condition, however, is difficult to achieve experimentally and bears little resemblance to that of a hydrated exchanger contacting a soil solution. Moreover, the chemical potential of a homoionic exchanger in the Standard State (which in the present case is the same as the Reference State) no longer could be measured directly in a solubility experiment. For example, the measured values of μ^0 for homoionic montmorillonites, such as might be employed to calculate K_{dis} in Eq. 3.18, would be invalid if they were determined routinely with the help of solubility data. For these reasons, it is necessary to include a careful statement concerning the aqueous solution phase when defining the Reference State (or the Standard State) of an adsorbate component.[2]

It may seem enough to define the Reference State simply as the homoionic exchanger, at 298.15 K and 1 atm pressure, in contact with an aqueous solution containing a soluble salt of the exchanging ion. That this statement will not suffice may be seen as follows. Consider a soil exchanger that is saturated completely with Na^+ and that is in equilibrium with a 100 mol m^{-3} solution of NaCl at 298.15 K under a pressure of 1 atm. In this situation, $\tilde{\mu}(Na^+, ex) = \tilde{\mu}(Na^+, aq)$, according to the discussion in Special Topic 2 (Chapter 2), where $\tilde{\mu}$ is an electrochemical potential and "ex" refers to the soil exchanger. This equality will provide a value of the Reference State electrochemical potential of Na^+ on the exchanger if $\tilde{\mu}(Na^+, aq)$ can be measured. Now suppose that the same soil exchanger is put into contact with a solution of 50 mol m^{-3} NaCl at the same temperature and pressure. By hypothesis, the exchanger is still in its Reference State. At equilibrium, the electrochemical potentials of Na^+ on the exchanger and in aqueous solution again must be equal, but this time they cannot have the same values as when the concentration of NaCl was 100 mol m^{-3}. In both cases, then, the exchanger supposedly was in its Reference State, with unit activity coefficient, yet the value of $\tilde{\mu}(Na^+, ex)$ is different in each case.

This paradox can be avoided by acknowledging that something specific must be said about the composition of the aqueous solution phase when the Reference State of an exchanger is defined. The prescription that is most widely accepted (in principle, if not in practice) is that given by Gaines and Thomas.[3] It specifies that an adsorbate component is in the Reference State if it is at unit mole fraction, is at 298.15 K under 1 atm pressure, and is in equilibrium with an *infinitely dilute* solution containing a salt of the exchangeable ion. In ion exchange experiments, the Gaines-Thomas definition of the Reference State may be realized by measuring K_{exc} at fixed exchanger composition for several ionic strengths, and then extrapolating to zero ionic strength to obtain a value of K_{exc} based on the Infinite Dilution Reference State.[4] Unfortunately, this procedure, or one equivalent to it, very seldom is followed in published ion exchange experiments.

The definition given here applies to the Standard State as well, once it is specified additionally that the infinitely dilute solution must contain salts of all the ions in the exchanger structure. With respect to the determination of μ^0 for solid exchangers by solubility measurements, the Gaines-Thomas convention may be adopted by determining the ion activity product (IAP) for the exchanger at several ionic strengths, and then extrapolating the data to obtain the IAP at zero ionic strength. In this limit, the IAP equals the solubility product constant for the exchanger, since the activity of the exchanger now has unit value, by definition.

Exchange isotherms, like activity-ratio diagrams and other graphical representations of chemical equilibria, can provide a qualitative picture of ion exchange selectivity on a comparative basis. Broadly speaking, the greater the convexity of the isotherm, the more selective is the exchanger for the ion whose charge fractions are plotted.[1] This general rule is given a more quantitative foundation in chemical thermodynamics after the *non-preference exchange isotherm* is defined. An exchanger is said to exhibit "no relative preference" for either ion in an exchange reaction if $K_{ex} = 1.0$ (i.e., $\Delta_r G^0 = 0$) and if the adsorbed ions show ideal solid solution behavior [i.e., $f_A = f_B = 1$ or $f_C = f_D = 1$;

cf. Section 3.3]. The implication of this definition for Eq. 5.9 is as follows:

$$1 = x_B^a (A^{a+})^b / x_A^b (B^{b+})^a \tag{5.11a}$$

The exchange isotherm implicit in this equation can be derived after introducing the relationships between charge fractions and mole fractions or molalities for a binary exchange process:

$$E_A = ax_A/(ax_A + bx_B) \qquad E_B = bx_B/(ax_A + bx_B) \tag{5.12}$$

$$\tilde{E}_A = am_A/(am_A + bm_B) \qquad \tilde{E}_B = bm_B/(am_A + bm_B) \tag{5.12b}$$

These equations are just special cases of Eq. 5.7. If they are solved for the mole fractions or molalities and then substituted into Eq. 5.11a, the result is

$$1 = E_B^a (bE_A + aE_B)^{b-a} \Gamma \tilde{E}_A^b / E_A^b \tilde{E}_B^a \tilde{Q}_{ab} \tag{5.11b}$$

where

$$\Gamma \equiv \gamma_A^b / \gamma_B^a \qquad \tilde{Q}_{ab} \equiv (\tilde{Q}/ab)^{a-b} \tag{5.13}$$

Equation 5.11b can be expressed entirely in the charge fractions for either exchanging cation, since $E_A + E_B = \tilde{E}_A + \tilde{E}_B \equiv 1$. For example, if cation A is of interest, Eq. 5.11b becomes

$$E_A^b \frac{[a + (b - a)E_A]^{a-b}}{(1 - E_A)^a} = \frac{\Gamma}{\tilde{Q}_{ab}} \frac{\tilde{E}_A^b}{(1 - \tilde{E}_A)^a} \tag{5.14}$$

For homovalent cation exchange, a = b and Eq. 5.14 reduces to the simple expression

$$\frac{E_A}{(1 - E_A)} = \frac{\tilde{E}_A}{(1 - \tilde{E}_A)} \tag{5.15}$$

which is readily demonstrated to be a linear relationship between E_A and $\tilde{E}_A$ that plots a straight-line exchange isotherm making a 45° angle with either axis. [In deriving Eq. 5.15, the Davies equation (Eq. 1.21) has been assumed to describe γ_A and γ_B.] Thus, for a binary adsorbate, the nonpreference isotherm for homovalent ion exchange reactions is a straight line. This behavior is approximated in the exchange isotherm for Ca → Mg exchange in Fig. 5.1, although some selectivity for Ca is indicated by the slight concavity of the graph.

For heterovalent cation exchange, the nonpreference isotherm generally will not be a straight line, nor will it be independent of $\tilde{Q}$ (or ionic strength) as in the case of homovalent exchange. At any fixed value of E_A, a variation in $\tilde{Q}$ cannot change the ratio on the left side of Eq. 5.14. The parameter Γ will usually shift only slightly as $\tilde{Q}$ is varied, whereas $\tilde{Q}_{ab}$ will change as some power of $\tilde{Q}$ (Eq. 5.13). Thus at a given value of E_A, a variation in $\tilde{Q}$ may cause a significant change in $\tilde{Q}_{ab}$ and a large, compensating shift in $\tilde{E}_A$ to maintain the constancy of the left side of Eq. 5.14. This conclusion can be illustrated readily for the case a = 1, b = 2 (uni-bivalent cation exchange). In this case, Eq. 5.14 becomes

$$\frac{E_A^2(1+E_A)^{-1}}{1-E_A} = \frac{1}{2}\Gamma\tilde{Q}\,\frac{\tilde{E}_A^2}{1-\tilde{E}_A} \qquad (5.16a)$$

and the nonpreference exchange isotherm is

$$E_A = \left[1 + \frac{2}{\Gamma\tilde{Q}}\left(\frac{1}{\tilde{E}_A^2} - \frac{1}{\tilde{E}_A}\right)\right]^{1/2} \qquad (5.16b)$$

Equation 5.16b is plotted in Fig. 5.2 for values of $\tilde{Q}$ equal to 10^3, 10^2, and 10 mol m^{-3}. (The parameter Γ was calculated with Eq. 1.21.) It is evident that the nonpreference isotherm for uni-bivalent exchange will approach a straight line only at large values of $\tilde{Q}$. The isotherms at lower values of $\tilde{Q}$ are decidedly concave and therefore would give the impression that the exchanger is selective for B^{2+}. However, in this case, the cations A^+ and B^{2+} are preferred *equally* by the exchanger because of the imposed conditions that $K_{ex} = 1$ and that the adsorbate form an ideal solid solution. Therefore no selectivity exists for B^{2+} over A^+, insofar as selectivity is defined thermodynamically by K_{ex} and the rational activity coefficients. This result shows clearly that *the magnitude of* K_{ex} *cannot be inferred simply from the appearance of an exchange isotherm.*

5.2 Binary Ion Exchange Equilibria

The ion exchange reactions in Eqs. 5.1 and 5.2 involve pairs of cations or anions, respectively, but this choice of the number of reacting ionic species is not unique, nor does it imply that the exchange has only two ions in the adsorbate. As the form of the general adsorption–desorption reaction in Eq. 4.3 makes apparent, ion exchange reactions can exhibit several ionic species, and it is only by analogy with complexation reactions that two reacting species are

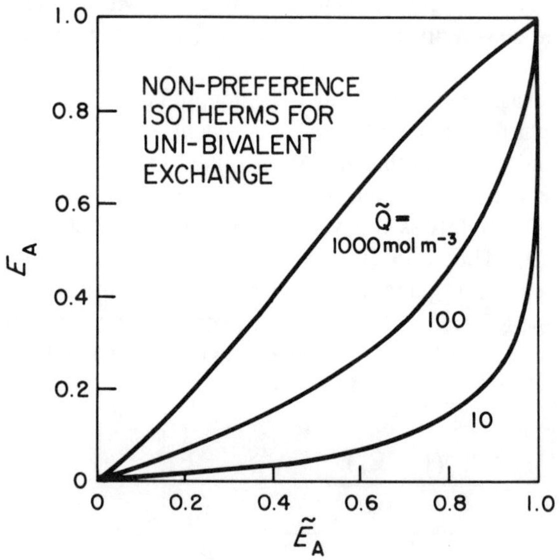

FIG. 5.2. Nonpreference exchange isotherms calculated with Eq. 5.16b for three values of the total moles of adsorptive charge, $\tilde{Q}$ (Eq. 5.7b).

featured instead of more. The principal advantages of this approach are its evident simplicity and the possibility that ion exchange in multicomponent exchange reactions (i.e., all natural soils) can be described in terms of a set of independent, two-species reactions.

A large body of experimental research exists concerning two-component ion exchangers, whose behavior is described by Eq. 5.1 or 5.2.[5] These systems thus exhibit *binary* ion exchange equilibria. The central problem in applying chemical thermodynamics to them is to derive equations that permit the calculation of K_{ex} and the activity coefficients of the two adsorbate species.[6] Several approaches have been taken to solve this problem, each of which reflects a particular notion of how exchanger composition data can be utilized most effectively to calculate thermodynamic quantities.

One conceptually simple approach is to express all binary ion exchange reactions as combinations of the special case of Eq. 4.3, in which the species Q is deleted and SR′ ≡ SR, with the possibility that more than 1 mol of SR may combine with adsorptive ions to form the adsorbate species. This approach portrays ion adsorption formally as a complexation reaction and builds ion exchange reactions as combinations of these reactions.[7] Since the adsorbate species may be formed from both cations and anions (cf. Eq. 4.3), ion exchange reactions involving charged complexes [e.g., $CaCl^+(aq)$] as well as monatomic ions [e.g., $Na^+(aq)$] can be described. If the further simplification is made that adsorbate species activity coefficients do not depend on exchanger composition, then equilibrium speciation calculations can be performed exactly as described in Section 2.4. Given the values of the equilibrium constants K_{ads} for each component ion adsorption reaction, a convergent speciation calculation will provide a set of compatible species concentrations in both the aqueous solution phase and the adsorbate. Computations of this kind can simulate ion exchange isotherms accurately.[7] As pointed out in Section 1.2, however, the species concentrations and the models adopted for their activity coefficients are connected inextricably, *neither having chemical significance without the other.* In the present example, the assumption of constant adsorbate species activity coefficients is a critical factor in determining the predicted species concentrations, whose accuracy must then be evaluated by independent experiments (e.g., spectroscopic studies).

It is evident from Eq. 5.9, for example, that the exchanger-composition dependence of the conditional exchange equilibrium constant derives from that of the adsorbate species activity coefficients. Measurements of K_{exc} show typically a significant exchanger-composition dependence,[5] such that constant adsorbate species activity coefficients are the exception, not the rule. A straightforward approach to obtaining a systematic view of how K_{exc} depends on exchanger adsorbate composition is to use model expressions for the rational activity coefficients as functions of the mole fraction of one of the two exchanger components. For example, suppose that $\ln f_A$ and $\ln f_B$, pertaining to the exchangeable cations, A and B, respectively, are expressed by the MacLaurin expansions:

$$\ln f_A = a_1 x_B + a_2 x_B^2 + a_3 x_B^3 + a_4 x_B^4 + \cdots + a_i x_B^i \tag{5.17a}$$

$$\ln f_B = b_1 x_A + b_2 x_A^2 + b_3 x_A^3 + b_4 x_A^4 + \cdots + b_i x_A^i \tag{5.17b}$$

where the a_i and b_i are empirical coefficients. Equation 5.17 is termed a *Margules equation* for the activity coefficients.[8] Not all of these empirical coefficients are nonzero and independent, however, because the Gibbs-Duhem equation (Eq. 4.9b), applied to the exchanger at fixed T and P, imposes a constraint on the variation of the two activity coefficients with adsorbate composition. In the case of an expansion terminating at third order, for example, the activity coefficients in Eq. 5.17 take the form[8]

$$\ln f_A = a_2 x_B^2 + a_3 x_B^3 \tag{5.18a}$$

$$\ln f_B = \left(a_2 + \frac{3}{2} a_3\right) x_A^2 - a_3 x_A^3 \tag{5.18b}$$

Note that imposition of Eq. 4.9b requires that the first-order terms vanish; that there be differing coefficients for the second-order terms; and that there be equal but opposite coefficients for the third-order terms. Moreover, if the infinite-dilution limits of the activity coefficients, that is,

$$\ln f_{AB}^{\infty} \equiv \lim_{x_A \downarrow 0} f_A \qquad \ln f_{BA}^{\infty} \equiv \lim_{x_B \downarrow 0} f_B \tag{5.19}$$

are evaluated with Eq. 5.18, the empirical coefficients are constrained further by the expressions[8]

$$\ln f_{AB}^{\infty} = a_2 + a_3 \qquad \ln f_{BA}^{\infty} = a_2 + \frac{1}{2} a_3 \tag{5.20a}$$

or

$$a_2 = 2 \ln f_{BA}^{\infty} - \ln f_{AB}^{\infty} \qquad a_3 = 2(\ln f_{AB}^{\infty} - \ln f_{BA}^{\infty}) \tag{5.20b}$$

upon solving for a_2 and a_3. Thus the empirical coefficients in the Margules equation are related to the infinite-dilution limits of the activity coefficients for the adsorbate species in a binary exchange system. In applications, however, Eq. 5.17 is introduced into Eq. 5.9 and the coefficients a_i ($i = 1, \ldots, n$) are estimated from, for example, least-squares fitting of the resulting model equation to the observed composition dependence of K_{exc}.[1]

Since the Margules expansions represent a convergent power series in the mole fractions,[8] they can be "summed selectively" to yield closed-form model equations for the adsorbate species activity coefficients. A variety of two-parameter models can be constructed in this way by imposing a constraint on the empirical coefficients in addition to the Gibbs-Duhem equation. For example, a simple interpolation equation that connects the two limiting values of f (f^{∞} at infinite dilution and $f = 1.0$ in the Reference State) can be derived after imposing the scaling constraint

$$\frac{\ln f_A}{\ln f_B} = \frac{\ln f_{BA}^{\infty}}{\ln f_{AB}^{\infty}} \left(\frac{x_B}{x_A}\right)^2 \tag{5.21}$$

along with Eq. 4.9b. The resulting activity coefficient expressions constitute the *van Laar model*:[9]

$$\ln f_A = \ln f_{AB}^{\infty} \frac{(x_B \ln f_{BA}^{\infty})^2}{[x_A \ln f_{AB}^{\infty} + x_B \ln f_{BA}^{\infty}]^2} \tag{5.22b}$$

Equation 5.21 implies that the right sides of Eqs. 5.17a and 5.17b are identical once they are divided by the scale factors, $x_B^2 \ln f_{BA}^{\infty}$ or $x_A^2 \ln f_{AB}^{\infty}$, respectively.

This factoring of ln f produces a "scale-invariant" part that serves to interpolate smoothly between the value of f^{∞} and f = 1.0. Two well-known special cases of the van Laar model are the *regular solution model*, which results by setting $f_{AB}^{\infty} = f_{BA}^{\infty}$, and the *Vanselow model*, which results by setting $f_{AB}^{\infty} = f_{BA}^{\infty} \equiv 1.0$.[1] The Vanselow model thus corresponds to an ideal solid solution ($f = f^{id} \equiv 1.0$; cf. Section 3.3).

Another two-parameter model of widespread application results from imposing the constraint that the activity of an adsorbate species can be a power-law expression in its exchanger charge fraction (cf. Eq. 5.10):

$$f_A = E_A^q / x_A \qquad f_B = E_B^p / x_B \tag{5.23}$$

where E_A and E_B are defined in Eq. 5.12a (taking cations as an example). The special case, $q = a/\beta$, $p = b/\beta$, defines the one-parameter *Rothmund-Kornfeld model*,[6,7] where a and b are cation valences (Eq. 5.9) and β is the model parameter. The further restriction, $\beta = 1$, then produces the *Gapon model*,[6] which has no adjustable parameters. Similarly, the choice, q = p = 1, produces the *Gaines-Thomas model*,[6] with no adjustable parameters. This last model can also be pictured as an analog of the Vanselow model, in which charge fractions replace mole fractions in order to account for adsorbate nonideality.

The methodological approach underlying these models of the adsorbate species activity coefficients is the idea of supplementing the Gibbs-Duhem equation (Eq. 4.9b) with one other equation that links f_A and f_B (or their differentials), thereby creating two equations in the two f-variables that can be solved uniquely, given the Reference State as described in Section 5.1. Equation 5.21 or 5.23 can serve as the second constraint on the activity coefficients, but the obvious ad hoc nature of these expressions is less than fully satisfactory. Instead of an assumed constraining equation, the condition that the composition dependence of K_{exc} is the result of that of the activity coefficients, mentioned in Section 5.1, can be exploited by forming the differential of ln K_{exc} based, for example, on Eq. 5.9:

$$d \ln K_{exc} = b\, d\ln f_A - a\, d\ln f_B \tag{5.24}$$

This equation can be combined with Eq. 4.9b following the method described in Section 4.1 to derive expressions analogous to Eq. 4.11:[10]

$$b \ln f_A = E_B \ln K_{exc} - \int_0^{E_B} \ln K_{exc}\, dE_B' \tag{5.25a}$$

$$a \ln f_B = -(1 - E_B) \ln K_{exc} + \int_{E_B}^{1} \ln K_{exc}\, dE_B' \tag{5.25b}$$

Thus measurements of K_{exc} as a function of E_B in a binary exchange system are sufficient to calculate the adsorbate species activity coefficients at any exchanger composition.

An example of the application of Eq. 5.25 to the cation exchange reaction

$$MgX_2(s) + 2K^+(aq) = 2\ KX(s) + Mg^{2+}(aq) \tag{5.26}$$

on an Entisol[11] is shown in Table 5.1. The activity coefficients listed in the third and fourth columns of the table were calculated with the help of the data in the first and second columns, as well as Eqs. 5.25a (with a = 1, A = Mg) and 5.25b (with b = 2, B = K). The generally small effect on f_{Mg} and f_K of a tenfold drop in ionic strength is evident, suggesting that extrapolation of K_{exc} to infinite dilution is not necessary within experimental precision. The values of K_{ex} calculated at the two ionic strengths and at each E_K according to the equation

$$K_{ex} = f_K^2 K_{exc}/f_{Mg} \tag{5.27}$$

Table 5.1 Thermodynamic Data for $Mg^{2+} \rightarrow K^+$ Exchange on an Entisol

E_K	K_{exc}	f_{Mg}	f_K	K_{ex}
		I = 10 mol m⁻³		
0.0	142.0	1.000	0.400	22.7
0.1	107.6	0.985	0.460	23.1
0.2	51.8	0.886	0.624	22.8
0.3	28.3	0.760	0.785	23.0
0.4	20.4	0.677	0.873	23.0
0.5	15.7	0.603	0.938	22.9
0.6	13.1	0.543	0.979	23.1
0.7	11.9	0.512	0.995	23.0
0.8	11.3	0.496	0.999	22.7
0.9	11.4	0.496	0.999	23.0
1.0	11.2	0.493	1.000	22.7
		I = 1 mol m⁻³		
0.0	149.0	1.000	0.393	23.0
0.1	113.2	0.986	0.446	22.8
0.2	54.6	0.884	0.609	22.9
0.3	31.2	0.769	0.752	22.9
0.4	20.8	0.666	0.859	23.0
0.5	15.6	0.584	0.931	23.2
0.6	13.4	0.538	0.962	23.1
0.7	11.3	0.482	0.992	23.1
0.8	10.3	0.450	1.000	22.9
0.9	10.7	0.465	0.999	23.0
1.0	10.3	0.448	1.000	23.0

Source: H.E. Jenson and K.L. Babcock.[11]

are listed in the fifth column of the table. The mean values of K_{ex} at the two ionic strengths (22.9 ± 0.1 and 22.99 ± 0.09) are not significantly different.

In light of the generality of Eq. 5.25, Eqs. 5.21 and 5.23 can be regarded as special models describing how K_{exc} is to vary with exchanger composition. For example, Eq. 5.21 is tantamount to the model expression

$$\ln K_{exc} = \ln K_{ex} + \ln f_{AB}^{\infty} \ln f_{BA}^{\infty} \frac{[bx_B^2 \ln f_{BA}^{\infty} - ax_A^2 \ln f_{AB}^{\infty}]}{[x_A \ln f_{AB}^{\infty} + x_B \ln f_{BA}]^2} \tag{5.28}$$

as can be deduced by substitution of Eq. 5.22 into Eq. 5.9.

The introduction of Eq. 5.25 into Eq. 5.9 converted to logarithmic form produces an expression analogous to Eq. 4.12:[10]

$$\ln K_{ex} = \int_0^1 \ln K_{exc} \, dE_B \tag{5.25c}$$

It follows from this equation that the adsorbed species activity coefficients must satisfy the integral condition

$$\int_0^1 \ln(f_A^b / f_B^a) \, dE_B \equiv 0 \tag{5.25d}$$

(Express Eq. 5.9 in logarithmic form and integrate both sides over the domain of E_B while noting Eq. 5.25c.) Equation 5.25d can be applied to examine calculated values or model expressions for the adsorbate species activity coefficients to ensure their consistency with chemical thermodynamics as embodied in Eqs. 4.96 and 5.24.

Another kind of experimental consistency check with Eq. 5.25 can be made if K_{ex} is determined for three or more binary exchange reactions. Consider, for example, the three following cation exchange reactions:

$$bAX_a(s) + aB^{b+}(aq) = aBX_b(s) + bA^{a+}(aq) \tag{5.29a}$$

$$eBX_b(s) + bE^{e+}(aq) = bEX_e(s) + eB^{b+}(aq) \tag{5.29b}$$

$$eAX_a(s) + aE^{e+}(aq) = aEX_e(s) + eA^{a+}(aq) \tag{5.29c}$$

Since these reactions are coupled, only two can be independent (cf. Section 1.5), and the dependence of the remaining reaction is reflected in a constraint on the three equilibrium constants

$$K_{ex}^{(1)} = (\mathbf{BX}_b)^a (\mathbf{A}^{a+})^b / (\mathbf{AX}_a)^b (\mathbf{B}^{b+})^a \tag{5.30a}$$

$$K_{ex}^{(2)} = (\mathbf{EX}_e)^b (\mathbf{B}^{b+})^e / (\mathbf{BX}_b)^e (\mathbf{E}^{e+})^b \tag{5.30b}$$

$$K_{ex}^{(3)} = (\mathbf{EX}_e)^a (\mathbf{A}^{a+})^e / (\mathbf{AX}_a)^e (\mathbf{E}^{e+})^a \tag{5.30c}$$

It is evident from an inspection of Eq. 5.30 that this constraint takes the form

$$[K_{ex}^{(3)}]^b = [K_{ex}^{(2)}]^a [K_{ex}^{(3)}]^e \tag{5.31a}$$

or, in terms of $\Delta_r G^0$ (Eq. s1.8 in Chapter 1),

$$b\ \Delta_r G^0(3) = a\ \Delta_r G^0(2) + e\ \Delta_r G^0(1) \tag{5.31b}$$

Equation 5.31 provides the basis for a test of both experimental precision and thermodynamic self-consistency when the $\Delta_r G^0$ values are calculated using values of ln K_{ex} determined with the help of Eq. 5.25c. Table 5.2 shows some typical examples of this test applied to cation exchange reactions on three montmorillonites.[12] The values of b $\Delta_r G^0(3)$ expected according to Eq. 5.31b are given in the eighth column of the table. In each case the measured value of b $\Delta_r G^0(3)$ in column 7 is in fair agreement with the theoretical value based on the requirement of self-consistency, the mean deviation between measured and theoretical values being ± 0.62 kJ mol^{-1}.

Table 5.2 Consistency Tests According to Eq. 5.31b for Cation Exchange Reactions on Three Montmorillonites. ($\Delta_r G^0$ values in kJ mol^{-1})

Montmorillonite	A^{a+}	B^{b+}	E^{e+}	$e\Delta_r G^0(1)$	$a\Delta_r G^0(2)$	$b\Delta_r G^0(3)$	$b\Delta_r G^0(3)_{theor}$ [a]
Wyoming	Na^+	K^+		-1.28			
		K^+	Li^+		+2.05		
	Na^+		Li^+			+0.20	+0.77
Chambers (25°C)	Na^+	Cs^+		-7.89			
		Cs^+	Rb^+		+3.38		
	Na^+		Rb^+			-5.64	-4.51
Chambers (30°C)	Na^+	Cs^+		-18.00			
		Cs^+	Ba^{2+}		+15.38		
	Na^+		Ba^{2+}			-2.08	-2.62
Camp-Berteau	Na^+	NH_4^+		-8.54			
		NH_4^+	Ba^{2+}		+7.87		
	Na^+		Ba^{2+}			-0.42	-0.67

[a] $b\Delta_r G^0(3)_{theor} \equiv a\Delta_r G^0(2) + e\Delta_r G^0(1)$

5.3 Multicomponent Ion Exchange Equilibria

Most of the ion exchange reactions that occur in soils involve three or more principal ions. For example, in soils at pH < 4, the trio $\{H^+, Ca^{2+}, Al^{3+}\}$ might be considered, and, in soils at pH > 7, the set $\{Na^+, Mg^{2+}, Ca^{2+}\}$ is important.[1] Multicomponent cation exchange is the norm for reactions on humic substances[13] and the same is true for anion exchange on metal oxides (e.g., Cl^-, NO_3^-, and SO_4^{2-} on iron or aluminum oxyhydroxides).[14]

In a multicomponent ion exchange system, the reactions between exchanging ions still can be expressed as in Eqs. 5.1 and 5.2, which are binary processes, but the description of exchanger composition in terms of mole or charge fractions requires a generalization of expressions like Eq. 5.12:

$$x_i = n_i^{(w)}/\Sigma_k n_k^{(w)} \qquad E_i = |Z_i|n_i^{(w)}/\Sigma_k|Z_k|n_k^{(w)} \tag{5.32}$$

where $n_i^{(w)}$ is the relative surface excess of an ion i with valence Z_i (Eq. 4.2) and the sum is over all exchanging ions. Quantitation of the adsorbate species, even in the case of a ternary or a quaternary ion exchange system, can impose significant data requirements.[15] The application of chemical thermodynamics to describe the isothermal, isobaric equilibrium of multicomponent ion exchange systems also requires a generalization of the approach based on Eqs. 4.9b and 5.25.[16] The generalization is straightforward, just as it is in Eq. 5.32, but formidable data requirements are implicit in the resulting equations of the activity coefficients and exchange equilibrium constants.

As in Section 4.1, it is convenient to label exchanging ions with numerical indices in describing multicomponent ion exchange equilibria. The thermodynamic approach will be illustrated for a ternary cation exchange system (e.g., Na^+, Mg^{2+}, Ca^{2+}), but the extension to an N-ary system is direct. This is evident, for example, in the Gibbs-Duhem equation for a ternary system:

$$x_1 d\ln f_1 + x_2 d\ln f_2 + x_3 d\ln f_3 = 0 \qquad (T, P \text{ constant}) \tag{5.33}$$

For each additional exchanging ion, one simply adds a term, $x_i d \ln f_i$, with the understanding that the sum of the mole fractions x_i ($i = 1, 2, 3, \ldots$) must always equal 1.0. Similarly, for each possible cation exchange reaction (Eq. 5.1), one defines a conditional exchange equilibrium constant

$$K_{ijc} \equiv K_{ij}(f_i^{Z_j}/f_j^{Z_i}) = 1/K_{jic} \tag{5.34}$$

Strictly speaking, both K_{ijc} and the adsorbate species activity coefficients should carry a designation of the number of exchanging cations (e.g., T for ternary) to emphasize their composition dependence. (This designation is not necessary for the exchange equilibrium constant K_{ij} because it is independent of composition.) The generalization of the constraint in Eq. 5.24 is simply to replicate it for each cation exchange reaction:

$$\ln K_{12} = Z_1 \ln f_2 - Z_2 \ln f_1 + \ln K_{12c} \tag{5.35a}$$

$$\ln K_{23} = Z_2 \ln f_3 - Z_3 \ln f_2 + \ln K_{23c} \tag{5.35b}$$

$$\ln K_{31} = Z_3 \ln f_1 - Z_1 \ln f_3 + \ln K_{31c} \tag{5.35c}$$

and

$$Z_1 d \ln f_2 - Z_2 d \ln f_1 = -d \ln K_{12e} \tag{5.36a}$$

$$Z_2 d \ln f_3 - Z_3 d \ln f_2 = -d \ln K_{23c} \tag{5.36b}$$

$$Z_3 d \ln f_1 - Z_1 d \ln f_3 = -d \ln K_{31c} \tag{5.36c}$$

Equations 5.33 and 5.36 constitute four differential equations in the three dependent variables, f_1, f_2, f_3 (the mole fractions and the conditional equilibrium constants are assumed to be known through measurement). If there are more than three exchanging ions, for each term added to the Gibbs-Duhem equation, there will be additional constraints like Eq. 5.36, so that enough equations always will be generated to express the activity coefficients in terms of the conditional equilibrium constants, as in Eq. 5.25.

In fact, there is an apparent excess of constraint equations for a ternary system: *four* equations in the *three* unknown activity coefficients. The excess is removed, however, by noting that the "closure relation" in Eq. 5.31 can be applied to the three replicates of Eq. 5.34:

$$\begin{aligned} Z_1 \ln K_{23} &+ Z_2 \ln K_{31} + Z_3 \ln K_{12} \\ &= Z_1 \ln K_{23c} + Z_2 \ln K_{31c} + Z_3 \ln K_{12c} \equiv 0 \end{aligned} \tag{5.37}$$

Thus only two of Eq. 5.35 or 5.36 are independent and can be applied to constrain the Gibbs-Duhem equation. Equation 5.37 also can be used to calculate one of the K_{ijc} or K_{ij}, thereby reducing data requirements.

The combination of Eqs. 5.33, 5.36a, and 5.36b leads to the differential expressions

$$Z_2 Z_3 d \ln f_1 = Z_3(1 - E_1) d \ln K_{12c} + Z_1 E_3 d \ln K_{23c} \tag{5.38}$$

$$Z_1 Z_2 d \ln f_3 = -Z_1(1 - E_3) d \ln K_{23c} - Z_3 E_1 d \ln k_{12c} \tag{5.39}$$

The left sides of Eqs. 5.38 and 5.39 are total differentials. Equation 5.38 is integrated along a path between the points (E_1, E_2, E_3) and (1, 0, 0); one finds, after integrating by parts (cf. Eq. 4.11),

$$-Z_2Z_3 \ln f_1 = -Z_3(1 - E_1) \ln K_{12c} + Z_3 \int_{(E_1,E_2,E_3)}^{(1,0,0)} \ln K_{12c}\, dE_1$$
$$-Z_1E_3 \ln K_{23c}^{T} - Z_1 \int_{(E_1,E_2,E_3)}^{(1,0,0)} \ln K_{23c}^{T}\, dE_3 \qquad (5.40a)$$

Similarly, after integrating Eq. 5.39 from (E_1, E_2, E_3) to (0, 0, 1), one obtains

$$-Z_1Z_2 \ln f_3 = Z_1(1-E_3) \ln K_{23c} - Z_1 \int_{(E_1,E_2,E_3)}^{(0,0,1)} \ln K_{23c}\, dE_3$$
$$+ Z_3E_1 \ln K_{12c} + Z_3 \int_{E_1,E_2,E_3)}^{(0,0,1)} \ln K_{12c}\, dE_1 \qquad (5.40b)$$

Equation 5.40 is a generalization of Eq. 5.25. A comparable expression for ln f_2 can be derived after combining Eqs. 5.33, 5.36a, and 5.36c.

To derive an equation for the exchange equilibrium constant, one forms the difference between Eqs. 5.40b and 5.40a:

$$Z_1Z_2 \ln f_3 - Z_2Z_3 \ln f_1 = Z_3 \ln K_{12c} + Z_1 \ln K_{23c} + Z_3 \int_{(0,0,1)}^{(1,0,0)} \ln K_{12c}\, dE_1$$
$$+ Z_1 \int_{(1,0,0)}^{(0,0,1)} \ln K_{23c}\, dE_3 \qquad (5.41a)$$

and introduces in it the sum of Eqs. 5.35a and 5.35b:

$$Z_1Z_2 \ln f_3 - Z_2Z_3 \ln f_1 + Z_3 \ln K_{12c} + Z_1 \ln K_{23c} = Z_3 \ln K_{12} + Z_1 \ln K_{23} \qquad (5.41b)$$

to find the result

$$Z_3 \ln K_{12} + Z_1 \ln K_{23} = \int_{(0,0,1)}^{(1,0,0)} (Z_3 \ln K_{12c}\, dE_1 - Z_1 \ln K_{23c}\, dE_3) \qquad (5.41c)$$

Finally, the closure relation in Eq. 5.37 and the compositional constraint, $dE_3 = -(dE_1 + dE_2)$ are used in Eq. 5.41c to obtain an equation for K_{31}:

$$Z_2 \ln K_{31} = \int_{(0,0,1)}^{(1,0,0)} (Z_2 \ln K_{31c}\, dE_1 - Z_1 \ln K_{23c}\, dE_2) \qquad (5.41d)$$

where the integration now is along *any path* between the points (0,0,1) and (1,0,0).

The procedure just described can be repeated to yield general differential equations for the activity coefficients (which can then be integrated as in Eq. 5.40) and general integral expressions for the exchange equilibrium constants:[16]

$$Z_iZ_j\, d\ln f_k = -Z_i(1 - E_k)\, d\ln K_{jkc} + Z_kE_i\, d\ln K_{jic} \qquad (5.42a)$$

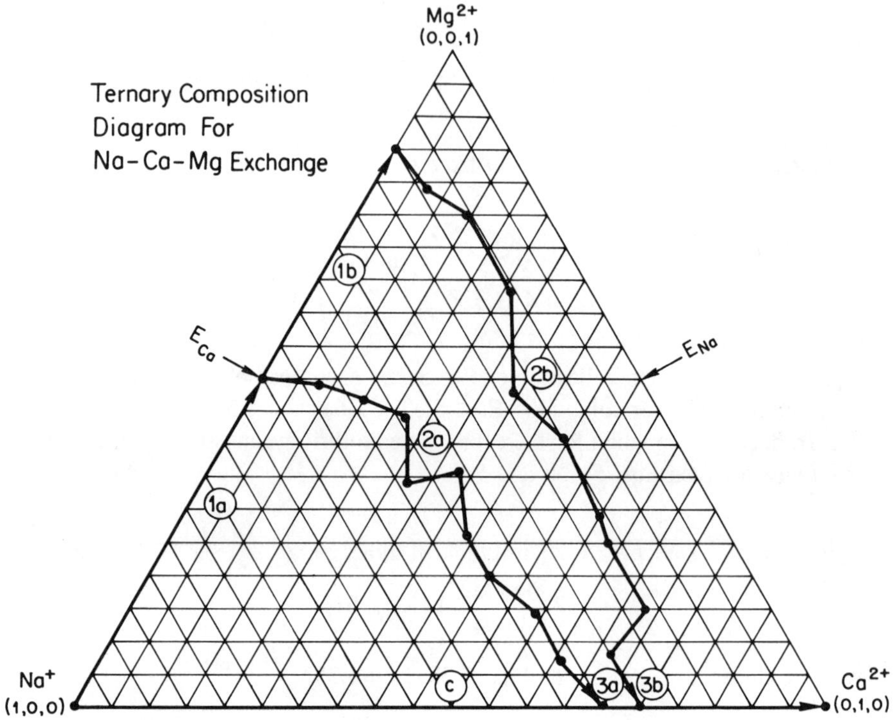

FIG. 5.3. Three pathways of integration for Eq. 5.43, shown in a ternary exchanger composition diagram, for Na–Ca–Mg exchange reactions on montmorillonite. Note that integration path c, a *binary* pathway from Na- to Ca-montmorillonite, is simply the bottom edge of the ternary diagram.

$$\ln K_{ij} = \int_{(E_i = 1, E_j = 0, E_k = 0)}^{(E_i = 0, E_j = 1, E_k = 0)} \left(\ln K_{ijc} \, dE_j - \frac{Z_j}{Z_k} \ln K_{kic} \, dE_k \right) \tag{5.42b}$$

where {ijk} = {123}, {312}, or {231}. These equations reduce to Eq. 5.25 when the system is binary. For the ternary system they illustrate how interactions (as expressed through the conditional equilibrium constants) *other* than that between a given pair of exchanging ions must be considered in order to calculate thermodynamic properties for the pair. Thus K_{31} in Eq. 5.41d depends not only on K_{31c}, as in a binary system, but also on the interaction between ion 3 and ion 2, as expressed through K_{23c}.

An application of Eq. 5.42b is shown in Fig. 5.3, which is a ternary composition diagram for exchange reactions between Na^+ (ion 1), Ca^{2+} (ion 2), and Mg^{2+} (ion 3) on montmorillonite suspended in 50 mol m^{-3} perchlorate solution at pH 7 and 25°C.[17] A single point in the diagram is prescribed by the set $\{E_1, E_2, E_3\}$, subject to $E_1 + E_2 + E_3 = 1.0$. Thus the diagram comprises points representing the exchanger composition at equilibrium, and paths connecting these points represent a sequence of equilibrium compositions as determined in a series of cation exchange experiments. Equation 5.42b can be applied to any of the paths to calculate the exchange equilibrium constant K_{ij}.

For example, the equilibrium constant K_{12} for $Na^+ \rightarrow Ca^{2+}$ exchange can be calculated from data obtained along any path on which the charge fractions of Ca^{2+} and Mg^{2+} are varied:

$$\ln K_{12} = \int_{(1,0,0)}^{(0,1,0)} \ln K_{12c}\, dE_2 + \int_{(1,0,0)}^{(0,1,0)} \ln K_{13c}\, dE_3 \qquad (5.43)$$

The path a in Fig. 5.3, with "legs" 1a, 2a, and 3a, represents the sequence of exchanger compositions: (1, 0, 0) → (0.5, 0, 0.5) → (0.3, 0.7, 0) → (0, 1, 0). The sequence begins at the left corner of the triangle (Na-montmorillonite) and moves up the left side, then turns to follow a somewhat tortuous route to the base of the triangle, and finally travels along the base to the right corner of the triangle (Ca-montmorillonite). For this path, $E_2 = 0$ along the first leg and only an integral of $\ln K_{13c}$ with respect to E_3 in a *binary* system must be evaluated. This was done by calculating K_{13c} with Na → Mg exchange data that permitted representing the conditional exchange constant with the regression equation

$$\ln K_{13c} = -0.7118 + 1.7726E_3 \qquad (r^2 = 0.90) \qquad (5.44)$$

and inserting Eq. 5.44 into the second term on the right side of Eq. 5.43. Along the second leg of the path of integration, values of $\ln K_{12c}$ and $\ln K_{13c}$ for the ternary system were fit to polynomial expressions, but examination of the correlation coefficients for the regression indicated that the conditional exchange constants showed no significant dependence on exchanger composition. Therefore the mean values

$$\ln K_{12c} = 0.55 \pm 0.14 \qquad \ln K_{13c} = 0.47 \pm 0.29$$

were introduced into Eq. 5.43 for the second leg. For the third leg of the path of integration, $E_3 = 0$ and only an integral of $\ln K_{12c}$ in a *binary* system is involved. In this case a polynomial expression was again inserted into the first term on the right side of Eq. 5.43 to complete the calculation. The final result for $\ln K_{12}$ was

$$\ln K_{12} = \underset{(\text{leg 1})}{-\,0.134} + \underset{(\text{leg 2})}{0.153} - \underset{(\text{leg 3})}{0.075} = -\,0.056 \qquad [\text{path a}]$$

which is the same as $K_{12} = 1.06$. This procedure was repeated for the integration path b with legs 16, 26, 36, representing the sequence: (1,0,0) → (0.15, 0, 0.85) → (0.25, 0.75, 0) → (0,1,0). This sequence extends the path along the left side of the triangle to a point nearer the apex (Mg-montmorillonite) with a consequent lengthening of the tortuous route down to the right corner. The final result for $\ln K_{12}$ was

$$\ln K_{12} = \underset{(\text{leg } 1)}{0.035} - \underset{(\text{leg } 2)}{0.112} - \underset{(\text{leg } 3)}{0.050} = -0.127 \quad [\text{path b}]$$

The result found is the same as $K_{12} = 0.88$. Finally, $\ln K_{12}$ was computed entirely in a binary system (path c in Fig. 5.3, along the base of the triangle) by setting $E_3 = 0$ in Eq. 5.43 and representing $\ln K_{12c}$ by a polynomial expression in E_2 (cf. Eq. 5.25c):

$$\ln K_{12} = \int_0^1 \ln K_{12c}\, dE_2 \tag{5.45a}$$

where[17]

$$\ln K_{12c} = 1.0001 - 1.5371\,E_2 - 8.0645\,E_2^2 + 32.4619\,E_2^3 - 74.2044\,E_2^4 + 84.821\,E_2^5 - 34.408\,E_2^6 \quad (r^2 = 0.98) \tag{5.45b}$$

The result of this calculation was $\ln K_{12} = 0.039$, which is the same as $K_{12} = 1.04$ [path c]. These three values for K_{12} agree, within the precision of the data, substantiating the path independence of Eq. 5.43.

This example is also a concrete illustration of a basic hypothesis underlying the myriad of experimental studies on binary exchange systems: *The thermodynamic characteristics of ion exchange in a multicomponent system can be evaluated solely with data from binary systems.* Thus any exchange equilibrium constant applicable to a ternary system can be calculated solely with data obtained in a binary exchange system (cf. Eqs. 5.43 and 5.45). That the same relationship can be found to apply to cation exchange isotherms is exemplified in Fig. 5.4, which shows exchange isotherms for $Ca^{2+} \rightarrow Mg^{2+}$ exchange on specimen or soil clay minerals suspended in 50 mol m^{-3} perchlorate solution at pH 7.[17,18] Within experimental precision, no effect of the charge fraction of Na^+ on these isotherms is apparent. Thus a $Ca^{2+} \rightarrow Mg^{2+}$ exchange isotherm measured in a binary system can be used to predict Ca → Mg in a ternary ($Na^+ - Ca^{2+} - Mg^{2+}$) system if the charge fractions in the latter system are calculated with Eq. 5.1 instead of Eq. 5.32 (i.e., after "renormalization" to an equivalent binary system).

It is important to emphasize that N-ary ion exchange relationships which do not enjoy thermodynamic status must be examined case by case to determine whether they can be "built up" from binary exchange data. For example, there is no reason to expect any conditional equilibrium constant K_{ijc} to remain invariant under changes in the composition of an ion exchange system from binary to ternary. As another example, one can pose the question as to whether binary exchange data for adsorbed species activity coefficients can be used, in

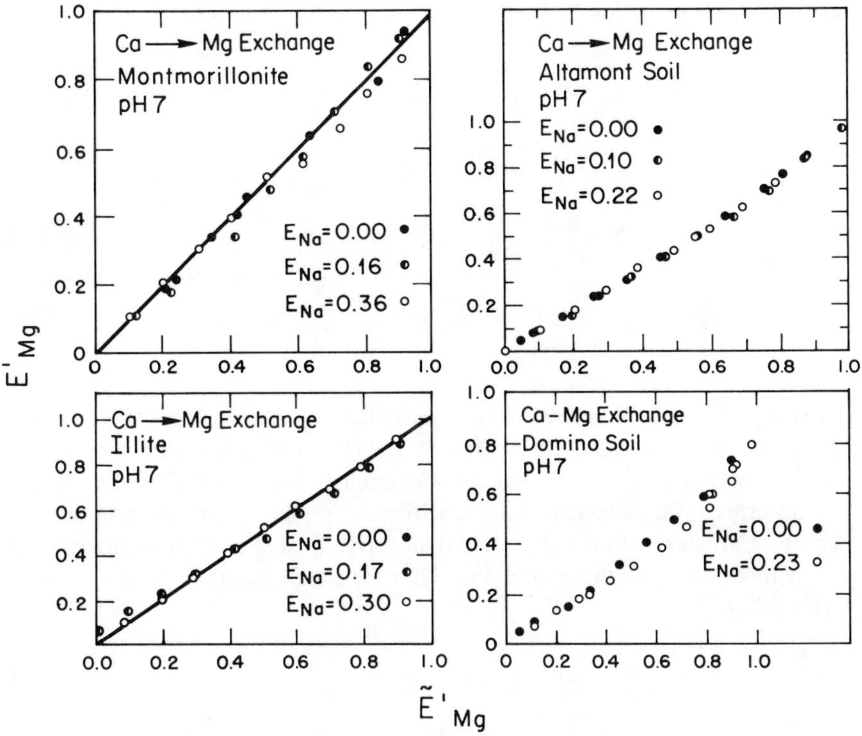

FIG. 5.4. Exchange isotherms corresponding to the reaction in Eq. 5.8b for montmorillonite and a montmorillonitic soil (Altamont series), and for illite and an illitic soil (Domino series) at pH 7, 298 K, and $\tilde{Q}$ = 50 $mol_c\ m^{-3}$. In each graph, the charge fraction of adsorbed Na is a fixed parameter and $E'_{Mg} \equiv n^{(w)}_{Mg}/(n^{(w)}_{Mg} + n^{(w)}_{Ca})$, $\tilde{E}' \equiv m_{Mg}/(m_{Mg} + m_{Ca})$ are "renormalized" Mg charge fractions. (Data from Sposito et al.[17,18] and Fletcher et al.[18])

general, to predict those of these species in an N-ary system (cf. 5.34). The answer to this question provided by chemical thermodynamics is *no, not in general.*[19]

To see this point in detail, one can express the activity coefficients in Eq. 5.33 with a third-order Margules expansion, as in Eq. 5.18:[20]

$$\ln f_i = \sum_{j,k \neq i} c^{(i)}_{jk} x_j x_k + \sum_{j,k,\ell \neq i} d^{(i)}_{jk\ell} x_j x_k x_\ell \tag{5.46}$$

where each mole fraction refers to an ion different from ion i, and the binary-system coefficients $c^{(i)}_{jj}$ and $d^{(i)}_{jjj}$ (equal to a_2 and a_3, respectively, in Eq. 5.18a) are related to the infinite-dilution limit of f_i:

$$\ln f^{\infty}_{ij} = c^{(i)}_{jj} + d^{(i)}_{jjj} \tag{5.47}$$

The introduction of Eqs. 5.46 and 5.47 into Eq. 5.33 leads to the equations[20]

$$d_{jjj}^{(i)} = -d_{iii}^{(j)} \tag{5.48a}$$

$$2c_{jj}^{(i)} - 2c_{ii}^{(j)} + 3d_{jjj}^{(i)} = 0 \tag{5.48b}$$

$$d_{jjk}^{(i)} = -c_{jk}^{(i)} - c_{ki}^{(j)} + 2c_{ii}^{(j)} \tag{5.48c}$$

$$c_{jk}^{(i)} - c_{ki}^{(j)} = 2(\ln f_{ji}^{\infty} + \ln f_{ki}^{\infty}) - 2(\ln f_{kj}^{\infty} + \ln f_{ij}^{\infty}) \tag{5.48d}$$

These equations can be solved for the coefficients $c_{jj}^{(i)}$, $d_{jjj}^{(i)}$, $c_{jk}^{(i)}$, and $d_{jjk}^{(i)}$ in terms of the infinite-dilution activity coefficients (which are binary-system properties), *but the solution will not be unique*. Equation 5.48d, connecting the ternary second-order Margules expansion coefficients to the binary infinite-dilution activity coefficients, shows that a constant (say, c_0) can be added to any $c_{jk}^{(i)}$ and $c_{ki}^{(j)}$, without changing this expression. Thus the complete solutions of Eqs. 5.47 and 5.48 are[19]

$$c_{jj}^{(i)} = 2\ln f_{ji}^{\infty} - \ln f_{ij}^{\infty} \tag{5.49a}$$

$$d_{jjj}^{(i)} = 2\ln f_{ij}^{\infty} - 2\ln f_{ji}^{\infty} \tag{5.49b}$$

$$\begin{aligned} c_{jk}^{(i)} = c_0 &+ \frac{3}{2}(\ln f_{ji}^{\infty} + \ln f_{ki}^{\infty}) - \frac{1}{2}(\ln f_{ij}^{\infty} + \ln f_{ik}^{\infty}) \\ &- \frac{1}{2}(\ln f_{jk}^{\infty} + \ln f_{kj}^{\infty}) \qquad (j \neq k) \end{aligned} \tag{5.49c}$$

$$\begin{aligned} d_{jjk}^{(i)} = -2c_0 &+ 3(\ln f_{ij}^{\infty} - \ln f_{ji}^{\infty}) + (\ln f_{ik}^{\infty} - \ln f_{ki}^{\infty}) \\ &+ (\ln f_{jk}^{\infty} - \ln f_{kj}^{\infty}) \qquad (j \neq k) \end{aligned} \tag{5.49d}$$

where c_0 is a constant parameter arising uniquely because of the ternary nature of the system. Equations 5.49c and 5.49d demonstrate that the activity coefficient of ion i (i = 1, 2, or 3) cannot be calculated solely from data obtained in a binary exchange system as epitomized in the infinite-dilution activity coefficients. The "ternary constant" c_0 must be calculated with data obtained in the ternary system.[19] In some cases c_0 will be found equal to zero experimentally, but this cannot be assumed a priori.

5.4 Ion Exchange Kinetics

Given the interpretation of the ion exchange reactions in Eqs. 5.1 and 5.2 as examples of the adsorption–desorption reaction in Eq. 4.3 (suitably generalized for heterovalent exchange), it follows that the kinetics of ion exchange can, in principle, be described in terms of the concepts developed in Chapter 4 for adsorption–desorption kinetics. In this sense what is understood concerning the rate laws for adsorption–desorption processes can be applied to study the kinetics of the exchange reaction in Eq. 5.6. Essential to this approach is the choice made concerning mechanism—in particular, the decision as to whether the rate-controlling step is diffusive transport or surface reaction, as discussed in Section 4.5. A variety of data suggests that *rates of ion exchange processes often are transport controlled, not reaction controlled.*[21] This is especially likely if the ions involved form solvated adsorbate species, the extreme case being two mobile ions that merely replace one another in the diffuse ion swarm of the electrical double layer. Thus *readily exchangeable* ions (as defined in Section 5.1) most probably engage in reactions whose rates are transport controlled, whereas *specifically adsorbed* ions may participate in reactions that are surface controlled (or at least are primarily so, in the sense of Eq. 4.63). Ion exchange reactions that involve both solvated and unsolvated adsorbate species (e.g., the exchange of Na^+ in the diffuse ion swarm for K^+ in an inner-sphere complex on a 2:1 layer-type clay mineral[1]) could exhibit rates that are influenced by both diffusion and surface complexation processes (cf. Eq. 4.67).

Even with the selection of diffusion as the rate-limiting process for most ion exchange reactions, there remains a need to choose between film diffusion and intraparticle diffusion (Eq. 4.62 versus Eq. 4.72). This can be done experimentally by the "interruption test."[21,22] Once an ion exchange reaction has been initiated, the exchanger particles are separated physically from the aqueous-solution phase, and then reimmersed in it after a short time interval. If film diffusion is the rate-limiting step, no significant effect of this interruption on the kinetics should be observed. If intraparticle diffusion is the rate-limiting step, the concentration gradient driving the exchange process should relax to near zero during the interruption, and the rate of ion exchange should increase after the exchanger particles are reimmersed in the aqueous-solution phase and a steep gradient is reestablished. In general, exchanger particles with large specific surface areas should favor the film diffusion mechanism, whereas those with significant microporosity should favor the intraparticle diffusion mechanism.

A generic rate law for ion exchange controlled by film diffusion can be developed by applying Eq. 4.62 to both ions in Eqs. 5.1 or 5.2. Taking cations, just to be concrete, one can consider the two Fick rate laws:[23]

$$j_i = (D_i/\delta)\,([i]_{bulk} - [i]_{surf}) \qquad (i = 1, 2) \tag{5.50}$$

where D_i (i = 1, 2) is a diffusion coefficient for ion i and δ is the thickness of the "Nernst film" through which the ion must diffuse from the (well-stirred) bulk aqueous-solution phase to the exchanger surface. The rate laws in Eq. 5.50 are subject to the constraint of charge balance implicit in Eq. 5.1 (let A ≡ 1, B ≡ 2, as in Section 5.3):

$$Z_1 \frac{dn_1^{(w)}}{dt} + Z_2 \frac{dn_2^{(w)}}{dt} = 0 \qquad (5.51)$$

where Z_i (i = 1, 2) is the valence and $n_i^{(w)}$ is the surface excess of ion i. The rate, $dn_i^{(w)}/dt$ (i = 1, 2), is proportional to the adsorptive flux, j_i (i = 1, 2):

$$\frac{dn_i^{(w)}}{dt} = Vc_s a_s j_i \qquad (i = 1, 2) \qquad (5.52)$$

where c_s is the concentration of exchanger particles in suspension, a_s is their specific surface area, and V is the suspension volume. Equations 5.50–5.52 provide the basis for a complete mathematical description of the kinetics of binary cation exchange governed by film diffusion.

A useful model expression that combines Eqs. 5.50–5.52 into a single equation is the *Bunzl rate law*.[24] This equation can be derived by forming the product of $Z_j D_j[j]_{surf}$ with $Z_i dn_i^{(w)}/dt$ for i = 1 and 2, j ≠ i, and then subtracting the result for i = 1, j = 2 from that for i = 2, j = 1:

$$Z_2 D_2 [2]_{surf} Z_1 \frac{dn_1^{(w)}}{dt} - Z_1 D_1 [1]_{surf} Z_2 \frac{dn_2^{(w)}}{dt}$$

$$= (Z_1 Z_2 D_2 D_2 V c_s a_s/\delta)\{[2]_{surf}([1]_{bulk} - [1]_{surf}) - [1]_{surf}([2]_{bulk} - [2]_{surf})\}$$

$$= (Z_1 Z_2 D_1 D_2 V c_s a_s/\delta)([2]_{surf}[1]_{bulk} - [1]_{surf}[2]_{bulk}) \qquad (5.53a)$$

where Eq. 5.52 is introduced to obtain the second step. The left side of Eq. 5.53a can be transformed into the expression

$$Z_2 D_2 [2]_{surf} Z_1 \frac{dn_1^{(w)}}{dt} - Z_1 D_1 [1]_{surf} Z_2 \frac{dn_2^{(w)}}{dt} = (Z_2 D_2 [2]_{surf} + Z_1 D_1 [1]_{surf}) Z_1 \frac{dn_1^{(w)}}{dt} \qquad (5.53b)$$

after introducing Eq. 5.51. Equation 5.53 is the same as the rate equation

$$Z_1 \frac{dn_1^{(w)}}{dt} = \frac{(Z_1 Z_2 D_1 D_2 V c_s a_s/\delta)([2]_{surf}[1]_{bulk} - [1]_{surf}[2]_{bulk})}{(Z_2 D_2 [2]_{surf} + Z_1 D_1 [1]_{surf})} \qquad (5.54)$$

Finally, Eq. 5.54 becomes the Bunzl rate law upon incorporating the definition of the *exchange separation factor*:[25,26]

$$\alpha_{ij} \equiv n_j^{(w)}[i]_{surf}/n_i^{(w)}[j]_{surf} \qquad (5.55)$$

and dividing both sides by Q, the total moles of adsorbed charge (Eq. 5.7b):

$$\frac{dE_1}{dt} = \left(\frac{D_1D_2Vc_sa_s}{\delta Q}\right)\frac{(\alpha_{21}Z_1[1]_{bulk}E_2 - Z_2[2]_{bulk}E_1)}{\alpha_{21}D_2E_2 + D_1E_1} \qquad (5.56)$$

where E_i (i = 1, 2) is a charge fraction on the exchanger (Eq. 5.32). Equation 5.56 describes the rate of change with time of the charge fraction of adsorbate ion 1 during an exchange reaction with ion 2. The prefactor on the right side reflects the film diffusion mechanism, whereas the numerator it multiplies compares the "forward" and "backward" rates of the exchange reaction. At equilibrium this numerator vanishes and

$$\alpha_{21} \underset{\frac{dE_1}{dt}\to 0}{\sim} Z_2[2]_{bulk}E_1/Z_1[1]_{bulk}E_2 = n_{1eq}^{(w)}[2]_{bulk}/n_{2eq}^{(w)}[1]_{bulk} \equiv \alpha_{21}^{eq} \qquad (5.57)$$

Unlike the value of α_{21} in general (Eq. 5.55), the *equilibrium* value of α_{21} depends only on the *bulk* concentrations of ions 1 and 2 in aqueous solution. Aside from α_{21}, however, the Bunzl rate law itself displays only bulk ion concentrations, regardless of the extent of reaction.

The chemical significance of the Bunzl rate law can be appreciated by examining the roles played by the parameters it contains. Its principal dependence on the film diffusion mechanism, for example, is epitomized in the *film diffusion rate constant* (Eq. 4.66)

$$k_{idiff} \equiv D_ic_sa_s/\delta \qquad (i = 1, 2) \qquad (5.58)$$

to which the prefactor in Eq. 5.56 reduces (with i = 1) after multiplication of both sides of the equation by Q/V, in order to express the rate formally in concentration units (cf. Eq. 4.17), and then division by D_2 factored from the numerator to leave the ratio D_1/D_2 as the coefficient of E_1 in the second term. The first-order rate constant k_{idiff} increases as either the diffusion coefficient of ion i or the exchanger surface area per unit volume of suspension (c_sa_s) increases, and it decreases with increasing film thickness. The numerator that multiplies k_{idiff} in Eq. 5.56 has the *appearance* of a rate law for the cation exchange reaction in Eq. 5.1 based on surface reaction control of the rate-limiting process (cf. Eq. 4.17 and Problem 4 in Chapter 4). Indeed, with α_{21} considered as the ratio of an adsorption rate coefficient to a desorption rate coefficient for ion 1, the numerator has the same mathematical form as the difference between (second-order) "forward" and "backward" reaction rate laws for an exchanging ion, based on Eq. 5.1. The Bunzl rate law as a whole also is similar in *mathematical form* to the mixed transport–reaction control rate law for adsorption–desorption processes, presented in Eq. 4.67. These comparisons demonstrate once again the pitfalls in attempting to attribute a particular mechanism to a particular mathematical form of a rate law. Despite the appearance of the separation factor in Eq. 5.56, the observed rate of change in the charge fraction of ion 1 with time is in fact determined solely by the relative rates of diffusion of ions 1 and 2 through a boundary layer around the exchanger, as represented in Eqs. 5.50–5.52. The rate of the surface reaction is assumed to be so large that, on the diffusion time scale, it has no influence on the kinetics.

In order to apply the Bunzl rate law an experimental scenario must be envisioned. One possibility is *differential ion exchange*,[22,26] in which an exchanger at equilibrium is perturbed by the sudden addition of a small aliquot containing $\Delta\tilde{Q}V$ moles of charge of exchangeable cations in a solution whose composition is otherwise similar to that in which the exchanger is suspended. The exchanger and the aqueous-solution phase then will undergo a relaxation process, mediated by the cation exchange reaction in Eq. 5.1, until a new equilibrium composition is achieved. The rate of the relaxation process as it affects ion 1 will be described by Eq. 5.56, subject to charge-balance constraints. For example, if the added aliquot contains only ion 1,

$$\Delta\tilde{Q} = Z_1\,\Delta[1]_0 - Z_1\,\Delta[1]_t + Z_2\,\Delta[2]_t \qquad (t \geq 0) \tag{5.59a}$$

where $[\ ]_t$ is a *bulk* aqueous-solution concentration at time $t \geq 0$. Moreover, since the sudden addition of ion 1 must provoke the adsorption of this ion and the consequent ejection of ion 2 from the adsorbate, electrical neutrality also requires the condition

$$Q\,\Delta E_1(t) = VZ_2\,\Delta[2]_t \qquad (t > 0) \tag{5.59b}$$

neglecting the small change in suspension volume produced by addition of the aliquot of solution containing ion 1. Equation 5.59 and the exchanger charge-balance constraint, $\Delta E_1 + \Delta E_2 = 0$, are to be imposed on Eq. 5.56 throughout the relaxation of $E_1(t)$.

If the perturbation $\Delta\tilde{Q}$ is very small when compared to Q/V, the Bunzl rate law can be linearized by following the approach described in Section 4.3. The resulting approximate rate equation for $\Delta E_1(t)$ is, for $t \geq 0$,

$$\left(\frac{Q}{V}\right)\frac{d\Delta E_1}{dt} = k_{1diff}[\alpha_{21}^0 Z_1(\Delta[1]_t E_{20} + [1]_0\,\Delta E_2) - Z_2(\Delta[2]_t E_{10} + [2]_0\,\Delta E_1)]/[\alpha_{21}^0 E_{20} + (D_1/D_2)E_{10}]$$
$$= \{k_{1diff}\alpha_{21}^0\,\Delta\tilde{Q}(1-E_{10}) - k_{1diff}[\alpha_{21}^0[(Q/V)(1-E_{10}) + Z_1[1]_0] + (Q/V)E_{10} + Z_2[2]_0]\,\Delta E_1\}/[\alpha_{21}^0(1-E_{10}) + (D_1/D_2)E_{10}] \tag{5.60}$$

This unwieldy expression can be simplified by defining the parameters:

$$\tilde{Q}_0 \equiv Z_1[1]_0 + Z_2[2]_0 \tag{5.61a}$$

$$S_0 \equiv (Q/V)E_{10} + Z_2[2]_0 \tag{5.61b}$$

$$\Delta E_1^{eq} \equiv \alpha_{21}^0\,\Delta\tilde{Q}(1-E_{10})/\{\alpha_{21}^0[(Q/V)+\tilde{Q}_0 - S_0] + S_0\} \tag{5.61c}$$

$$K_1 \equiv k_{1diff}\{\alpha_{21}^0[(Q/V) + \tilde{Q}_0 - S_0] + S_0\}/\{(Q/V)[\alpha_{21}^0(1-E_{10}) + (D_1/D_2)E_{10}]\} \tag{5.61d}$$

The combination of Eqs. 5.60 and 5.61 then produces a rate law with the same mathematical form as Eq. 1.55:

$$\frac{d\Delta E_1}{dt} = -K_1\,\Delta E_1 + K_1\,\Delta E_1^{eq} \tag{5.62}$$

whose solution is therefore given by Eq. 1.56:

$$\Delta E_1(t) = \Delta E_1^{eq}[1 - \exp(-K_1 t)] \tag{5.63}$$

where $\Delta E_1(0) \equiv 0$ for a relaxation scenario as described by Eq. 5.59. The characteristic time scale of the first-order rate process in Eq. 5.62 is $0.693/K_1$ (see Note 7 in Chapter 2), with K_1 given by Eq. 5.61d. In general, then, the "relaxation time constant" (Section 4.3) will be determined by the film diffusion properties (k_{1diff}) and exchange selectivity (α_{21}^0) of the binary system, for a given set of initial conditions (Eqs. 5.61a and 5.61b). The equilibrium value of ΔE_1, however, is independent of the diffusion properties of the ions (after infinite time!) and instead is determined by the initial exchange selectivity and the initial conditions imposed (Eq. 5.61c).[22]

Illustrative kinetics data for the cation exchange reaction[27]

$$MX_2(s) + 2H^+(aq) \rightleftarrows 2HX(s) + M^{2+}(aq) \tag{5.64}$$

where M = Ca, Cd, or Cu, and X represents 1 mol of titratable charge on a sphagnum peat are shown in Fig. 5.5. The filled circles in each graph are data from an experiment at pH $\approx$ 4.5 in which $\Delta\tilde{Q}$ = 250 $\text{mmol}_c\ \text{m}^{-3}$, $\tilde{Q}_0 \approx 0$, Q = 0.11 $\pm$ 0.03 mol_c, V = 2×10^{-4} m^3, and E_{10} = 0 (H-peat initially). Since $\Delta\tilde{Q}/(Q/V) = 4.5 \times 10^{-4}$, Eq. 5.60 should be applicable, with Eqs. 5.61c and 5.61d reduced to the form

$$\Delta E_1^{eq} = \Delta\tilde{Q}/(Q/V) \tag{5.65a}$$

$$K_1 = k_{1diff} \tag{5.65b}$$

A qualitative exponential increase of the specific moles of adsorbed charge, q $\equiv Q\,\Delta E_1/Vc_s$ (c_s = 0.5 kg m^{-3}), with time is apparent in the graphs, as are the similar initial rates of adsorption, given in this case by the following expression (E_{10} = 0):

$$\text{initial rate} \equiv \lim_{t\downarrow 0}\left(\frac{d\Delta E_1}{dt}\right) = \frac{k_{1diff}\,\Delta\tilde{Q}}{Q/V} \tag{5.65c}$$

which should differ only slightly among the three metal cations. On the other hand, the "infinite-time" values of q (shown as tick marks on the right side of each graph above the curve with filled circles) are not the same for each metal and are smaller than $Q\,\Delta E_1^{eq}/Vc_s$ = 500 $\text{mmol}_c\ \text{kg}^{-1}$. This result could be simply an effect of finite equilibration time. The open circles in Fig. 5.5 are data from a desorption experiment in which an aliquot containing 0.05 mmol H^+ was added to a suspension of the equilibrated M/H-peat in distilled water. In this case the rates *are* expected to differ among the three metals, because $E_{10} \neq 0$ (cf. Eqs. 5.61c and 5.61d).

5.5 Heterogeneous Ion Exchange

Ion exchange reactions in soils almost always involve mixtures of inorganic and organic substances whose individual exchange reactions are described by different equilibrium constants. For example, in an agricultural soil the exchanger might consist of the clay minerals, montmorillonite and kaolinite; oxides of aluminum and iron; and humus, with each adsorbent reacting differently in Na^+–Ca^{2+} exchange. Montmorillonite and soil organic matter themselves possess more than one class of metal-complexing surface functional group, and therefore more than one exchange equilibrium constant is needed for them in order to account in detail for a given exchange reaction. Thus soil ion exchangers are *polyfunctional* ion exchangers. The description of this polyfunctionality requires an extension of the concepts developed in Sections 5.1 and 5.4.

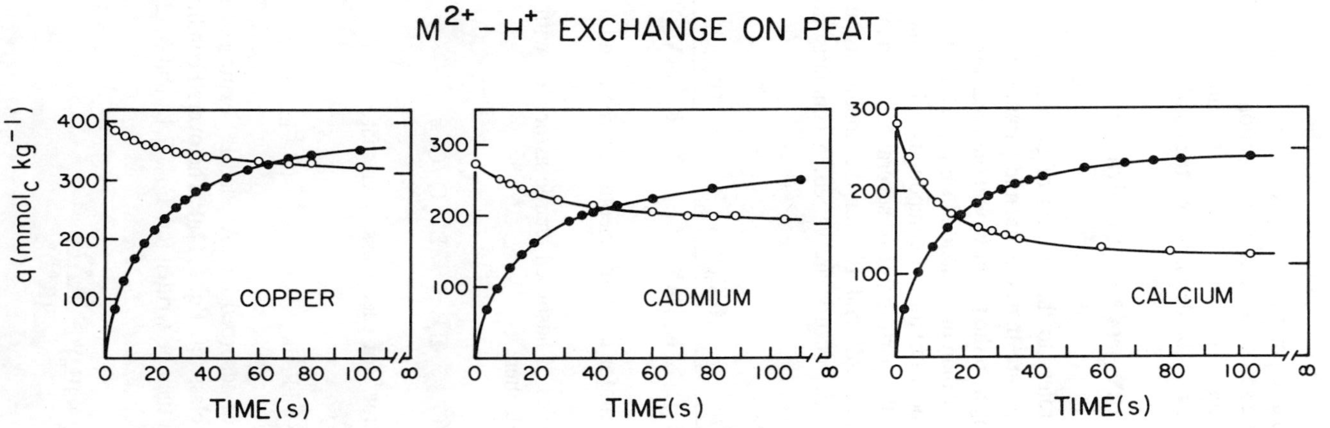

FIG. 5.5. Graphs of the specific moles of adsorbed charge q versus time for the exchange reaction in Eq. 5.64 on a sphagnum peat. Filled circles refer to the ← direction, whereas open circles refer to the → direction of the exchange reaction. (Data from Bunzl et al.[27])

To be concrete, the discussion will focus on cation exchangers participating in the reaction in Eq. 5.1. However, all the results to be derived apply just as well to anion exchangers participating in the reaction in Eq. 5.7. An overall equilibrium constant for a soil cation exchange reaction is given in Eqs. 5.9 and 5.30. If there are n classes of exchanger, each class undergoes a reaction like that in Eq. 5.1, and each may be described by the equilibrium constant

$$K_{exi} = (\mathbf{BX}_{ib})^a(\mathbf{A}^{a+})^b/(\mathbf{AX}_{ia})^b(\mathbf{B}^{b+})^a \tag{5.66}$$

where i = 1, . . . , n refers to the ith class of exchanger. It is important to understand that since they are thermodydnamic parameters, each distinct K_{exi} *defines* a class of exchanger in the mixture, without regard to chemical structure or chemical behavior other than the reaction in Eq. 5.1. Therefore a given class of exchanger may itself be a mixture of compounds (which happen to have the same value of K_{exi}), and the partitioning of several soil exchangers into classes may change as the cations A^{a+} and B^{b+}, or the conditions under which the exchange reaction in Eq. 5.1 occurs, are varied. A comparison of Eq. 30 with Eq. 5.66 indicates that

$$\frac{(\mathbf{B}^{b+})^a}{(\mathbf{A}^{a+})^b} = \frac{(\mathbf{BX}_b)^a}{K_{ex}(\mathbf{AX}_a)^b} = \frac{(\mathbf{BX}_{1b})^a}{K_{ex1}(\mathbf{AX}_{1a})^b} = \cdots = \frac{(\mathbf{BX}_{nb})^a}{K_{exn}(\mathbf{AX}_{nb})^b} \tag{5.67}$$

is an equation of constraint for a heterogeneous adsorbent and for each class of exchanger.

If Q_i is the total moles of adsorbed charge associated with the ith class of exchanger (cf. Eq. 5.7b), then

$$p_i \equiv Q_i/Q \qquad \Sigma_{i=1}^n p_i = \Sigma_{i=1}^n Q_i/Q \equiv 1 \tag{5.68a}$$

defines a "weighting factor" for each class, such that

$$E_A = \Sigma_{i=1}^n p_i E_{Ai} \qquad E_B = \Sigma_{i=1}^n p_i E_{Bi} \tag{5.68b}$$

are the charge fractions of adsorbed ions A^{a+} and B^{b+} expressed in terms of the charge fractions in each class (cf. Eq. 5.12a). The overall conditional exchange equilibrium constant can then be written (cf. Eqs. 5.11 and 5.12)

$$K_{exc} = \frac{a^a E_B^a}{b^b E_A^b}(bE_A + aE_B)^{b-a}\frac{(\mathbf{A}^{a+})^b}{(\mathbf{B}^{b+})^a}$$

$$= \frac{a^a(\Sigma_i p_i E_{Bi})^a}{b^b(\Sigma_i p_i E_{Ai})^b}(b\Sigma_i p_i E_{Ai} + a\Sigma_i p_i E_{Bi})^{b-a}\frac{(\mathbf{A}^{a+})^b}{(\mathbf{B}^{b+})^a} \tag{5.69}$$

This complicated equation can be transformed further by invoking the constraint in Eq. 5.67, but its principal implication can be appreciated more simply by examining the example of homovalent exchange (a = b ≡ 1):

$$K_{exc} = \frac{\Sigma_i^n p_i E_{Bi}}{\Sigma_i^n p_i E_{Ai}} \frac{(\mathbf{A}^+)}{(\mathbf{B}^+)} = \frac{\Sigma_i^n p_i E_{Bi} (\mathbf{A}^+)/(\mathbf{B}^+)}{\Sigma_i^n p_i [E_{Bi} (\mathbf{A}^+)/K_{exci} (\mathbf{B}^+)]}$$

$$= \Sigma_i^n p_i E_{Bi} / \Sigma_i^n p_i (E_{Bi}/K_{exci}) \qquad (5.70a)$$

or

$$K_{exc}^{-1} = \frac{\Sigma_i^n p_i (E_{Bi}/K_{exci})}{\Sigma_i^n p_i E_{Bi}} \qquad (5.70b)$$

where

$$K_{exci} = E_{Bi} (\mathbf{A}^+)/E_{Ai} (\mathbf{B}^+) \qquad (5.71)$$

in this example. Equation 5.70 shows that K_{exc}^{-1} is a weighted average of the set of K_{exci}^{-1}. The weighting factors, $p_i E_{Bi}/E_B$, will vary with E_B, and therefore K_{exc} *will be a function of exchanger composition even if each* K_{exci} *is constant* (ideal behavior, cf. Section 3.3).

This point can be seen directly in the special case of Eq. 5.70a when n = 2:

$$K_{exc} = \frac{(p_1 E_{B1} + p_2 E_{B2})}{(p_1 E_{A1} + p_2 E_{A2})} \frac{(\mathbf{A}^+)}{(\mathbf{B}^+)}$$

$$= \frac{(p_1 K_{exc1} E_{A1} + p_2 K_{exc2} E_{A2})}{(p_1 E_{A1} + p_2 E_{A2})} = \frac{p_1 E_{A1} K_{exc1} + p_2 E_{A2} K_{exc2}}{E_A} \qquad (5.72)$$

This equation shows once again that K_{exc} is a weighted average of the K_{exc1} with composition-dependent weights. Since $E_{A1} = 1 - E_{B1}$, Eq. 5.72 may be rewritten in the form

$$K_{exc} = \frac{p_1 K_{exc1} (1 - E_{B1}) + p_2 K_{exc2} (1 - E_{B2})}{1 - E_B}$$

$$= \frac{p_1 K_{exc1} + p_2 K_{exc2} - p_1 K_{exc1} E_{B1} - p_2 K_{exc2} E_{B2}}{1 - E_B} \qquad (5.73)$$

Now Eq. 5.73 is multiplied on both sides by K_{exc} and the identity

$$p_1K_{exc1}E_{B1} + p_2K_{exc2}E_{B2} \equiv E_B(K_{exc1} + K_{exc2}) - (p_1E_{B1}K_{exc2} + p_2E_{B2}K_{exc1}) \tag{5.74}$$

is noted. The result of this manipulation is

$$K_{exc}^2 = \frac{[p_1K_{exc1} + p_2K_{exc2} - E_B(K_{exc1} + K_{exc2})]}{1 - E_B}K_{exc} + \frac{(p_1E_{B1}K_{exc2} + p_2E_{B2}K_{exc1})}{1 - E_B}K_{exc} \tag{5.75}$$

According to Eq. 5.70a,

$$\begin{aligned} K_{exc} &= E_B/[p_1(E_{B1}/K_{exc1}) + p_2(E_{B2}/K_{exc2})] \\ &= K_{exc1}K_{exc2}E_B/[p_1E_{B1}K_{exc2} + p_2E_{B2}K_{exc1}] \end{aligned} \tag{5.76}$$

The combination of Eqs. 5.75 and 5.76 produces the quadratic equation

$$K_{exc}^2 - \frac{[p_1K_{exc1} + p_2K_{exc2} - E_B(K_{exc1} + K_{exc2})]}{1 - E_B}K_{exc} - \frac{K_{exc1}K_{exc2}E_B}{1 - E_B} = 0 \tag{5.77}$$

The solution of Eq. 5.77 for K_{exc} as a function of E_{B1}, K_{exc1}, and K_{exc2} is straightforward. If classes 1 and 2 are ideal mixtures of $AX_i(s)$ and $BX_i(s)$ ($i = 1, 2$), then K_{exc1} and K_{exc2} are constant parameters, and the composition dependence of K_{exc} may be calculated numerically without difficulty. Sample calculations of this type[28] show that the variation of K_{exc} with E_B is more pronounced, the greater the difference between K_{exc1} and K_{exc2}. In addition, if $K_{exc1} > K_{exc2}$, then the variation of K_{exc} with E_B occurs primarily at low charge fractions of BX(s) if p_1 is small, and primarily at high fractions of BX(s) if p_1 is large. For the general case, expressed in Eq. 5.69, the analysis of K_{exc} pertaining to a heterogeneous exchanger system can be made somewhat easier to carry out if the K_{exc1} can be developed in the power series or closed-form equations in the E_{Bi} along the lines of what was described in Section 5.2. Even with relatively simple equations for the K_{exci}, it is evident that the composition dependence of K_{exc} can be complicated. In particular, K_{exc} may display maxima or minima as E_B is varied.[29] As a general rule, this lack of monotonicity in the composition dependence of K_{exc} may be taken as a sign of polyfunctionality.

Transport-controlled, heterogeneous cation exchange kinetics can be described by a generalization of Eq. 5.56.[30] For each class the Bunzl rate law is

$$\frac{dE_{1i}}{dt} = \left(\frac{Vk^{i}_{1diff}}{Q_i}\right) \frac{(\alpha_{21i}Z_1[1]_{bulk}E_{2i} - Z_2[2]_{bulk}E_{1i})}{\alpha_{21i}E_{2i} + (D_1/D_2)E_{1i}} \tag{5.78}$$

where 1 = ion A and 2 = ion B, as in Section 5.4. The n equations having the form of Eq. 5.78 must be solved simultaneously under given initial conditions and the constraint of charge balance (cf. Eq. 5.59b)

$$VZ_2([2]_t - [2]_0) = Q[E_1(t) - E_1(0)] = \Sigma_i^n Q_i[E_{1i}(t) - E_{1i}(0)] \tag{5.79}$$

An example of this approach based on a numerical solution of Eq. 5.78 is shown in Fig. 5.6 for the case n = 2 and the parameter values:[30] $Q_1 = Q_2 = 1$ mmol$_c$, $V = 2 \times 10^{-4}$ m^3, $\alpha_{211} = 10$, $\alpha_{212} = 0.1$, $Z_1 = Z_2 = 1$, $k^1_{1diff} = k^2_{1diff} = 3 \times 10^{-4}$ s^{-1}, and $D_1/D_2 = 0.5$. The initial conditions imposed are $E_{11}(0) = E_{12}(0) = 0$, $[1]_0 = 10$ mol m^{-3}, and $[2]_0 = 0$. Plotted in the figure is the ratio $E_{1i}(t)/E^{eq}_{1i}$ for each class of exchanger, with $E^{eq}_{11} = 0.91$, $E^{eq}_{12} = 0.09$, as calculated from setting the left side of Eq. 5.78 equal to zero and solving the resulting set of two algebraic equations under the given parameter values.[30] Because the two exchangers differ only in their exchange separation factors (assumed not to vary with exchanger composition), the rate of ion exchange is initially *the same* for both under the given initial conditions. [The rate of increase of $E_{1i}(t)/E^{eq}_{1i}$ (i = 1, 2) is, correspondingly, ten times larger for exchanger 2 than it is for exchanger 1,

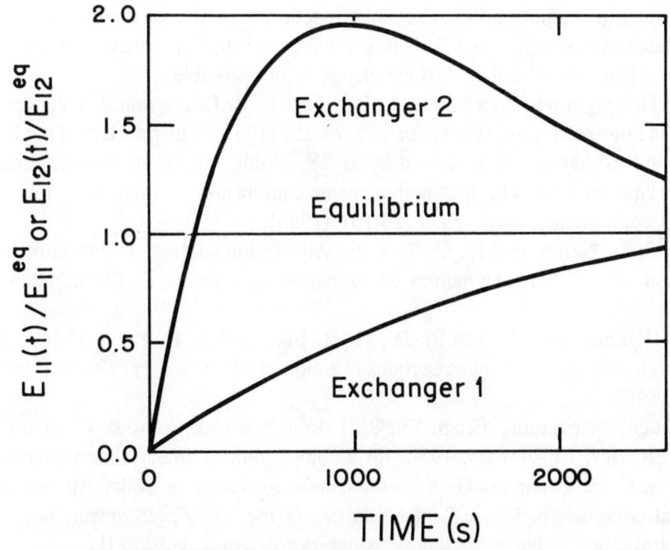

FIG. 5.6. Graphs of numerical solutions of Eq. 5.78 for two exchangers with different separation factors that react with two monovalent exchangeable cations with differing diffusion coefficient. (Data from Bunzl.[30])

because E_{11}^{eq} is ten times larger than E_{12}^{eq}. This difference is evident in Fig. 5.6.] The early, diffusion-driven rate of ion exchange allows exchanger 2 to adsorb more of ion 1 than its selectivity for that ion warrants, with the result that $E_{12}(t)$ overshoots its equilibrium value, as shown in Fig. 5.6. Eventually, however, the influence of the difference in separation factors on the rate of ion exchange becomes dominant, and the excess of ion 1 adsorbed by exchanger 2 is claimed by exchanger 1. Thus $E_{12}(t)/E_{12}^{eq}$ exhibits a maximum during the equilibration period, while $E_{11}(t)/E_{11}^{eq}$ climbs monotonically toward 1.0. This example illustrates the interplay of diffusion and selectivity that will govern any competition for ions between two exchanges of differing properties. Similar results are found if exchangers having the same selectivity but different film diffusion rate constants (e.g., with differing specific surface area) are studied using Eq. 5.78.[30] They make clear the need to ensure that full equilibration has occurred before exchange separation factors are inferred from an application of Eq. 5.57.

NOTES

1. For an introductory discussion of exchangeable ions and exchange isotherms, see Chaps. 7–9 in G. Sposito, *The Chemistry of Soils*, Oxford University Press, New York, 1989. Fundamental to a chemical thermodynamics description of the reactions in Eqs. 5.1 and 5.2 as applied to soils is the assumption that X^- and Y^+ can be associated with a macroscopic, chemically homogeneous solid, insofar as ion exchange is concerned. This assumption, which requires the structural integrity of the exchanger to be maintained throughout an ion exchange reaction, must be verified experimentally in each application of Eq. 5.1 or 5.2. If it is not accurate, then a representation of the solid exchanger as merely 1 mol of surface charge is not possible.

2. This point was raised by K. L. Babcock, L. E. Davis, and R. Overstreet, Ionic activities in ion-exchange systems, *Soil Sci.* **72**:253 (1951). The problem of the Reference State for ion exchangers is reviewed by L. W. Holm, On the thermodynamics of ion exchange equilibria. I. The thermodynamical equilibrium in relation to the reference states and components, *Aark. Kem.* **10**:151 (1956).

3. G. L. Gaines and H. C. Thomas, Adsorption studies on clay minerals. II. A formulation of the thermodynamics of exchange adsorption. *J. Chem. Phys.* **21**:714 (1953)

4. H. Laudelout, R. van Bladel, G. H. Bolt, and A. L. Page, Thermodynamics of heterovalent cation exchange reactions in a montmorillonite clay, *Trans. Faraday Soc.* **64**:1477 (1968).

5. See, for example, Chap. 5 in G. H. Bolt, *Soil Chemistry. B. Physico-Chemical Models*, Elsevier, Amsterdam, 1982, for a compilation of binary cation exchange data.

6. See, for example, G. Sposito, Cation exchange in soils: An historical and theoretical perspective, Chap. 2 in *Chemistry in the Soil Environment*, ed. by R. H. Dowdy et al., Soil Science Society of America, Madison, WI, 1981.

7. P. Fletcher and G. Sposito, The chemical modelling of clay–electrolyte interactions for montmorillonite, *Clay Minerals* **24**:375 (1989). See also R. S. Mansell, S. A. Bloom, and W. J. Bond, A tool for evaluating a need for variable selectivities in

cation transport in soil, *Water Resour. Res.* **29**:1855 (1993).

8. J. Grover, Chemical mixing in multicomponent solutions. An introduction to the use of Margules and other thermodynamics excess functions to represent nonideal behavior, pp. 67–97 in *Thermodynamics in Geology*, ed. by D. G. Fraser, D. Reidel, Dordrecht, The Netherlands, 1977. It follows from Eqs. 5.17 and 5.19 that, in general, $\ln f_{AB}^{\infty} = \Sigma_i a_i$ and $\ln f_{BA}^{\infty} = \Sigma_i b_i$. Equation 5.20a is a special case of this relation for a third-order Margules expansion.

9. See, for example, Chap. 6 in M. B. King, *Phase Equilibrium in Mixtures*, Pergamon Press, Oxford, 1969, for a discussion of this model.

10. The generalization of Eqs. 4.96 and 5.24 to include the possibility of imbibed water in an exchanger (thus making it a three-component mixture) is described in Chap. 5 of G. Sposito, *The Thermodynamics of Soil Solutions*, Clarendon Press, Oxford, 1981. The presence of charge fractions in Eq. 5.25 instead of mole fractions, as in Eq. 4.11, derives from the possible inequality of the stoichiometric coefficients, a and b, in the cation exchange reaction (cf. Eqs. 4.6 and 5.9). For *homovalent* exchange reactions, only mole fractions appear in the expressions for the adsorbate species activity coefficients.

11. H. E. Jensen and K. L. Babcock, Cation-exchange equilibria on a Yolo loam, *Hilgardia* **41**:475 (1973).

12. R. G. Gast, Standard free energies of exchange for alkali metal cations on Wyoming bentonite, *Soil Sci. Soc. Am. J.* **33**:37 (1969); R. G. Gast, Alkali metal cation exchange on Chambers montmorillonite, *Soil Sci. Soc. Am. J.* **36**:14 (1972); R. J. Lewis and H. C. Thomas, Adsorption studies on clay minerals. VIII. A consistency test of exchange sorption in the systems sodium–cesium–barium montmorillonite, *J. Phys. Chem.* **67**:1781 (1963); and H. Laudelout, R. van Bladel, M. Gilbert, and A. Cremers, Physical chemistry of cation exchange in clays, *Trans. 9th Int. Congress Soil Sci.* **1**:565 (1968).

13. D. S. Gamble, J. A. Marinsky, and C. H. Langford, Humic-trace metal ion equilibria in natural waters, *Ion Exchange and Solvent Extraction* **9**:373 (1985).

14. L. Charlet and G. Sposito, Monovalent ion adsorption by an Oxisol, *Soil Sci. Soc. Am. J.* **53**:691 (1989).

15. See, for example, C. Thellier and G. Sposito, Influence of electrolyte concentration on quaternary cation exchange by Silver Hill illite, *Soil Sci. Soc. Am. J.* **53**:705 (1989).

16. V. S. Soldatov and V. A. Bichkova, Ternary ion-exchange equilibria, *Separation Sci. Technol.* **15**:89 (1980); S.-Y. Chu and G. Sposito, The thermodynamics of ternary cation exchange systems and the subregular model, *Soil Sci. Soc. Am. J.* **45**:1084 (1981).

17. G. Sposito, C. Jouany, K. M. Holtzclaw, and C. S. LeVesque, Calcium–magnesium exchange on Wyoming bentonite in the presence of adsorbed sodium, *Soil Sci. Soc. Am. J.* **47**:1081 (1983).

18. P. Fletcher, K. M. Holtzclaw, C. Jouany, G. Sposito, and C. S. LeVesque, Sodium–calcium–magnesium exchange reactions on a montmorillonitic soil: II. Ternary exchange reactions, *Soil Sci. Soc. Am. J.* **48**:1022 (1984); G. Sposito, C. S. LeVesque, and D. Hesterberg, Calcium–magnesium exchange on illite in the presence of adsorbed sodium, *Soil Sci. Soc. Am. J.* **50**:905 (1986). See also S. Feigenbaum, A. Bar-Tal, Rita Portnoy, and D. L. Sparks, Binary and ternary exchange of potassium on calcareous montmorillonitic soils, *Soil Sci. Soc. Am. J.* **55**:49 (1991).

19. See S.-Y. Chu and G. Sposito, op. cit.,[16] for the correct thermodynamic analysis leading to this conclusion. See also D. J. Anderson and D. H. Lindsley, A valid

Margules formulation for an asymmetric ternary solution: Revision of the olivine–ilmenite thermometer, with applications, *Geochim. Cosmochim. Acta* **45**:847 (1981).

20. Equation 5.46 was analyzed incorrectly by K. L. Currie and L. W. Curtis, An application of multicomponent solution theory to geodetic pyroxenes, *J. Geol.* **84**:179 (1976) and by J.-J. Gruffat and J.-L. Bouchardon, Coefficients d'activité d'une solution subrégulière déduits d'un modèle d'interaction par triplets, *Compt. Rend. Acad. Sci. Paris*, **300**:259 (1985). The error lay in assuming $c_0 \equiv 0$ (see Eq. 5.49). Otherwise, the calculations in these two papers are correct.

21. See, for example, Chap. 6 in F. Helfferich, *Ion Exchange*, McGraw-Hill, New York, 1962, and Chap. 4 in D. L. Sparks and D. L. Suarez, *Rates of Soil Chemical Processes,* Soil Science Society of America, Madison, WI, 1991.

22. K. Bunzl, Kinetics of ion exchange in soil organic matter. III. Differential ion exchange reactions of Pb^{2+}-ions in humic acid and peat, *J. Soil Sci.* **25**:517 (1974). For a general discussion, see K. Bunzl, Kinetics of differential ion-exchange processes in a finite solution volume, *J. Chromatogr.* **102**:169 (1974).

23. The diffusion coefficient in Eq. 5.50 actually is a composite of diffusion tensor elements defined in a frame of reference fixed in the solvent phase (i.e., water). For a discussion of the ways in which the diffusion coefficient can be defined, see J. B. Brady, Reference frames and diffusion coefficients, *Am. J. Sci.* **275**:954 (1975).

24. K. Bunzl, Reactions of ion exchangers with salts of low solubility, *Zeit. Phys. Chemie (neue Folge)* **75**:118 (1971).

25. See, for example, p. 15 in F. Helfferich, op. cit.[21]

26. K. Bunzl and W. Schimmack, Kinetics of ion sorption on humic substances, Chap. 5 in D. L. Sparks and D. L. Suarez, op. cit.[21]

27. K. Bunzl, W. Schmidt, and B. Sansoni, Kinetics of ion exchange in soil organic matter. IV. Adsorption and desorption of Pb^{2+}, Cu^{2+}, Cd^{2+}, Zn^{2+}, and Ca^{2+} by peat, *J. Soil Sci.* **27**:32 (1976).

28. H. E. Jensen, Selectivity coefficients of mixtures of ideal cation-exchangers, *Agrochimica* **XIX**:247 (1975). See also K. Bunzl and W. Schultz, Distribution coefficients of ^{137}Cs and ^{85}Sr by mixtures of clay and humic material, *J. Radioanalyti. Nucl. Chem.* **90**(1):23 (1985).

29. A discussion of heterogeneous exchange systems that show extrema of this type is given by R. M. Barrer and J. Klinowski. Ion exchange involving several groups of homogeneous sites, *J. Chem. Soc. Faraday I* **68**:73 (1972).

30. K. Bunzl, Competitive ion exchange in mixed cation exchanger systems: Kinetics and equilibria, *J. Inorg. Nucl. Chem.* **39**:1049 (1977). See also K. Bunzl, Kinetics of ion exchange in polydisperse systems, *Anal. Chem.* **50**:258 (1978) for an application to a set of exchangers with different specific surface areas.

FOR FURTHER READING

Elprince, A. M., *Chemistry of Soil Solutions*, Van Nostrand Reinhold, New York, 1986. Part II of this anthology contains reprints of a dozen seminal research papers on ion exchange reactions in soils.

Greenland, D. J., and M. H. B. Hayes, *The Chemistry of Soil Processes*, Wiley, Chichester, UK, 1981. Chapters 4 and 5 of this compendium of soil chemistry offer useful surveys of ion exchange processes.

Grover, J., Chemical mixing in multicomponent solutions: An introduction to the use of Margules and other thermodynamic excess functions to represent non-ideal behavior, pp. 67–97 in *Thermodynamics in Geology*, ed. by D. G. Fraser, D. Reidel, Dordrecht, The Netherlands, 1977. This review article provides a fine introduction to the thermodynamic theory of mixtures underlying the Margules expansion for adsorbate-species activity coefficients.

Helfferich, F., *Ion Exchange*, McGraw-Hill, New York, 1962. For soil chemists, still the most readable experimentalist-oriented introduction to the pure chemistry of readily exchangeable ions.

Grant, S. A., and P. Fletcher, Chemical thermodynamics of cation exchange reactions, *Ion Exchange and Solvent Extraction* **11**:1 (1993). A useful survey of the chemical modeling of multicomponent cation exchange systems.

Sparks, D. L., and D. L. Suarez, *Rates of Soil Chemical Processes*, Soil Science Society of America, Madison, WI, 1991. Chapters 4 and 5 of this conference proceedings volume describe methodologies in the kinetics of ion exchange.

PROBLEMS

1. The cation exchange reaction in Eq. 5.1, for the case A = Na and B = Cs on a montmorillonitic Aridisol, was described by the fourth-order Margules expansions:

$$\ln f_{Na} = -1.145\, x_{Cs}^2 + 4.02\, x_{Cs}^3 - 3.015\, x_{Cs}^4$$
$$\ln f_{Cs} = -0.14 + 2.29\, x_{Cs} - 7.175\, x_{Cs}^2 + 8.04\, x_{Cs}^3 - 3.015\, x_{Cs}^4$$

With these two expressions, (a) calculate K_{exc} for $Na^+ \rightarrow Cs^+$ exchange and (b) examine the applicability of the van Laar model. (*Answer*: $K_{exc} = 32.7$ and $f_{NaCs}^{\infty} = f_{CsNa}^{\infty} = 0.869$, thus requiring $\ln f_{Na} = (x_{Cs}/x_{Na})^2 \ln f_{Cs}$.)

2. The exchange of Na^+ for H^+ on a peat was described well by the composition-dependent conditional equilibrium constant ("three-parameter model"):

$$\log K_{exc} = 4.06 x_{Na} + 2.09(1 - x_{Na}) + 0.82 x_{Na}(1 - x_{Na})$$

as applied to Eq. 5.1 with A = Na, B = H. Derive Margules expansions for the adsorbate species activity coefficients and calculate their infinite-dilution limits as well as the exchange equilibrium constant. (*Answer*: $\log f_{Na} = -0.575 x_H^2 - 0.547 x_H^3$; $\log f_H = -0.848 + 1.15 x_H + 0.245 x_H^2 - 0.547 x_H^3 = -1.396 x_{Na}^2 + 0.547 x_{Na}^3$;

$f^{\infty}_{NaH} = 0.0755$, $f^{\infty}_{HNa} = 0.142$; $K_{exc} = 1.63 \times 10^3$. Note that despite the appearance of three constants in the expression for log K_{exc}, there are only *two* independent parameters, $a_2 = -0.575$ and $a_3 = -0.547$, in the context of Eq. 5.18.)

3. Derive Eq. 5.22 by introducing Eq. 5.21 into Eq. 4.9b. Test the van Laar model with the data in Table 5.1. (*Hint*: Express Eq. 5.21 in the differential form

$$x_A^2 \, d\ln f_A - k x_B^2 \, d\ln f_B = -2(x_A \ln f_A + k x_B \ln f_B) \, dx_A$$

where $k \equiv \ln f^{\infty}_{BA} / \ln f^{\infty}_{AB}$ and combine it with Eq. 4.9b to derive a differential expression that relates $d(\ln f_A)$ to dx_A. Then apply the result

$$\int \frac{dx}{(a + bx)(a' + b'x)} = (ab' - a'b)^{-1} \ln \left(\frac{a' + b'x}{a + bx} \right)$$

to the factor in dx_A in order to deduce Eq. 5.21, after choosing an appropriate constant of integration. In respect to the data in Table 5.1, test the scaling relationship in Eq. 5.21.)

4. Derive Eq. 5.37 and apply it to the experimental data in Table 5.3 for the ternary exchange system, Na-Ca-Mg-montmorillonite, to examine them for self-consistency. (*Answer*: The mean value of the right side of Eq. 5.37 is 0.0029 ± 0.0039.)

Table 5.3 Conditional Exchange Constants and Charge Fractions

$\ln K_{12c}$	$\ln K_{13c}$	$\ln K_{23c}$	E_1	E_2	E_3
0.376	-0.006	-0.76	0.427	0.080	0.493
0.477	0.245	-0.46	0.384	0.146	0.470
0.643	0.439	-0.40	0.340	0.224	0.436
0.318	0.113	-0.40	0.397	0.265	0.338
0.708	0.784	0.152	0.316	0.328	0.357
0.510	0.522	0.024	0.357	0.385	0.258
0.541	0.579	0.076	0.344	0.453	0.202
0.675	0.736	0.122	0.322	0.536	0.143
0.722	0.778	0.112	0.316	0.612	0.072

1 = Na, 2 = Ca, 3 = Mg; E_i (i=1, 2, 3) is a charge fraction on the clay mineral.

5. In Table 5.4 on the next page are data at pH 7.2 on cation exchange in the ternary system, Na-Ca-Mg, for an Aridisol. Use these data to calculate K_{32c} (1 = Na, 2 = Ca, 3 = Mg) as a function of E_{Ca}; then estimate K_{32} by applying Eq.

5.42b. (*Hint*: Use Eq. 1.21 to compute aqueous-phase activity coefficients. Note that $E_{Na} = 0.23$ is constant in the data set. The computed values of $\ln K_{32c}$ can be integrated either numerically with available software or analytically after fitting them to a power series in E_{Ca}.)

6. Sphagnum peat, proton-saturated and suspended in distilled water at pH 4, is made to undergo the cation exchange reaction in Eq. 5.64 by adding 20 μmol_c Pb^{2+} ions. The initial rate of Pb^{2+} adsorption, $(Q/V)(d\,\Delta E_{Pb}/dt)_{t=0} = 2.8$ $mmol_c\ m^{-3}\ s^{-1}$, where $Q = 0.13\ mmol_c$ and $V = 0.2\ dm^3$. Calculate the characteristic time scale for the $H^+ \rightarrow Pb^{2+}$ exchange process. (*Hint*: Apply the linearized Bunzl rate law, after verifying $\Delta\tilde{Q} << (Q/V)$, to estimate $0.693/k_{diff} \approx 25$ s.)

7. A Mollisol undergoes the exchange reaction in Eq. 5.1 for H^+ and Cs^+. Measurements of the separation factor show its composition dependence to be

$$\alpha_{HCs}^{eq} = 2.53 - 1.22E_{Cs} - 0.62E_{Cs}^2$$

Use this result to calculate the initial rate of Cs^+ adsorption, at pH 5 and $[Cs^+]_0 = 0.1$ mol m^{-3}, for $E_{Cs}(0) = 0.1$ or 0.9, given $k_{diff} = 3.5 \times 10^{-3}\ s^{-1}$ for Cs^+ and $D_H/D_{Cs} = 5$. (*Answer*: $(Q/V)(d\Delta E_{Cs}/dt)_{t=0} = 3.3$ mmol $m^{-3}\ s^{-1}$ at $E_{Cs} = 0.1$, $\alpha_{HCs}^{eq} = 2.4$; and 3.5 $\mu mol\ m^{-3}\ s^{-1}$ at $E_{Cs} = 0.9$, $\alpha_{HCs}^{eq} = 0.93$, according to the Bunzl rate law.)

Table 5.4 Cation Exchange Composition Data

Expt.	$[Na]_e$	$[Ca]_e$	[Mg]	q_{Na}[a]	q_{Ca}	q_{Mg}
		mol m^{-3}			$mol_c\ kg^{-1}$	
1	41.5	0.97	2.49	0.074	0.142	0.122
2	41.4	0.68	2.83	0.078	0.120	0.141
3	41.4	0.36	3.20	0.078	0.093	0.169
4	41.4	0.074	3.51	0.086	0.054	0.212
5	41.7	3.40	0.01	0.074	0.260	0.002
6	41.6	3.08	0.37	0.077	0.244	0.0193
7	41.8	2.73	0.677	0.073	0.229	0.0341
8	41.7	2.33	0.981	0.079	0.216	0.050
9	41.7	1.990	1.42	0.073	0.202	0.0656
10	41.5	1.689	1.77	0.079	0.188	0.0820
11	41.5	1.313	2.14	0.079	0.167	0.102
12	43.2	0.305	3.00	0.06	0.067	0.164
13	43.1	0.580	2.63	0.13	0.096	0.142
14	43.1	2.102	1.02	0.07	0.19	0.046
15	43.0	0.34	2.97	0.07	0.075	0.172
16	42.8	0.60	2.65	0.058	0.100	0.144
17	42.7	2.11	1.03	0.060	0.208	0.052

[a] $q_i \equiv Z_i n_i^{(w)}/Vc_s$ (i = Na, Ca, or Mg)

8. The results in Problem 4 of Chapter 4 imply that a kinetics analysis of the ion exchange reaction in Eq. 5.6 for binary, homovalent exchange leads to the adsorption isotherm equation:

$$[(SR')_{|Z_c|}C]_e = \frac{SR_T\, A\,([C]_e/[Q]_e)^b}{1 + A([C]_e/[Q]_e)^b} \qquad (Z_C = Z_Q)$$

if the rate of the reaction is *surface controlled*. Use this result to show that the isotherm equation leads to the *Rothmund-Kornfeld model* of the adsorbate species activity coefficients (Eq. 5.23), with $\beta \equiv b$. Thus the parameter β is an index of the breadth of the distribution of k_{ads} and k_{des} in Eq. 5.6. (*Hint*: Show that the equation above implies $(E_C/E_Q)^{1/b}\ [Q]_e/[C]_e = \text{constant}$ where $E \equiv [(SR')_{|Z_c|}C]_e/SR_T$. Then consider Eqs. 5.9 and 5.10 to derive a model expression for f_C/f_Q. Note that for cation exchange, the valences, a and b, in Eq. 5.9 are the same as Z_Q and Z_C.)

9. Follow the approach in Section 4.2 to derive a rate law for a homovalent cation exchange reaction described by Eq. 5.6 under the assumption that the kinetics are *surface controlled*. Compare your result with the Bunzl rate law in Eq. 5.54. (*Hint*: Show that Eq. 4.26 generalizes to the rate law:

$$\frac{d[(SR')_{Z_C}C]}{dt} = k_{ads}[C][(SR')_{Z_Q}Q] - k_{des}[Q][(SR')_{Z_C}C]$$

for a homogeneous exchanger, where $Z_C = Z_Q$. The left side of this rate law is the same as $V^{-1}\,dn_1^{(w)}/dt$ in Eq. 5.54, with $Z_1 = Z_2$. Show that the denominator on the right side of Eq. 5.54 is constant if $D_1 = D_2$, and that therefore the correspondence

$$k_{ads} \leftrightarrow Vk_{diff}/Q \leftrightarrow k_{des}$$

is possible in the two rate laws.)

10. Measurements of the conditional equilibrium constant for Mg → Ca exchange on a Vertisol showed the composition dependence

$$K_{exc} = 1.4218 - 0.5156\,E_{Ca} + 0.6094E_{Ca}^2 - 0.4223\,E_{Ca}^3$$

Test the applicability of the two-component model in Eq. 5.77 to represent the composition dependence of K_{exc} in terms of an ideal mixture of two exchangers (e.g., montmorillonite and humus), one of which shows no preference for Ca versus Mg. Why is it possible to apply Eq. 5.77 to *bivalent* cation exchange? (*Hint*: Consider Eq. 5.69 for the case a = b and derive Eq. 5.77. Then show that this equation has the limits

$$\lim_{E_{Ca}\downarrow 0} K_{exc} = p_1 K_{exc1} + p_2 K_{exc2}$$

$$\lim_{E_{Ca}\uparrow 1} K_{exc}^{-1} = p_1 K_{exc1}^{-1} + p_2 K_{exc2}^{-1}$$

and use these expressions to compute p_1 and K_{exc1} given $K_{exc2} \equiv 1.0$. With the results, $p_1 = 0.106$, $K_{exc1} = 4.972$, Eq. 5.77 can be solved for K_{exc} at any E_{Ca} to compare with the measured value.)

6

COLLOIDAL PROCESSES

6.1 Flocculation Pathways

The process through which soil colloids (i.e., insoluble solid particles whose diameter falls between 0.01 and 10 μm) in suspension coalesce and undergo gravitational settling to form bulky, porous masses is termed *flocculation*.[1] This process necessarily restricts the mobility of colloids and therefore is critical in determining whether soil erosion or the transport of adsorbates (either nutrients or pollutants) that may be bound to colloids will occur.[2] The soil particles formed during flocculation and thereby removed from suspension are themselves candidates for further transformation into *aggregates*, the dense, organized solid masses that figure essentially in the structure, permeability, and fertility of soil.[3]

Flocculation processes are complicated phenomena because of the varieties of both particle morphology and chemical reactions they encompass.[3,4] A few concepts of a general nature have emerged, however, and they will be the focus of this chapter. From the perspective of kinetics, perhaps the most important of these broad generalizations is the distinction that can be made between *transport-controlled* and *reaction-controlled flocculation*, parallel to the classification of adsorption processes described in Section 4.5. Flocculation kinetics are said to exhibit transport control if the rate-limiting step is the movement of two (or more) particles toward one another prior to their close encounter and subsequent combination into a larger particle. Reaction control occurs if it is particle combination instead of particle movement (toward collision) that limits the rate of flocculation.

Transport control of flocculation is realized in an especially direct way in the process known as *diffusion-limited cluster–cluster aggregation*[5] (*aggregation* as used in this term means "flocculation" in the present chapter). In this process, which is straightforward to simulate and visualize on a computer, particles undergo Brownian motion (i.e., diffusion) until they come together in close proximity, after which they coalesce instantaneously and irreversibly to form floccules (or "clusters"). The "clusters" then diffuse until they contact one another and combine to form larger "clusters," and so on, until gravitational settling dominates and the "clusters" are removed from suspension. Figure 6.1 typifies a two-dimensional computer simulation of this process.[6] In this simulation, 2000 "primary particles" were placed randomly on a square lattice. Of these, 110 had neighboring particles on one or more of the nearest-neighbor sites, thus qualifying them to be regarded as binary (or higher-order) clusters initially. The 1890 initial "primary particles" (either single particles or "clusters") in the "suspension" were then moved about on the lattice through random selection followed by translation over one unit of distance on the lattice in a random direction. After a "cluster" had been moved, it was examined to see whether it had any nearest neighbors. If it did, it was combined with them at once to form a larger "cluster," which then became a candidate for additional random translation. Figure 6.1 shows the initial state of the "suspension" and three subsequent states comprising smaller numbers of "clusters" (but the same number of "primary particles"), the third exhibiting just ten "clusters." The open, dendritic structure of these final "clusters" is evident. Detailed examination of the successive states of the "suspension" indicates that the clusters tend to form more by the combination of two particle units of comparable size than by large particle units coalescing with small ones.

Diffusion-limited cluster–cluster flocculation is expected to be a very rapid process. The diffusion coefficient of a colloid generally lies between 10^{-14} and 10^{-11} m^2 s^{-1}, implying a diffusive time scale [(colloid diameter)^2/diffusion coefficient] in the range 10 ms to 10 s.[1] Therefore, *diffusion-controlled flocculation should correspond to rapid flocculation*, and laboratory experiments that realize conditions for rapid flocculation should produce floccules with structures like those of the "clusters" in Fig. 6.1. Examples of floccules that were created in this way are shown in Fig. 6.2. The "primary particles" in the floccules are gold spheres of diameter 15 $\pm$ 2 nm that were prepared in aqueous suspension by reduction of $NaAuCl_4$ with $Na_3C_6H_5O_7$ (Na citrate).[7] These spheres bear a negative surface charge arising from adsorbed citrate ions and repel one another strongly, thus forming a very stable suspension (no observable flocculation over a few weeks time). If, however, uncharged pyridine molecules are added to the suspension in small quantities, they displace the adsorbed citrate ions quickly and increase the surface charge on the gold spheres toward a zero value. The discharged spheres are observed to flocculate on a time scale as short as minutes, depending on the amount of pyridine added. The structures in Fig. 6.2 were produced by rapid flocculation of about 4700 Au spheres initially separated by approximately 1 μm in aqueous suspension. The open, complicated structure of the floccules is apparent, consistent with the computer simulation results in Fig. 6.1. During the course of diffusion-limited flocculation, floccules of particles collide and coalesce instantaneously, leaving no opportunity for them to optimize their packing into a close-fitting, organized structure. Thus the arrangement of particles in a floccule whose formation is controlled by transport processes is expected to be highly convoluted, as in Fig. 6.2.

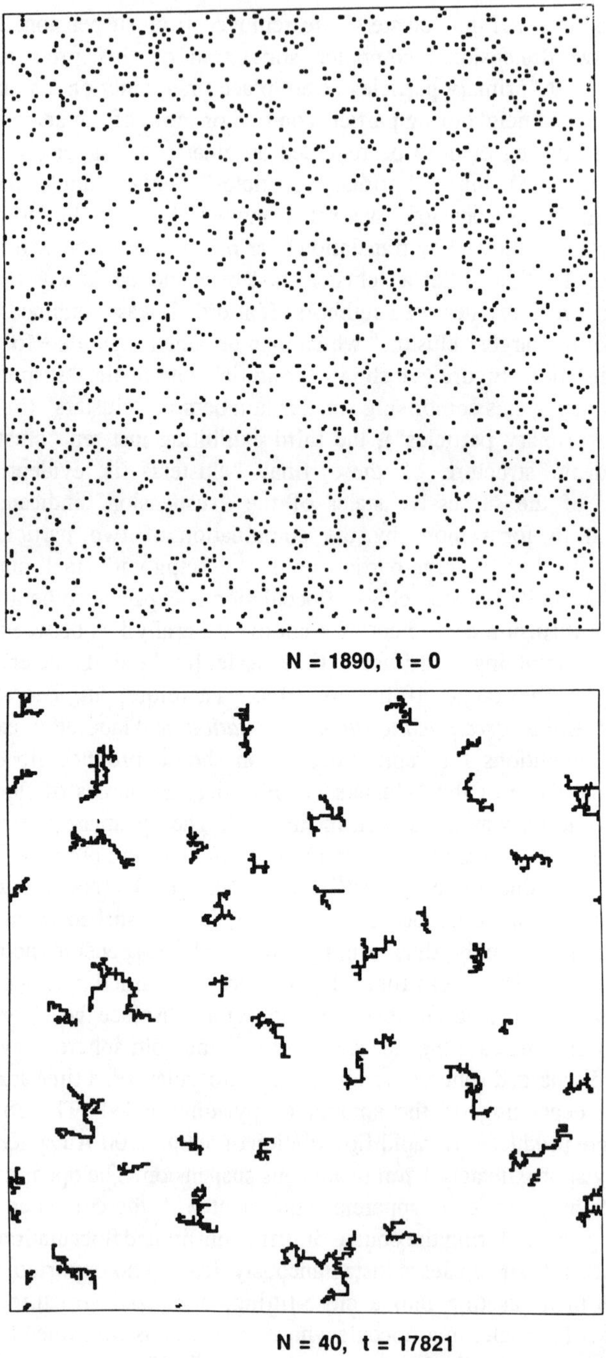

Fig. 6.1. Visualization sequence of diffusion-limited cluster–cluster flocculation, based on a two-dimensional computer simulation of the random encounter and instantaneous combination of 2000

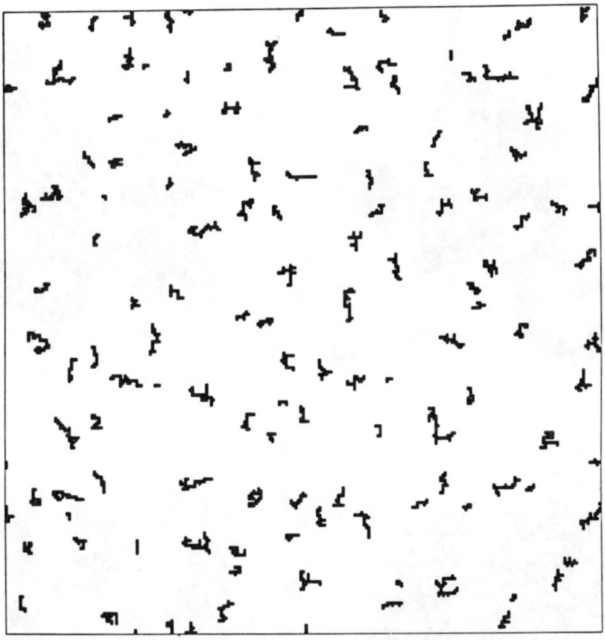

N = 125, t = 1845

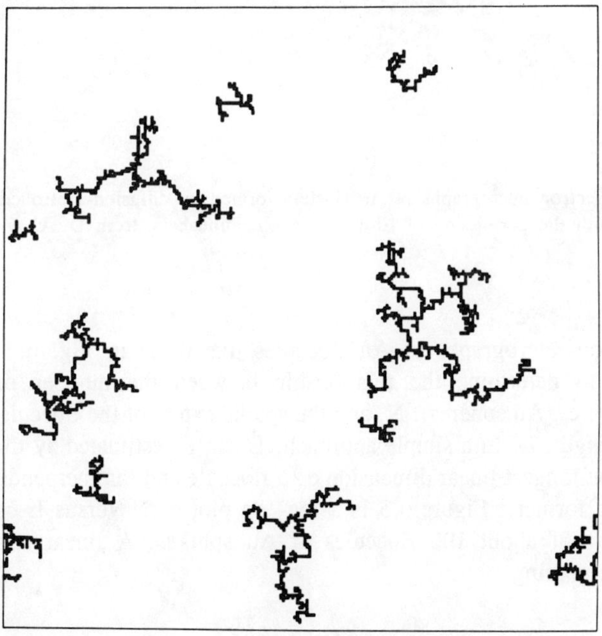

N = 10, t = 203585

"primary particles" on a square lattice. (Adapted, with permission, from P. Meakin.[6] Copyright © by the American Geophysical Union.)

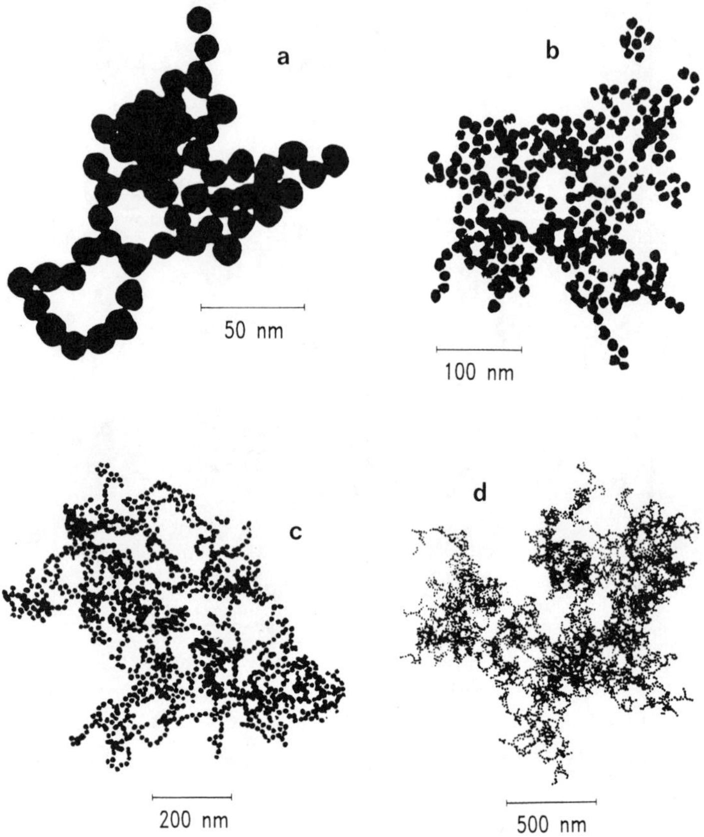

FIG. 6.2. Electron micrographs of Au clusters formed by diffusion-controlled flocculation. (Adapted, with the permission of Elsevier Science Publishers, from D. A. Weitz and J. S. Huang.[7])

Electron micrographs of Au floccules like those in Fig. 6.2 have been examined to determine the relationship between the number of "primary particles" (i.e., Au spheres), N, and the spatial extent of the floccule, expressed as some length, L. In a simple approach, L can be estimated by the geometric mean of the longest linear dimension of a floccule and that perpendicular to the axis of the former.[7] Figure 6.3 is a log–log plot of N versus L based on the examination of about 100 floccules of Au spheres. A linear relationship is apparent, implying

$$N(L) = AL^D \tag{6.1}$$

where A and D are positive parameters. For the data in Fig. 6.3, $D = 1.7 \pm 0.1$.[7] This kind of power-law relationship between floccule size and spatial extent has also been observed in computer simulations of diffusion-limited cluster–cluster flocculation in three spatial dimensions.[8] In this case the computed value of D is 1.78 ± 0.04, in remarkable agreement with the results in Fig. 6.3.

The power-law relationship in Eq. 6.1 also has implications for measurements of floccule size and dimension during the flocculation process itself. If the principal contributor to floccule growth is collisional encounters between particle units of comparable size, the increase in N per encounter will be equal approximately to N itself. Moreover, if diffusionally mediated collisions are the cause of these encounters, the kinetics of collision will be described by a second-order rate coefficient:[9]

$$\begin{aligned} K_D &= 4k_BT/3\eta \\ &= 6.2 \times 10^{-18}\ \mathrm{m^3\ s^{-1}} \qquad (T = 298.15\ \mathrm{K}) \end{aligned} \tag{6.2}$$

where k_B is the Boltzmann constant, T is absolute temperature, and η is the coefficient of viscosity of water (assuming an aqueous suspension of colloids). The characteristic time scale is then $(K_D\rho)^{-1}$, where ρ is the number of floccules per cubic meter. The rate of floccule growth can be expressed as[9]

$$\frac{dN}{dt} \approx \frac{\text{increase in N per encounter}}{\text{time scale for encounter}} \approx NK_D\rho = K_D\rho_0 = \text{constant} \tag{6.3}$$

where $\rho_0 = N\rho$ is the initial, "primary particle" number density. Integration of Eq. 6.3 leads to the equation

$$N(t) = N(0) + K_D\rho_0 t \sim K_D\rho_0 t \quad [t >> N(0)/K_D\rho_0] \tag{6.4}$$

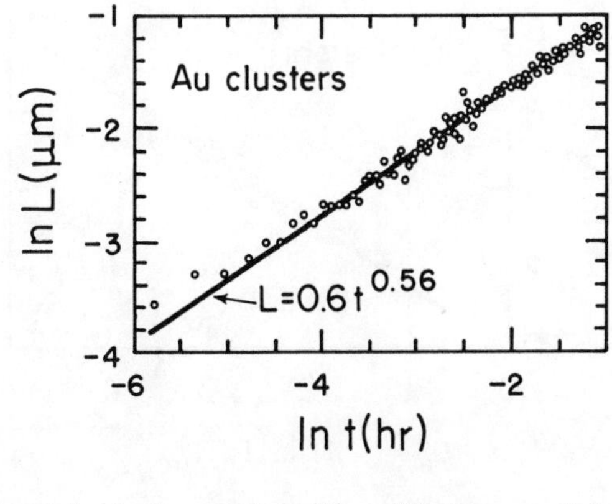

FIG. 6.3. Test of Eq. 6.1 for Au floccules formed by a diffusion-controlled process. (Data from D. A. Weitz and M. Oliveria.[7])

which, in combination with Eq. 6.1, yields a relation between cluster dimension and time:

$$L(t) \sim (K_D \rho_0 / A)^{1/D} \, t^{1/D} \quad [t >> N(0)/K_D \rho_0] \tag{6.5}$$

Figure 6.4 is a log–log plot of the average floccule radius (as measured by light-scattering techniques) versus time for a suspension of Au spheres where initial number density was 10^{18} m^{-3}. The line through the data points conforms to the equation[10]

$$L(t) = 0.60 \, t^{0.56} \tag{6.6}$$

from which it can be inferred that D = 1/0.56 = 1.79, in excellent agreement with the value of D as obtained directly from Fig. 6.3. Similar results, with D = 1.75 ± 0.05, have been found in studies of the rapid flocculation of silica and goethite colloids suspended in 1–1 electrolyte solutions.[11]

The power-law relation in Eq. 6.1 can be interpreted physically as indicative of a *cluster fractal*.[12] The exponent D is then termed the *cluster fractal dimension*. Some basic concepts about cluster fractals are introduced in Special Topic 3 at the end of this chapter. Suffice it to say here that Eq. 6.1 can be pictured as a generalization of the geometric relation between the *number* of "primary particles" in a cluster that is d-dimensional (d = 1, 2, or 3) and the d-dimensional *size* of the cluster. For example, if a cluster is one-dimensional (d = 1), it can be portrayed as a straight chain of, say, circular "primary particles" of diameter L_0. The number of particles in a chain of length L is

$$N(L) = L/L_0 \equiv AL^1 \tag{6.7a}$$

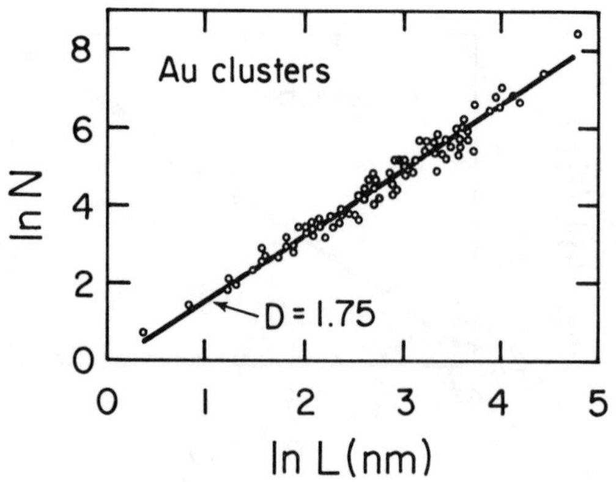

FIG. 6.4. Test of Eq. 6.5 for Au floccules formed by diffusion-controlled flocculation. (Data from D. A. Weitz et al.[10])

where $A \equiv 1/L_0$ in this case. Equation 6.7a has the appearance of Eq. 6.1 with $D = 1$. If the cluster is two-dimensional, it can be represented by a parquet of circular "primary particles" packed together so that they touch. The number of particles in the cluster will then be

$$N(L) = g(L/L_0)^2 \equiv AL^2 \tag{6.7b}$$

where now $A \equiv g/L_0^2$ and g is a geometric factor whose value depends on exactly how the circular "primary particles" have been packed to form the parquet cluster. In this case Eq. 6.1 is recovered if $D = 2$. (Strictly speaking, Eq. 6.7 applies in a limiting sense as $L_0/L \downarrow 0$.) Evidently, Eq. 6.1, with $1 < D < 2$, represents a cluster of dimension D whose character is intermediate between that of a chain and that of a parquet. In particular, if the cluster has the form of a highly convoluted (as opposed to straight) chain of particles, one that winds about irregularly in space but does not fill it completely, then it is not unreasonable to suppose that the size-dimension relation in Eq. 6.1 could describe it with a noninteger value of D. Indeed, the floccules or clusters of gold colloids in Fig. 6.3 have an irregular, convoluted-chain form that does not fill the photographic plane in which it is viewed, and it can be described by Eq. 6.1 with noninteger D. Thus it qualifies as a cluster fractal. Its ability to fill the plane of view, irrespective of its shape, is quantified by how closely D approaches the value 2.0.

Reaction control of flocculation can be simulated and visualized on a computer by a straightforward extension of the algorithm used to model diffusion-limited cluster–cluster flocculation.[5] Instead of permitting a pair of floccules to coalesce at once when they are in close proximity, one can introduce a "sticking probability" that represents the influence of repulsive particle interactions and the multiplicity of configurations possible for two combined floccules. Zero value of the "sticking probability" would evidently prevent flocculation entirely, whereas a value sufficiently large would lead again to transport control of the flocculation process. In a computer simulation, once two floccules are in close enough proximity to combine, a "sticking probability" can be included simply by enumerating all possible distinct configurations of the combined floccules, and then selecting one at random. This process completely "washes out" any influence of the transport of the floccules to the site of their encounter if the probability of combining is small enough. Simulations of cluster–cluster flocculation with this algorithm result in floccules that are also fractal (i.e., are described by Eq. 6.1), but they are relatively dense, three-dimensional structures whose fractal dimension is approximately equal to 2.[5,6] The greater density and larger fractal dimension, as compared to the floccules whose growth is transport controlled, are results of the greater opportunity the floccules have to combine in close-packed arrangements when their coalescence is not instantaneous upon collision.

Reaction-controlled flocculation is expected to be a much slower process than transport-controlled flocculation because of the intervention of particle repulsion. In the context of the gold colloid flocculation experiments described in conjunction with Fig. 6.2, particle repulsion can be introduced by reducing the amount of pyridine added, so that the Au spheres continue to bear negative charge (from adsorbed citrate ions) and the time scale for flocculation is lengthened from minutes to days or weeks.[7,12] The floccules formed under these conditions are illustrated in Fig. 6.5, which also shows examples of both computer simulations and experimental floccules typifying either transport ("DLCA") or reaction ("RLCA") control of flocculation.[6] The relatively dense nature of the floccules in the latter case is readily apparent and is characterized by $D \approx 2$ in applying Eq. 6.1.[12] For silica or goethite colloids that flocculate slowly in 1–1 electrolyte solutions,[11] the same results are observed, in respect to the visual appearance of the floccules (Fig. 6.5) and to the application of Eq. 6.1, but with the interesting feature that rapidly flocculated silica floccules apparently can restructure over the time scale of minutes to hours to form the more dense floccule that is characteristic of slow flocculation.

6.2 The von Smoluchowski Rate Law

Inherent to the flocculation pathways described in Section 6.1 is the notion of two floccules combining to form a larger floccule, which then commences on a trajectory that eventually brings it closer to another floccule with which it, too, may combine to form yet a larger unit, and so on, until flocculation is complete. This encounter-and-combine picture lends itself readily to a formal representation of flocculation processes in terms of chemical reactions. If m denotes a floccule comprising m "primary particles" and {m + n} denotes a floccule created by the combination of two floccules, m and n, then a flocculation process can be pictured in terms of the sequential reaction

$$
\begin{aligned}
m + n \; &\underset{k_{m+n}}{\overset{k_{m,n}}{\rightleftarrows}} \; \{m + n\} \\
\{m + n\} + p \; &\underset{k_{m+n+p}}{\overset{k_{m+n,p}}{\rightleftarrows}} \; \{m + n + p\}
\end{aligned}
\tag{6.8a}
$$

In this reaction scheme the floccule {m + n} eventually combines with a floccule of p "primary particles" to form the floccule {m + n + p}. No distinction is made between the floccules {m + n} and q, where $q \equiv m + n$; thus Eq. 6.8a also can be expressed more economically as the reaction sequence

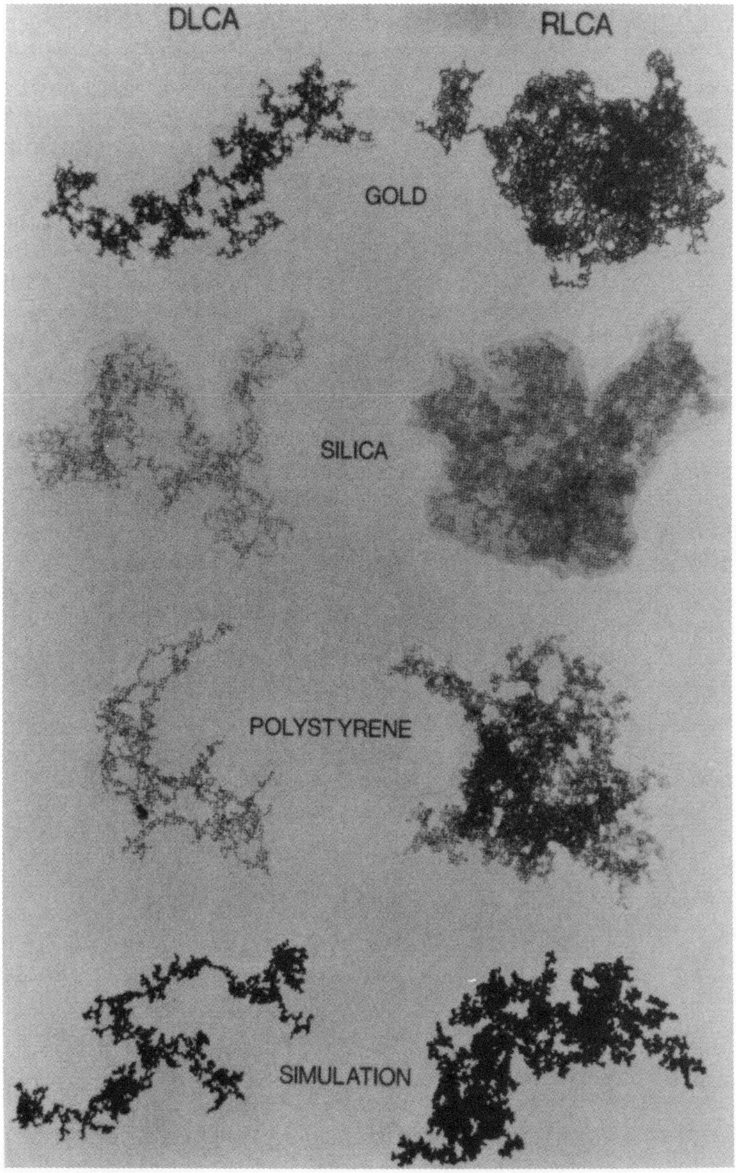

Fig. 6.5. Comparison between floccules formed by transport-controlled ("DLCA") or reaction-controlled ("RLCA") flocculation and the results of computer simulations of these two processes (bottom row). (Adapted, with permission, from P. Meakin.[6] Copyright © by the American Geophysical Union.)

$$
\begin{aligned}
&m + n \underset{k_q}{\overset{k_{mn}}{\rightleftarrows}} q \qquad (q \equiv m + n) \\
&q + p \underset{k_r}{\overset{k_{qp}}{\rightleftarrows}} r \qquad (r \equiv q + p)
\end{aligned}
\tag{6.8b}
$$

In Eq. 6.8b the fact that a flocculation process is a realization of the abstract sequential reaction scheme in Eq. 1.51 becomes transparent, and the rate law

$$\frac{d\rho_q}{dt} = k_{mn}\rho_m\rho_n - k_q\rho_q - k_{qp}\rho_q\rho_p + k_r\rho_r \tag{6.9}$$

follows at once from Eq. 1.53b, where ρ_q is the number density of floccule q in a suspension (number of such floccules per unit volume of suspension). Equation 6.9 can be simplified by assuming that floccule formation is "irreversible" ($k_q = k_r \equiv 0$); it also must be generalized to allow independent, parallel reactions of floccule formation by summing over all possible combinations of m and n that lead to q, or of q and p that lead to r:

$$\frac{d\rho_q}{dt} = \frac{1}{2}\sum_m \sum_n\limits_{(m+n=q)} k_{mn}\rho_m\rho_n - \rho_q \sum_p k_{qp}\rho_p \tag{6.10}$$

where the factor 1/2 is inserted to eliminate "overcounting" associated with the combinations m + n and n + m, both yielding the same q (i.e., $k_{mn} = k_{nm}$). Equation 6.10 is the *von Smoluchowski rate law*.[13] Note that the rate coefficients that appear in this equation are second-order rate coefficients for (binary) floccule *formation*. Thus the von Smoluchowski rate law is a model for the kinetics of floccule growth dominated by the binary encounters characteristic of dilute suspensions.

Full solution of Eq. 6.10 for all values of q, given a set of initial conditions, yields a mathematical description of the distribution of floccule size in a suspension as time passes. From the point of view of experimentation, information about the temporal evolution of floccule size often is obtained by measurement of the *q-moments*:[14]

$$M_\alpha(t) \equiv \sum_q q^\alpha \rho_q(t) \qquad (\alpha = 0, 1, 2, \ldots) \tag{6.11}$$

A differential equation for $M_\alpha(t)$ is derived after multiplying both sides of Eq. 6.10 by q^α and summing over q:

$$\begin{aligned}\frac{dM_\alpha}{dt} &= \frac{1}{2}\sum_q \sum_m \sum_n q^\alpha k_{mn}\rho_m\rho_n - \sum_q q^\alpha \rho_q \sum_p k_{qp}\rho_p \\ &\qquad\qquad (m+n=q) \\ &= \frac{1}{2}\sum_n \sum_m (m+n)^\alpha k_{mn}\rho_m\rho_n - \frac{1}{2}\sum_q \sum_p (q^\alpha + p^\alpha) k_{qp}\rho_q\rho_p \\ &= \frac{1}{2}\sum_m \sum_n [(m+n)^\alpha - m^\alpha - n^\alpha] k_{mn}\rho_m\rho_n \end{aligned} \tag{6.12}$$

where the second step comes from applying the constraint, $m + n = q$, to the sum over q, and then subtracting $\Sigma_q \Sigma_p p^\alpha k_{qp}\rho_q\rho_p$ from the second term, correcting for the "overcounting" with the factor 1/2; the last step results from changing the dummy indices q, p to the indices m, n in the second term.[15]

Equation 6.12 implies that the zeroth moment M_0 is the solution of the differential equation

$$\frac{dM_0}{dt} = -\frac{1}{2}\sum_m \sum_n k_{mn}\rho_m\rho_n \tag{6.13a}$$

This moment is equal to the *total number of floccules per unit volume*, and Eq. 6.13a shows that it must always decline with increasing time if $k_{mn} > 0$. The first moment M_1 is equal to the *total number of "primary particles" per unit volume*, which should be constant according to the definition of the flocculation process and to Eq. 6.12: $dM_1/dt = 0$.[15] The *number-average size* is the ratio of M_1 to M_0:[16]

$$\bar{M}_N(t) \equiv M_1/M_0 = \sum_q q\rho_q(t) / \sum_q \rho_q(t) \tag{6.14a}$$

This parameter can be measured in experiments on the colligative properties of a suspension. The constancy of M_1 and Eq. 6.13 together imply that $\bar{M}_N$ increases with time during a flocculation process.

The second moment M_2 satisfies the differential equation

$$\frac{dM_2}{dt} = \sum_m \sum_n mn\, k_{mn}\rho_m\rho_n \tag{6.13b}$$

The *mass-average size* is the ratio of M_2 to M_1:[16]

$$\bar{M}_M(t) \equiv M_2/M_1 = \sum_q q^2\rho_q(t) / \sum_q q\rho_q(t) \tag{6.14b}$$

This parameter, which can be measured in light-scattering experiments (cf. Fig. 6.4), is often taken as a convenient estimate of the average floccule size. Like $\bar{M}_N(t)$, it increases with time, but at a greater rate.

The second-order rate coefficient k_{mn} in Eq. 6.10 represents in parametric form whatever may be the detailed mechanism of the process limiting the rate of floccule formation via binary encounter and irreversible combination. If this process is Brownian motion, then k_{mn} is equal to the ratio of the diffusional rate of one floccule (say, n) toward another (say, m), divided by the product of the number densities of the two.[17] The diffusional rate, in turn, is the product of the diffusional-flux collision frequency and the number density ρ_m of the floccule designated as the "target" for an "incoming" floccule. This collision frequency, CF, at the surface of a sphere of radius r surrounding a "target" is (cf. Eq. 4.71)

$$\text{CF} \equiv \text{diffusional flux} \times \text{sphere area} = \mathbf{D}\,\frac{d\rho}{dr} \times 4\pi r^2 \qquad (6.15a)$$

where **D** is the diffusion coefficient for the Brownian motion of an "incoming" floccule, relative to the Brownian motion of the "target," and ρ is the number density of the "incoming" floccule on the surface of the sphere of total area $4\pi r^2$. The collision frequency must actually be independent of the size of the enclosing sphere, so a convenient evaluation is obtained by separating variables in Eq. 6.15a and performing the integrations

$$\text{CF} \int_{R_m + R_n}^{\infty} \frac{dr}{4\pi r^2} = \mathbf{D} \int_0^{\rho_n} d\rho = \mathbf{D}\rho_n$$

$$= \frac{\text{CF}}{4\pi(R_m + R_n)} \qquad (6.15b)$$

where R_m and R_n are the radii of "target" and "incoming" floccules, respectively, and note is taken of the vanishing of the "incoming" floccule number density at the distance of closest approach, $R_m + R_n$. Thus, by Eq. 6.15,

$$\text{diffusional rate} \equiv \text{CF}\rho_m = 4\pi\mathbf{D}(R_m + R_n)\rho_m\rho_n \qquad (6.15c)$$

and

$$k^{\mathbf{D}}_{mn} \equiv \text{diffusional rate} / \rho_m\rho_n = 4\pi\mathbf{D}(R_m + R_n) \qquad (6.16a)$$

Under the assumption that the "target" and "incoming" floccules engage in Brownian motion independently,[18] $\mathbf{D} = \mathbf{D}_m + \mathbf{D}_n$, and the rate coefficient becomes

$$k^{\mathbf{D}}_{mn} = 4\pi\,(\mathbf{D}_m + \mathbf{D}_n)\,(R_m + R_n) \qquad (6.16b)$$

Equation 6.16b indicates that the rate of cluster formation increases with the relative mobility of the colliding clusters (as expressed by $\mathbf{D}_m$ or $\mathbf{D}_n$) and with their spatial extent (as expressed by R_m or R_n). It is usually a good approximation to assume that the product of these two factors (e.g., $\mathbf{D}_m R_m$) is independent of cluster properties.[19] Under this assumption, Eq. 6.16b takes the simpler form ($\mathbf{D}_1 R_1 = \mathbf{D}_m R_m = \mathbf{D}_n R_n$):

$$k_{mn}^{D} = 4\pi \mathbf{D}_1 R_1 \left(2 + \frac{R_m}{R_n} + \frac{R_n}{R_m} \right) \tag{6.16c}$$

where $\mathbf{D}_1$ is the diffusion coefficient of a "primary particle" of radius R_1. With the substitution $R_m = R_n(1 + \epsilon_{mn})$ and expansion of the resulting binomials to second order in $\epsilon_{mn} \equiv (R_m - R_n) / R_n$, Eq. 6.16c then can be approximated by

$$\begin{aligned} k_{mn}^{D} &= 4\pi \mathbf{D}_1 R_1 \left[2 + (1 + \epsilon_{mn}) + (1 + \epsilon_{mn})^{-1}\right] \\ &\approx 16\pi \mathbf{D}_1 R_1 + O(\epsilon_{mn}^2) \end{aligned} \tag{6.16d}$$

Equation 6.16d should be a good estimate of the floccule formation rate coefficient if $\mathbf{D}$ scales as R^{-1} and if collisions between floccules of approximately equal size dominate.[20]

The approximate rate constant in Eq. 6.16d is independent of floccule size, which results in a considerable simplification of the von Smoluchowski rate law:

$$\frac{d\rho_q}{dt} = 2K_D\left(\frac{1}{2}\sum_m \rho_m \rho_{q-m} - \rho_q \sum_p \rho_p\right) \tag{6.17}$$

where $K_D \equiv 8\pi \mathbf{D}_1 R_1$ is the same as the second-order rate coefficient that appears in Eq. 6.2.[9] With the initial condition of a fully dispersed suspension of "primary particles" (i.e., $\rho_q(0) = \rho_0$ if $q = 1$ and equals zero if $q > 1$), the solution of Eq. 6.17 is:[21]

$$\rho_q(t) = \rho_0 \frac{(K_D \rho_0 t)^{q-1}}{[1 + K_D \rho_0 t]^{q+1}} \qquad (q \geq 1) \tag{6.18}$$

Equation 6.18 is graphed in Fig. 6.6 for the cases $q = 1, 2, 3$. The number density of "primary particles," $\rho_1(t)$, decreases monotonically with time as these particles are consumed in the formation of floccules. The number densities of the floccules, on the other hand, rise from zero to a maximum at $t = (q - 1)/2K_D\rho_0$, and then decline. This mathematical behavior reflects creation of a floccule of given size from smaller floccules, followed by a period of dominance, and finally consumption to form yet larger particle units as time passes. Both experimental data and computer simulations, like that whose visualization appears in Fig. 6.1, are in excellent qualitative agreement with Eq. 6.18 when they are used to calculate the $\rho_q(t)$.[13,14] Thus the von Smoluchowski rate law with a uniform rate coefficient appears to capture the essential features of diffusion-controlled flocculation processes.

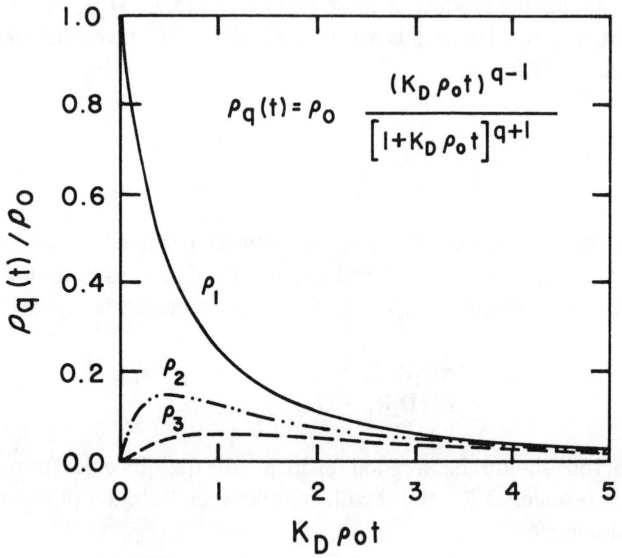

FIG. 6.6. Graphs of Eq. 6.18 for q = 1, 2, 3. Note that $\rho_1(t)$ is the number of "primary particles" in suspension.

The q-moments that correspond to Eq. 6.18 can be calculated recursively after inserting this latter equation into Eq. 6.11 and taking the time derivative of both sides:

$$\frac{dM_\alpha}{dt} = \sum_q q^\alpha \left\{\rho_0^2 \frac{K_D(q-1)(K_D\rho_0 t)^{q-2}}{(1+K_D\rho_0 t)^{q+1}} - \rho_0^2 K_D \frac{(q+1)(K_D\rho_0 t)^{q-1}}{(1+K_D\rho_0 t)^{q+2}}\right\}$$

$$= \sum_q q^\alpha \{[(q-1)/t]\rho_q(t) - [\rho_0 K_D(q+1)/(1+K_D\rho_0 t)]\rho_q(t)\}$$

$$= [(M_{\alpha+1}(t) - M_\alpha(t))/t - [\rho_0 K_D/(1+K_D\rho_0 t)](M_{\alpha+1}(t) + M_\alpha(t))$$

or, on solving for $M_{\alpha+1}(t)$,

$$M_{\alpha+1}(t) = (1+2K_D\rho_0 t)M_\alpha(t) + t(1+K_D\rho_0 t)\frac{dM_\alpha}{dt} \qquad (6.19)$$

Once $M_0(t)$ is known, all other q-moments can be calculated using Eq. 6.19. Equation 6.13a shows that $M_0(t)$ satisfies the differential equation

$$\frac{dM_0}{dt} = -K_D M_0^2 \qquad (6.20a)$$

whose solution is

$$M_0(t) = M_0(0) \,/\, [1 + K_D M_0(0)t] \tag{6.20b}$$

Thus the total number of clusters per unit volume declines monotonically to zero with a characteristic time scale, $1/K_D M_0(0)$ (for second-order kinetics), where $M_0(0) = \rho_0$ according to Eqs. 6.11 and 6.18. The combination of Eq. 6.20 and 6.19 then yields

$$M_1(t) = \frac{(1 + 2K_D\rho_0 t)\,\rho_0}{(1 + K_D\rho_0 t)} - \frac{t\,(1 + K_D\rho_0 t)\,\rho_0^2 K_D}{(1 + K_D\rho_0 t)^2} = \rho_0 \tag{6.21}$$

which is the same as $M_0(0)$, as expected from the basic definition of the first moment. The number-average size is then

$$\bar{M}_N(t) = 1 + K_D\rho_0 t \tag{6.22}$$

according to Eqs. 6.14a, 6.20b, and 6.21. This property of the flocculating suspension grows linearly with time.

The second q-moment follows readily from Eq. 6.19 after incorporation of Eq. 6.21:

$$M_2(t) = \rho_0(1 + 2K_D\rho_0 t) \tag{6.23}$$

The mass-average size is then

$$\bar{M}_M(t) = 1 + 2K_D\rho_0 t \tag{6.24}$$

according to Eqs. 6.14b and 6.21. This measure of floccule size grows at twice the rate of the number-average size. The two size parameters are related by the expression

$$2\bar{M}_N - \bar{M}_M = 1 \tag{6.25}$$

which can be applied to data on the colligative and light-scattering properties of a flocculating suspension in order to test the accuracy of Eq. 6.18. Data and computer simulations of the temporal evolution of $\bar{M}_N$ and $\bar{M}_M$ for a system of hard spheres undergoing Brownian motion with "irreversible" combination on collision tend to confirm the qualitative accuracy of Eqs. 6.22, 6.24, and 6.25.[13,14]

6.3 Scaling the von Smoluchowski Rate Law

Cluster fractals that are created by diffusion-limited flocculation processes are described mathematically by power-law relationships like those in Eqs. 6.1 and 6.5. These relationships are said to have a *scaling property* because they satisfy what in mathematics is termed a *homogeneity condition*:[22]

$$f(\lambda x) = \lambda^a f(x) \qquad (\lambda > 0) \tag{6.26}$$

where f(x) is thereby a homogeneous function of the independent variable x. Clearly, Eqs. 6.1, 6.5, and 6.7 exhibit scaling properties, with the exponent a taking on the values D, 1/D, 1, and 2, respectively. Scaling properties are inherent to cluster fractals, as discussed in Special Topic 3 at the end of this chapter.

If the von Smoluchowski rate law (Eq. 6.10) is to be consistent with the formation of cluster fractals, then it must in some way also exhibit scaling properties. These properties, in turn, have to be exhibited by its second-order rate coefficient k_{mn} since this parameter represents the flocculation mechanism, aside from the binary-encounter feature implicit in the sequential reaction in Eq. 6.8. The model expression for k_{mn} in Eq. 6.16b, for example, should have a scaling property. Indeed, if the assumption is made that $\mathbf{D}_m R_m$ (m = 1, 2, . . .) is constant, Eq. 6.16c applies, and if cluster fractals are formed, Eq. 6.1 can be used (with R replacing L) to put Eq. 6.16c into the form

$$\begin{aligned} k^{D}_{mn} &= \frac{1}{2} K_D \left[2 + \left(\frac{m}{n} \right)^{1/D} + \left(\frac{n}{m} \right)^{1/D} \right] \\ &= \frac{1}{2} K_D \left(m^{-1/D} + n^{-1/D} \right) \left(m^{1/D} + n^{1/D} \right) \end{aligned} \tag{6.27}$$

where note is taken of the fact that the indices m and n have the same physical meaning as N in Eq. 6.1. Equation 6.27 implies that this model rate constant has the scaling property

$$k^{D}_{\lambda m \lambda n} = k^{D}_{mn} \tag{6.28}$$

considering the indices m and n as the independent variables on which the rate constant depends. In this example the rate constant is *scale invariant*, as is evident also from Eq. 6.16c.

The special case of Eq. 6.27 that obtains where m = n, that is, $k^{D}_{mn} = 2K_D$, is trivially scale invariant, so this property ought to be implicit in the corresponding solution of the von Smoluchowski rate law, given in Eq. 6.18. That this is the case can be seen by noting the large-time limit of $\rho_q(t)$,

$$\rho_q(t) \underset{t \uparrow \infty}{\sim} \rho_0(K_D\rho_0 t)^{-2} \tag{6.29a}$$

and considering the product

$$(K_D\rho_0 t)^2 \rho_q(t) = \rho_0 \left[\frac{K_D\rho_0 t}{1 + K_D\rho_0 t} \right]^{q+1} \tag{6.29b}$$

The napierian logarithm of the right side of Eq. 6.29b has the large-time limit:

$$(q + 1) \ln\left[\frac{K_D\rho_0 t}{1 + K_D\rho_0 t} \right] \underset{t \uparrow \infty}{\sim} -(q + 1)/(1 + K_D\rho_0 t)$$

$$\approx -q/K_D\rho_0 t \qquad (t >> 1/K_D\rho_0) \tag{6.29c}$$

Therefore

$$\rho_q(t) \underset{t \uparrow \infty}{\sim} \rho_0(K_D\rho_0 t)^{-2} \exp(-q/K_D\rho_0 t) \tag{6.30a}$$

is the large-time limit of $\rho_q(t)$ in Eq. 6.18. This result can be expressed in the alternative form

$$\rho_q(t) \underset{t \uparrow \infty}{\sim} \rho_0 q^{-2}(q/K_D\rho_0 t)^2 \exp(-q/K_D\rho_0 t) \tag{6.30b}$$

which displays the ratio $(q/K_D\rho_0 t)$ more prominently. This asymptotic solution has the scaling property

$$\rho_{\lambda q}(\lambda t) = \lambda^{-2}\rho_q(t) \qquad (t >> 1/K_D\rho_0) \tag{6.31}$$

as can be demonstrated by replacing q with λq and t with λt in Eq. 6.30b (note that $\rho_q(t)$ depends on two independent variables and both must be scaled). The connection between Eqs. 6.31 and 6.28 is now made by scaling the von Smoluchowski rate law in Eq. 6.10:

$$d\rho_{\lambda q}/d(\lambda t) = \lambda^{-3}\, d\rho_q/dt$$

$$\rho_{\lambda m}\rho_{\lambda n} = \lambda^{-4}\rho_m\rho_n$$

$$\sum_{\lambda m}\sum_{\lambda n} {}_{(m+n=q)} k^D_{\lambda m \lambda n}\rho_{\lambda m}\rho_{\lambda n} = \sum_{\lambda m} k^D_{\lambda m \lambda(q-m)}\rho_{\lambda m}\rho_{\lambda(q-m)}$$

$$\approx \int k^{D}_{\lambda m \lambda(q-m)} \rho_{\lambda m} \rho_{\lambda(q-m)} d(\lambda m)$$

$$= \lambda \int k^{D}_{\lambda m \lambda(q-m)} \rho_{\lambda m} \rho_{\lambda(q-m)} dm \approx \lambda \sum_m k^{D}_{\lambda m \lambda(q-m)} \lambda^{-4} \rho_m \rho_{q-m}$$

$$= \lambda^{-3} \sum_m k^{D}_{\lambda m \lambda(q-m)} \rho_m \rho_{q-m}$$

so that

$$\lambda^{-3} \frac{d\rho_q}{dt} = \lambda^{-3} \left[\sum_m k^{D}_{\lambda m \lambda(q-m)} \rho_m \rho_{q-m} - \rho_q \sum_p k^{D}_{\lambda q \lambda p} \rho_p \right] \qquad (6.32)$$

The von Smoluchowski rate law will be *invariant in form* (i.e., scale invariant) under the scaling property in Eq. 6.31 only if Eq. 6.28 is true. Otherwise, additional factors of the parameter λ will appear on the right side of Eq. 6.32 that cannot be canceled by multiplying through the equation with λ^3. Thus the scaling property in Eq. 6.28 is implicit in the scaling property of the $\rho_q(t)$ at large times. That an asymptotic solution is necessary in order to demonstrate this point is consistent with the fact that the fractal properties in Eqs. 6.1 and 6.5 are also asymptotic characteristics that apply in the limit of large floccule size and large times, respectively (see Special Topic 3).

The scaling properties of the q-moments (Eq. 6.11) can be deduced from those of the $\rho_q(t)$:

$$M_\alpha(\lambda t) = \Sigma_{\lambda q} (\lambda q)^\alpha \rho_{\lambda q}(\lambda t)$$

$$\underset{t \uparrow \infty}{\sim} \lambda \Sigma_q \lambda^\alpha q^\alpha \lambda^{-2} \rho_q(t)$$

$$= \lambda^{\alpha - 1} M_\alpha(t) \qquad (t >> 1/K_D\rho_0) \qquad (6.33)$$

Equation 6.33 is consistent with the results in Eqs. 6.20b, 6.21, and 6.23 evaluated at large times. The same is true of Eq. 6.19, which is invariant in form under the scaling property in Eq. 6.31 if $t >> 1/K_D\rho_0$. A comparison of Eq. 6.30b with Eqs. 6.22 and 6.24 shows also that the asymptotic, large-time $\rho_q(t)$ can be expressed in terms of q and the average floccule size:

$$\rho_q(t) \underset{t \uparrow \infty}{\sim} \rho_0 q^{-2} [q/\bar{M}_N(t)]^2 \exp[-q/\bar{M}_N(t)] \qquad (6.34a)$$

$$\rho_q(t) \underset{t \uparrow \infty}{\sim} \rho_0 q^{-2} [2q/\bar{M}_M(t)]^2 \exp[-2q/\bar{M}_M(t)] \qquad (6.34b)$$

Usually it is Eq. 6.34b that is chosen to represent the asymptotic $\rho_q(t)$ because $\bar{M}_M(t)$ is more accessible to accurate measurement than is $\bar{M}_N$.[6] The physical significance of Eq. 6.34 is that flocculating clusters whose formation is described well by Eq. 6.17 (or, in principle, by Eq. 6.10 with a scale-invariant rate constant as defined by Eq. 6.28) exhibit no intrinsic length scale arising from the mechanism of flocculation. Therefore $\rho_q(t)$, at times large enough to "wash out" the early influence of the "primary particle" diameter, contains only some convenient measure of the average cluster size as a length-scale parameter.

Equation 6.28 represents the simplest effect of scaling on the second-order rate constant in the von Smoluchowski rate law: no effect whatsoever. More generally, if k_{mn} satisfies the homogeneity condition

$$k_{\lambda m \lambda n} = \lambda^{\theta} k_{mn} \qquad (\lambda > 0) \tag{6.35}$$

the scaling invariance of the von Smoluchowski rate law will imply large-time scaling properties of the $\rho_q(t)$ and the q-moments that differ from those in Eqs. 6.31 and 6.33. Broader scaling properties could arise, for example, if the diffusion coefficients in Eq. 6.16b were assumed to be proportional to a power of the cluster radius other than −1, such that[23]

$$k_{mn} = 2K\,(m^{\gamma/D} + n^{\gamma/D})\,(m^{1/D} + n^{1/D}) \tag{6.36}$$

where γ is an arbitrary parameter and K collects together all geometric factors and coefficients of proportionality not dependent on the number of "primary particles" in a cluster. The special case, $\gamma = -1$, $K = K_D$, reduces Eq. 6.36 to Eq. 6.27. Other possibilities include the model expression[24]

$$k_{mn} = 2K(mn^{\gamma})^{1/D} \tag{6.37}$$

to which Eq. 6.36 reduces if flocculation is dominated by encounters between very large and very small clusters ($m >> n$, $\gamma < 0$). The homogeneity condition satisfied by K_{mn} in Eqs. 6.36 and 6.37 is

$$k_{\lambda m \lambda n} = \lambda^{(\gamma + 1)/D} k_{mn} \tag{6.38}$$

which is a special case of Eq. 6.35. Computer simulations[6,23] of diffusion-limited cluster–cluster flocculations, carried out with $\mathbf{D}_m \propto R_m{}^{\gamma}$, are in excellent agreement with Eq. 6.36 for the case $D = 1.78$ when k_{mn} is calculated from the simulation results as the ratio of diffusional rate to the product $\rho_m \rho_n$ (the definition in Eq. 6.16a).

The more general homogeneity condition in Eq. 6.35 should permit a more general asymptotic form of $\rho_q(t)$ than what appears in Eq. 6.30b. This latter equation can be rewritten as the product function:

$$\rho_q(t) \underset{t \uparrow \infty}{\sim} \rho_0 q^{-2} f(q/K_D \rho_0 t) \tag{6.39}$$

where $f(y) \equiv y^2 \exp(y)$. An evident generalization of Eq. 6.39 is[25]

$$\rho_q(t) \underset{t\uparrow\infty}{\sim} \rho_0 q^{-\tau} g(q/t^z) \tag{6.40}$$

where τ and z are positive-valued parameters and g(y) is an arbitrary function of the ratio q/t^z. The physical significance of the parameters τ and z can be derived from introducing the model expression in Eq. 6.40 into Eq. 6.11 for the q-moments.[23,25] Consider first the q-moment $M_1(t)$, which must be constant because of "primary particle" mass conservation:

$$M_1(t) \underset{t\uparrow\infty}{\sim} \int q\rho_q(t)dq = \rho_0 \int q^{1-\tau} g(q/\tau^z)\, dq$$
$$= \rho_0\, t^{(2-\tau)z} \int y^{1-\tau} g(y)\, dy \equiv \rho_0 \tag{6.41}$$

It follows from Eq. 6.41 that $\tau = 2$ is a necessary condition for the constancy of $M_1(t)$. This value of τ already appears in Eq. 6.30b. Consider next the q-moment $M_0(t)$:

$$M_0(t) \underset{t\uparrow\infty}{\sim} \int \rho_q(t)\, dq = \rho_0 \int q^{-2} g(q/t^z)\, dq$$
$$= \rho_0\, t^{-z} \int y^{-2} g(y)\, dy \tag{6.42}$$

which demonstrates that the total number of floccules declines as t^{-z} at large times ($t \uparrow \infty$). Corresponding to this decline is the growth of $M_2(t)$:

$$M_2(t) \underset{t\uparrow\infty}{\sim} \int q^2\rho_q(t)\, dq = \rho_0 t^z \int g(y)\, dy \tag{6.43}$$

from which it can be deduced that the mass-average size has the asymptotic behavior (Eq. 6.14b):

$$\bar{M}_M(t) \underset{t\uparrow\infty}{\sim} [\int g(y)\, dy \,/\, \int y^{-1} g(y)\, dy] t^z \tag{6.44}$$

Equation 6.44 shows that the parameter z is a measure of how rapidly large floccules will be created. It also shows that the asymptotic form of $\rho_q(t)$ in Eq. 6.40 can be expressed as

$$\rho_q(t) \underset{t\uparrow\infty}{\sim} \rho_0 q^{-2} g[q/\bar{M}_M(t)] \tag{6.45}$$

just as is done in Eq. 6.34b for $\rho_q(t)$ in Eq. 6.30b. Thus the generalization in Eq. 6.40 does not produce an intrinsic length scale.

The implication of Eq. 6.40 for the more general homogeneity condition in Eq. 6.35 can be worked out by introducing that condition and the scaling property

$$\rho_{\lambda q}(\lambda^{1/z}t) = \lambda^{-2}\,\rho_q(t) \qquad (t \uparrow \infty) \tag{6.46}$$

into Eq. 6.10. Note that the time scaling in Eq. 6.46 differs from that in Eq. 6.31 because $g(q/t^z)$ must be invariant under any scaling transformation of its independent variables in order that the ratio q/t^z remain a meaningful combination. The analog of Eq. 6.32 is then the scaled rate law

$$\lambda^{-2-1/z}\,\frac{d\rho_q}{dt} = \lambda^{-3+\theta}\left[\sum_m k_{mm-q}\,\rho_m\,\rho_{m-q} - \rho_q \sum_p k_{qp}\rho_p\right] \tag{6.47}$$

The von Smoluchowski rate law is scale invariant if[23]

$$z = (1 - \theta)^{-1} \tag{6.48}$$

Since $z > 0$ for an increasing cluster size with time (Eq. 6.44), $\theta < 1$ in Eq. 6.48. Thus the homogeneity condition satisfied by the second-order rate coefficient determines the rate of floccule growth and the corresponding decline in the number of floccules (Eqs. 6.42 and 6.44).

Equation 6.45 can be written in the asymptotic *generic form*

$$\rho_q(t) \underset{t \uparrow \infty}{\sim} B\,[\bar{M}_M(t)]^{-2}\,h[q/\bar{M}_M(t)] \tag{6.49}$$

where $B \equiv \rho_0\,[\int g(y)dy \,/ \int y^{-1}g(y)\,dy\,]^2$ and $h(y) \equiv y^2 g(y)$, with the use of Eq. 6.44. Equation 6.49 implies that $[\bar{M}_M(t)]^2\rho_q(t)$ depends only on the ratio $q/\bar{M}_M(t)$ for any values of q and t large enough to make $\rho_q(t)$ take on its asymptotic form. This prediction has been verified experimentally in studies of the flocculation of Au colloids and polystyrene spheres.[6,26]

6.4 Fuchsian Kinetics

Reaction control of flocculation kinetics is described most simply by imagining that colliding floccules do not coalesce instantaneously (on the particle diffusion time scale), but instead must undergo numerous such encounters before they can combine to form larger units. In this way, interacting floccules can explore a variety of possible combinations before their coalescence occurs, leading, in general, to more compact structures than are formed by transport-controlled flocculation.[6] The effect of this exploratory process is expressed mathematically by a small "sticking probability" factor that may be introduced, for example, as a coefficient in the rate coefficient k_{mn} in Eq. 6.10. This factor evidently would reduce the value of the rate constant, as compared to the case of transport control, and thereby lengthen the characteristic time scale for flocculation.

A more mechanistic approach to reaction-controlled flocculation entails representing the "sticking probability" in terms of a Boltzmann factor that contains the potential energy of interaction between clusters: $\exp[-V(r)/k_BT]$, where V(r) is the potential energy of two clusters separated by the distance r. If $|V(r)| << k_BT$, the Boltzmann factor approaches 1.0 and flocculation becomes transport-controlled. If $|V(r)| >> k_BT$ and $V(r) > 0$, the Boltzmann factor approaches zero, the limiting situation in which no flocculation occurs. Computer simulation of reaction-controlled, cluster–cluster flocculation using this Boltzmann factor as a "sticking probability" shows that attractive cluster interactions [$V(r) < 0$] lead to more compact, dense structures than repulsive interactions [$V(r) > 0$].[27] The calculated dependence of the mass-average cluster size on time indicates that the growth of floccules also is more rapid if the interactions are attractive. The fractal dimension of the floccules as deduced from Eq. 6.1 (after permitting growth to go on long enough to obtain asymptotic results) appears to be near 2 if the interaction is short-ranged, regardless of its sign.

The effect of the potential energy V(r) on the second-order rate coefficient k_{mn} can be described mathematically in a simple way by modifying the expression for the diffusional flux in Eq. 6.15a to include the Boltzmann factor:

$$CF = \mathbf{D}\,\frac{d\rho}{dr}\exp\left[\frac{-V(r)}{k_BT}\right](4\pi r^2) \tag{6.50a}$$

Thus the collision frequency CF is reduced by the Boltzmann factor when cluster interactions are repulsive and is increased when they are attractive. The result corresponding to Eq. 6.15b is then

$$CF\int_{R_m+R_n}^{\infty}\exp\left[\frac{V(r)}{k_BT}\right]\frac{dr}{4\pi r^2} = \mathbf{D}\rho_n \tag{6.50b}$$

The integral on the left side of Eq. 6.50b defines the *stability ratio* W_{mn}:[28]

$$W_{mn} \equiv (R_m + R_n)\int_{R_m+R_n}^{\infty}\exp\left[\frac{V(r)}{k_BT}\right]\frac{dr}{r^2} \tag{6.51}$$

It follows from Eqs. 6.15c, 6.16, and 6.51 that the rate coefficient k_{mn} now becomes

$$k^F_{mn} \equiv 4\pi(\mathbf{D}_m + \mathbf{D}_n)(R_m + R_n)\,/\,W_{mn} \tag{6.52}$$

Equation 6.52 exhibits the inverse of the stability ratio playing the role of the "sticking probability" coefficient that reduces k_{mn} below its value for pure transport control of flocculation whenever cluster interactions are repulsive. (Note that $W_{mn} = 1$ when V(r) vanishes, according to Eq. 6.51.) Equation 6.52 is a model for k_{mn} known as *Fuchsian kinetics*.[28]

The stability ratio can be calculated once the potential energy V(r) has been specified. The investigation of floccule interactions is an active area of research[29] that has not yet resulted in a definitive mathematical expression for V(r), although there is consensus that both attractive and repulsive contributions exist. Irrespective of the exact form of V(r), if it is a continuous function of r, then the First Mean Value Theorem[30] can be applied to Eq. 6.51 to derive

$$W_{mn} = (R_m + R_n)\exp[V(\xi_{mn})/k_BT] \int_{R_m+R_n}^{\infty} \frac{dr}{r^2}$$

$$= \exp[V(\xi_{mn})/k_BT] \qquad [(R_m + R_n) < \xi_{mn} < \infty] \qquad (6.53)$$

Equation 6.53 implies that there is a value of r ($\equiv \xi_{mn}$) somewhere between the limits of the integral, such that W_{mn} is simply equal numerically to the exponential factor evaluated at ξ_{mn}. [The proof of Eq. 6.53 requires only the continuity of V(r) and the fact that $r^2 > 0$.[30]] If V(r) exhibits a positive maximum (corresponding to a "potential barrier"), then $V(\xi_{mn})$ is likely to be equal to it.[28]

Experiments on reaction-controlled flocculation kinetics using electron microscopy and light-scattering techniques[6,12,24,31,32] suggest that the time dependence of the average floccule radius during flocculation does not follow a power-law expression like Eq. 6.6, but instead can be described by the exponential relation

$$L(t) = L_0 \exp(Ct) \qquad (6.54)$$

where L_0 and C are positive-valued parameters. Moreover, although the generic scaling form for $\rho_q(t)$ in Eq. 6.49 is still applicable, the explicit q-dependence of the number density of cluster q is given by the power-law equation

$$\rho_q(t) \underset{t\uparrow\infty}{\sim} B[\bar{M}_M(t)]^{-2}[q/\bar{M}_M(t)]^{-3/2} \qquad (6.55)$$

Examples of the scaling behavior in Eq. 6.55 are shown in Fig. 6.7 for the flocculation of Au[31] and polystyrene (bearing carboxyl functional groups)[6] colloids.

Given that Eq. 6.1 (with $D \approx 2$) applies to reaction-controlled flocculation kinetics, Eq. 6.54 implies that $\bar{M}_M(t)$ [or $\bar{M}_N(t)$] must also exhibit an exponential growth with time. Therefore, by contrast with transport-controlled flocculation kinetics, a uniform value of the rate constant k_{mn} cannot be introduced into the von Smoluchowski rate law, as in Eq. 6.17, to derive a mathematical model of the number density $\rho_q(t)$. Equations 6.22 and 6.24 indicate clearly that a uniform k_{mn} leads to a *linear* time dependence in the

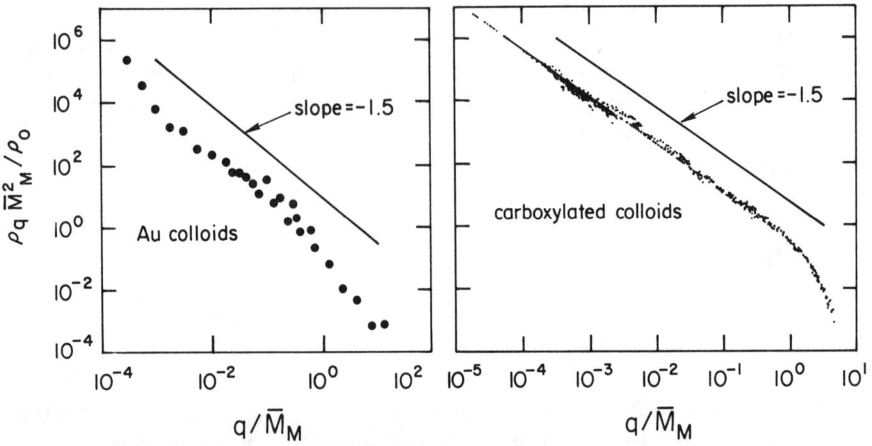

FIG. 6.7. Tests of the scaling relationship in Eq. 6.55 for Au (left, data from D. A. Weitz and M. Y. Lin[31]) and polystyrene (right, data from P. Meakin[6]) colloids. The line corresponds to $\tau = 1.5$.

number-average and mass-average cluster size. A model expression for k^F_{mn} that does lead to an exponential time dependence in these two latter functions is[32]

$$k^F_{mn} \approx k_F(m + n) \qquad (k_F > 0) \tag{6.56}$$

which is equivalent to postulating that the stability ratio varies inversely with floccule size (cf. Eqs. 6.16 and 6.52). Mechanistically speaking, this means that larger colliding floccules have a larger "sticking probability" than smaller colliding floccules, evidently because of a smaller "potential barrier" in V(r).

With the introduction of Eq. 6.56, Eq. 6.10 becomes

$$\frac{d\rho_q}{dt} = \frac{1}{2} k_F \sum_m q \rho_m \rho_{q-m} - k_F \rho_q \sum_p (q + p) \rho_p \tag{6.57}$$

The solution of Eq. 6.57 for the initial condition

$$\rho_q(0) = \begin{cases} \rho_0 & q = 1 \\ 0 & q > 1 \end{cases} \tag{6.58}$$

is given by the equation[14]

$$\rho_q(t) = \rho_0 [1 - u(t)] [qu(t)]^{q-1} \exp[-qu(t)]/q! \tag{6.59}$$

where

$$u(t) \equiv 1 - \exp(-k_F \rho_0 t) \tag{6.60}$$

and q! is the factorial function. This result is more complicated mathematically than Eq. 6.18, but its time dependence is qualitatively similar: The number density of "primary particles," $\rho_1(t)$, decreases monotonically with time, while the number densities of the floccules ($q > 1$) rise from zero to a maximum at $t = \ln q/2k_F\rho_0$, and then decline. Thus, once again, floccules of a given size are created from smaller floccules, enjoy a period of dominance, and then are consumed to form larger floccules than themselves.

The q-moments that correspond to Eq. 6.59 are solutions of Eq. 6.12 with Eq. 6.56 introduced:

$$\frac{dM_\alpha}{dt} = \frac{1}{2} k_F \sum_m \sum_n [(m+n)^\alpha - m^\alpha - n^\alpha](m+n)\rho_m \rho_n \tag{6.61}$$

In particular, the first three q-moments satisfy the differential equations (cf. Eq. 6.13):

$$\begin{aligned}\frac{dM_0}{dt} &= -\frac{1}{2} k_F \sum_m \sum_n (m+n)\rho_m \rho_n \\ &= -\frac{1}{2} k_F (M_1 \textstyle\sum_n \rho_n + M_1 \sum_m \rho_n) \\ &= -k_F M_1 M_0(t) \qquad [M_0(0) = \rho_0]\end{aligned} \tag{6.62a}$$

$$\frac{dM_1}{dt} \equiv 0 \qquad [M_1(0) = \rho_0] \tag{6.62b}$$

$$\begin{aligned}\frac{dM_2}{dt} &= k_F \sum_m \sum_n mn(m+n)\rho_m \rho_n \\ &= k_F \sum_m \sum_n (m^2 n + mn^2)\rho_m \rho_n \\ &= 2k_F M_1 M_2(t) \qquad [M_2(0) = \rho_0]\end{aligned} \tag{6.62c}$$

The solutions of these differential equations are[14]

$$M_0(t) = \rho_0 \exp(-k_F \rho_0 t) \tag{6.63a}$$

$$M_1(t) = \rho_0 \tag{6.63b}$$

$$M_2(t) = \rho_0 \exp(2k_F \rho_0 t) \tag{6.63c}$$

as can be verified by direct substitution into Eq. 6.62. The number-average and mass-average sizes now follow from Eq. 6.14:

$$\bar{M}_N(t) = \exp(k_F \rho_0 t) \tag{6.64a}$$

$$\bar{M}_M(t) = \exp(2k_F \rho_0 t) \tag{6.64b}$$

These two measures of average size show an exponential growth with time that is consistent with Eqs. 6.1 and 6.54. There is a corresponding exponential decay of $M_0(t)$, the total number of floccules at time t. (Evidently $k_F < 2K_D$ in order that Eqs. 6.59 and 6.63 describe a slower flocculation process than Eqs. 6.18, 6.20b, and 6.23.)

The scaling exponent $\theta = 1$ for the model rate constant in Eq. 6.56, as compared to $\theta = 0$ for the model expressed in Eq. 6.27. Thus k_{mn} is not scale invariant for reaction-controlled flocculation described by Eqs. 6.59 and 6.63. Moreover, the value of θ is not consistent with finite z in Eq. 6.48, indicating at once that a power-law time dependence of $M_0(t)$ at large time does not exist. This fact does not preclude the applicability of the generic scaling form of $\rho_q(t)$ in Eq. 6.49, however, since no explicit time dependence appears in the latter equation. To examine this possibility, Eq. 6.60 is transformed with the help of Eq. 6.64b:

$$u(t) = 1 - [\bar{M}_M(t)]^{-1/2} \tag{6.65}$$

and, similarly to Eq. 6.29c, the base e logarithm of the right side of Eq. 6.59 is calculated (neglecting ρ_0):

$$-\frac{1}{2} \ln \bar{M}_M(t) + (q-1) \ln q + (q-1) \ln\left\{ 1 - [\bar{M}_M(t)]^{-1/2} \right\}$$

$$-q \left\{ 1 - [\bar{M}_M(t)]^{-1/2} \right\} - \ln q!$$

At large time, $\ln\{1 - [\bar{M}_M(t)]^{-1/2}\} \approx -[\bar{M}_M(t)]^{-1/2}$, and, for large q-values,[33] $\ln q! \approx q \ln q - q + \frac{1}{2} \ln q$. With these two approximations, Eq. 6.59 has the asymptotic logarithmic form:

$$\ln [\rho_q(t)/\rho_0] \underset{t\uparrow\infty}{\sim} -\tfrac{1}{2} \ln \bar{M}_M(t) + (q-1) \ln q$$

$$- (q-1) [\bar{M}_M(t)]^{-1/2} - q\{1 - [\bar{M}_M(t)]^{-1/2}\}$$

$$-q \ln q + q - \tfrac{1}{2} \ln q = -\tfrac{1}{2} \ln \bar{M}_M(t)$$

$$-\frac{3}{2} \ln q + O([\bar{M}_M(t)]^{-1/2} \tag{6.66}$$

Equation 6.66 is equivalent to the asymptotic expression:

$$\rho_q(t) \underset{t \uparrow \infty}{\sim} \rho_0 \, [\, \bar{M}_M(t) \,]^{-2} \, [\, q/\bar{M}_M(t) \,]^{-3/2} \tag{6.67}$$

in agreement with Eq. 6.55 if $B \equiv \rho_0$. Therefore the solutions of Eq. 6.57 have the generic, asymptotic scaling form in Eq. 6.49 despite the exponential time dependence of the mass-average floccule size, $\bar{M}_M(t)$. This property is consistent with the fractal nature of the floccules as expressed in Eq. 6.1.

6.5 The Stability Ratio

Fuchsian kinetics lead to the model form of the von Smoluchowski rate law that is obtained by introducing Eq. 6.52 into Eq. 6.10:

$$\frac{d\rho_q}{dt} = \frac{1}{2} \sum_m \left(\frac{k^D_{mn}}{W_{mn}} \right) \rho_m \rho_{q-m} - \rho_q \sum_p \left(\frac{k^D_{qp}}{W_{qp}} \right) \rho_p \tag{6.68}$$

where k^D_{mn} is the model rate coefficient for transport-controlled flocculation kinetics that appears in Eq. 6.16. The corresponding differential equation for the total number of floccules per unit volume, $M_0(t)$, is

$$\left(\frac{dM_0}{dt} \right)_F \equiv -\frac{1}{2} \sum_m \sum_n \left(\frac{k^D_{mn}}{W_{mn}} \right) \rho_m \rho_n \tag{6.69}$$

according to Eq. 6.13a. The differential equation for the second moment, $M_2(t)$, is

$$\left(\frac{dM_2}{dt} \right)_F \equiv \sum_m \sum_n mn \left(\frac{k^D_{mn}}{W_{mn}} \right) \rho_m \rho_n \tag{6.69.b}$$

according to Eq. 6.13b. Both of these equations can be applied to define an *experimental stability ratio*:[34]

$$W^0_{exp}(t) \equiv \frac{(dM_0/dt)_D}{(dM_0/dt)_F} \tag{6.70a}$$

$$W^2_{exp}(t) \equiv \frac{(dM_2/dt)_D}{(dM_2/dt)_F} \tag{6.70b}$$

where $(\ )_D$ is equal to the right side of Eq. 6.13 with $k_{mn} = k^D_{mn}$(Eq. 6.16). Thus the experimental stability ratio is related to the rate of increase in either the number-average size (Eqs. 6.14a and 6.70a), or the mass-average size (Eqs. 6.14b and 6.70b), as compared to the rate of increase during a purely transport-controlled flocculation process. The former relationship depends on experimental methods to measure the number of floccules (direct particle counting or colligative properties), whereas the latter definition relates to light-scattering experiments that measure $\bar{M}_M(t)$.[16] Implicit in these definitions is the assumption that the same system of "primary particles" can be made to undergo both kinds of flocculation process, as in the case of the Au colloids described in Section 6.1.

The most common scenario for measuring the experimental stability ratio involves its *initial* value ($t = 0$). If the initial state of a suspension is arranged to comprise only "primary particles," then Eq. 6.58 applies and Eq. 6.70 reduces to the expression:

$$\lim_{t \downarrow 0} W^0_{exp}(t) = \lim_{t \downarrow 0} W^2_{exp}(t) = W_{11} \tag{6.71}$$

where W_{11} is given by Eq. 6.51 for $m = n = 1$. Model calculations of W_{11} suggest that Eq. 6.53 is an excellent approximation if $V(\xi_{11}) = V_{max}$, the (positive) maximum value of the potential energy of interaction between two "primary particles."[35,36] This latter condition is readily apparent after noting that because $V(r)$ is exponentiated in Eq. 6.51, a very small shift from its maximum value can produce order-of-magnitude decreases in W_{11}.[35,37]

The sensitivity of W_{11} to chemical factors can be appreciated through a simple model calculation based on Eq. 6.53. Perhaps the least complicated expression for the interparticle potential energy is the large-separation approximation.[38]

$$\varphi(d) \approx (64\, a^2/Z\kappa)cRT \exp(-Z\kappa d) - (A/12\pi d^2 k) \tag{6.72}$$

where $\varphi(d)$ is the potential energy of two *planar* particle surfaces per unit surface area, with the surfaces separated by the distance d, such that $Z\kappa d > 1$. In Eq. 6.72, the parameter a is related to the electric potential near the surface of a charged particle; $\kappa \equiv (\beta c)^{1/2}$, where $\beta = 1.084 \times 10^{16}$ m mol^{-1} at 298 K is a parameter dependent on the dielectric properties of water and c is the concentration of $Z - Z$ electrolyte solution in which the "primary particles" are suspended; R is the molar gas constant, T is absolute temperature, and A is the Hamaker constant.[38] The first term on the right side of Eq. 6.72 represents a screened electrostatic repulsion between two particles of like charge sign, whereas the second term represents particle attraction caused by van der Waals dispersion forces.[38] For natural colloids the parameter A lies in the range $0.1–5.0 \times 10^{-20}$ J.[39] Given the approximate expression for $\varphi(d)$, its maximum value can be calculated by imposing the condition $(\partial\varphi/\partial d)_{d = d_m} = 0$:

$$64\ a^2\ cRT\ \exp(-Z\kappa d_m) = (A/6\pi d_m^3) \tag{6.73}$$

The combination of Eqs. 6.72 and 6.73 leads to the expression

$$\begin{aligned}\varphi_{max}(d_m) &= (A/12\pi d_m^2)\ [(2/Z\kappa d_m) - 1] \\ &= (64\ a^2 cRT/Z\kappa)\ \exp(-Z\kappa d_m)\ [1 - (Z\kappa d_m/2)]\end{aligned} \tag{6.74}$$

which provides alternative equations for the maximum value of $\varphi(d)$. It follows from Eqs. 6.53 and 6.74 that

$$\begin{aligned}\ln W_{11} &\approx \varphi_{max}(d_m)/\kappa_B T \\ &= [(A/\kappa_B T)/12\pi d_m^2]\ [(2/Z\kappa d_m) - 1] \\ &= (64\ a^2\ cN_A/Z\kappa)\ \exp(-Z\kappa d_m)\ [1 - (Z\kappa d_m/2)]\end{aligned} \tag{6.75}$$

where N_A is the Avogadro constant. In Eq. 6.75 it is understood that the parameter d_m remains an implicit function of variables, such as c or T, and interaction parameters, such as a or A, because of Eq. 6.73.

The sensitivity of the stability ratio to chemical or particle interaction factors can be illustrated by an examination of the model expression for W_{11} in Eq. 6.75. For example, if temperature and the particle interaction parameters are fixed, then W_{11} will vary with the concentration, c (also included in κ), of Z–Z electrolyte. At low values of c, κ is also small, and the first equality in Eq. 6.75 indicates that W_{11} will take on its largest values. (Decreasing c also provokes an increase in d_m because of Eq. 6.73, but this effect is dominated by that of κ.[40]) Conversely, as c increases, the value of W_{11} will drop until it achieves its minimum, $W_{11} = 1.0$, when $Z\kappa d_m = 2$ (Eq. 6.75). At this concentration, termed the *critical coagulation concentration* (ccc), or *flocculation value*,[1] the flocculation process has become transport-controlled and therefore is rapid. Thus in general

$$\lim_{c \uparrow ccc} \ln W_{exp} = 0 \tag{6.76}$$

defines the critical coagulation concentration implicitly in terms of the experimental stability ratio (Fig. 6.8).

An explicit expression for ccc in the context of the model of $\varphi(d)$ in Eq. 6.72 is derived by combining the condition $Z\kappa d_m = 2$, which follows from applying Eq. 6.76 to Eq. 6.75, with the constraint on d_m in Eq. 6.73:

$$\frac{\kappa^3}{ccc} = \left(\frac{3072\,\pi}{e^2}\ \frac{a^2 RT}{A}\right) Z^{-3}$$

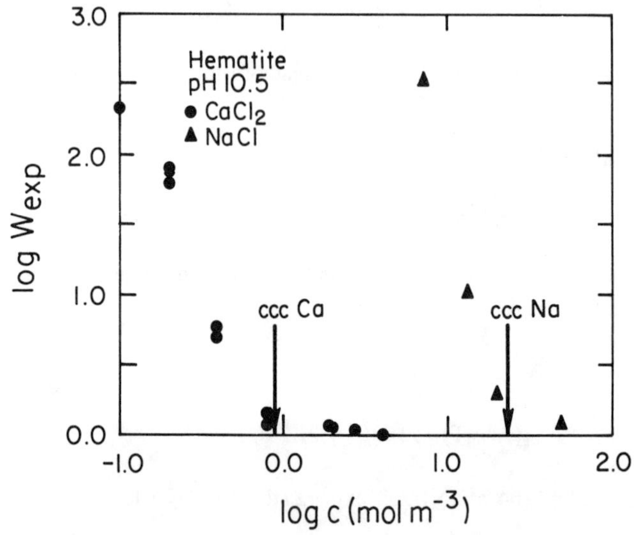

FIG. 6.8. Log-log plot of W_{exp} versus electrolyte concentration for hematite (α-Fe_2O_3) colloids suspended in either $CaCl_2$ or NaCl solution at pH 10.5. Arrows indicate critical coagulation concentrations [Eq. 6.76; data from L. Liang, Effects of surface chemistry on kinetics of coagulation of submicron iron oxide particles (α-Fe_2O_3) in water, Ph.D. dissertation, California Institute of Technology, Pasadena, CA, 1988. Environmental Quality Laboratory Report No. AC-5-88].

or

$$\text{ccc} = \left(\frac{3072\,\pi\, a^2 RT}{e^2 \beta^{3/2} A} \right) Z^{-6} \tag{6.77}$$

where e is the base of natural logarithms. Equation 6.77 predicts that the ccc will decrease significantly as the ion valence increases, in agreement with many experimental data (although not necessarily with the power −6 for Z).[39-41] This behavior is illustrated in Fig. 6.8 for the cases Z = 1 or 2. Note that the model also predicts that ccc will decrease as the strength of the attractive particle interaction increases (increasing value of A in Eq. 6.72). Calculations based on more detailed representations of the interparticle potential energy support this concept.[36,40]

The mechanism by which increasing electrolyte concentration decreases W_{exp} is essentially a screening of particle surface charge by an ion swarm.[39] This mechanism is represented in the factor $Z\kappa d_m$ in Eq. 6.75. Another mechanism for decreasing W_{exp} is represented in the factor a in Eq. 6.75, which is a function of the electric potential near the particle surface [$a = \tanh(ZF\psi_s/4RT)$ in the model equation for $\varphi(d)$, where ψ_s is a near-surface electric potential].[38]

Adsorption of ions by the interacting particles affects this electric potential, with inner-sphere surface complexation (i.e., charge neutralization) providing the most dramatic effect and adsorption in the diffuse ion swarm the smallest effect (i.e., charge screening). For example, in the case of H^+ or OH^-, strong adsorption by suspended particles can increase or decrease the near-surface electric potential significantly, with little or no change in κ, if other electrolyte ions are present at sufficient concentrations.[1] Thus, in principle, adjustment of the pH value can produce a zero value of the electric potential, and therefore particle charge, irrespective of the bulk electrolyte concentration. When this occurs, the screened electrostatic repulsion between particles also vanishes (cf. Eq. 6.72, for a = 0) and only attractive forces operate, causing rapid flocculation and $W_{exp} \leq 1$.[36] In general,

$$\lim_{\mathrm{pH} \to \mathrm{PZC}} \ln W_{exp} = 0 \tag{6.78}$$

defines the *point of zero charge* (PZC), the pH value at which rapid flocculation is produced at some fixed electrolyte concentration below the ccc. Clearly, the value of PZC will depend on the composition of the electrolyte solution and suspended particles.[1]

Surface electric potential control (or surface charge control) of the rate of flocculation is possible for any adsorptive that forms a surface complex with suspended particles, as discussed in Section 6.1 and in Chapter 4 (cf. Table 4.2). Among these adsorptives for soil colloids are oxyanions, such as phosphate or oxalate, and transition metal cations. An expression analogous to Eq. 6.78 can be developed to define points of zero charge for any such adsorptive, as illustrated in Fig. 6.9.[42]

Special Topic 3: Cluster Fractals

The term *fractal*, coined by Benoit Mandelbrot from the Latin adjective *fractus* ("broken"), refers to the limiting properties of mathematical objects that exhibit the three attributes of *similar structure* over a range of length (or time) scales; *intricate structure*, or complexity, that is scale independent; and *irregular structure* that cannot be captured entirely within the purview of classical (Euclidean) geometric concepts, necessitating, for example, the use of a spatial dimension that is not an integer.[43] Like the objects studied in classical geometry—circles, spheres—the objects studied in fractal geometry are idealizations that are only approximated in natural systems, but nonetheless are useful to their mathematical description.

The essence of a flocculation process as described in Section 6.1 is the combining of "primary particles" into floccules, followed by the combining of the floccules into larger floccules, and so on. That this process can lead to a fractal object is illustrated by the sequence of constructing clusters from a unit

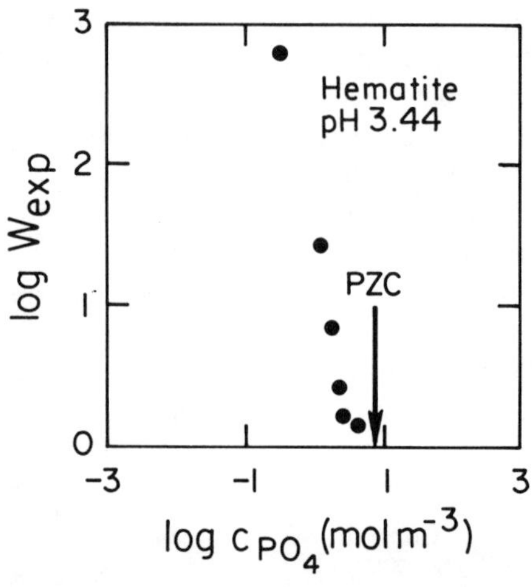

FIG. 6.9. Log-log plot of W_{exp}, for hematite (α-Fe_2O_3) colloids at pH 3.44, versus *o*-phosphate concentration. The arrow indicates the point of zero charge, defined operationally as the concentration of *o*-phosphate at which W_{exp} = 1.0 (data from L. Liang, Effects of surface chemistry on kinetics of coagulation of submicron iron oxide particles (α-Fe_2O_3) in water, Ph.D. dissertation, California Institute of Technology, Pasadena, CA, 1988. Environmental Quality Laboratory Report No. AC-5-88).

comprising five disks, each of diameter d_0 (Fig. 6.10).[44] The unit itself has a characteristic dimension (or diameter) equal to $3d_0$. If five of these units are combined to form a cluster with the same symmetry as a unit (i.e., each unit in the cluster arranged like a disk in the unit), the characteristic dimension grows to $9d_0$ (B in Fig. 6.10). If five of these clusters are then combined in a way that preserves the inherent symmetry (C in Fig. 6.10), the dimension increases to $27d_0$, and so on. The clusters formed in this simple process exhibit similar structure (in the sense of their symmetry), complexity, and irregularity (cf. D in Fig. 6.10). Therefore it would seem that they could qualify as fractal objects.

The size of each cluster in the sequence as expressed through the number N of primary particles it contains is 5, 25, 125, and 625 for the four examples shown in Fig. 6.10. Thus, $N = 5^n$, where n = 1, 2, ..., denotes the stage of cluster growth in the sequence. In the same vein the characteristic dimension, $L = 3^n d_0$, where n = 1, 2, . . . , once again. The relationship between these two properties, size and dimension, can be expressed mathematically by:[43]

$$N(L) = AL^D \qquad \text{(s3.1)}$$

where A and D are positive parameters. In the present example Eq. s3.1 has the form

$$5^n = A(3^n d_0)^D \quad (n = 1, 2, \ldots) \qquad \text{(s3.2)}$$

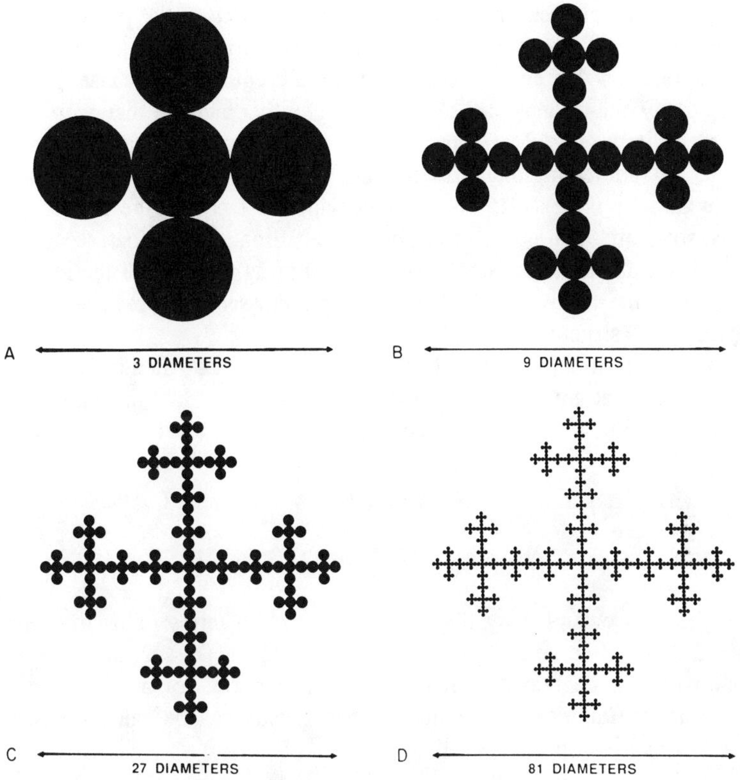

Fig. 6.10. A sequence of prefractal clusters created by combining five-particle units (A) in a way that preserves the inherent symmetry of the units (B–D). (Adapted, with permission, from P. Meakin[6]. Copyright © by the American Geophysical Union.)

from which it follows that $A = d_0^{-D}$ and

$$D = \ln 5/\ln 3 \approx 1.465 \tag{s3.3}$$

after substitution for A in Eq. s3.2 and solving the resulting expression for the exponent D. More generally, if r denotes the scale factor by which the characteristic dimension increases at each successive stage, then

$$D \equiv \lim_{n\uparrow\infty} \frac{1}{n} [\ln N(n)/\ln r] \tag{s3.4}$$

defines the *fractal dimension* of a cluster whose size is N(n) at any stage. For the cluster in Fig. 6.10, r = 3 and $N(n) = 5^n$, leading to a fractal dimension as given by Eq. s3.3.

The fractal dimension of a cluster is a numerical measure of its space-filling capability. The clusters in Fig. 6.10 occupy space in a plane (Euclidean dimension = 2). If they were constituted by disks arranged in a single row, the fractal dimension that characterizes them should equal 1.0 because they would be effectively one-dimensional objects. On the other hand, if they were compact structures comprising closely packed disks, their fractal dimension would be equal to 2.0, indicating their space-filling nature. The fractal dimension of the clusters actually is near 1.5, in keeping with the fact that the clusters have a porous structure that is not entirely space filling. (Note that, according to Eq. s3.4, the fractal dimension is strictly a limiting property of the infinite-size cluster, with the finite-size clusters in the growth sequence regarded therefore as "prefractal" structures.[43])

The porous structure of a cluster fractal can be quantified by estimating its number density at any stage of growth. For example, the "number density" of any cluster in Fig. 6.10 can be calculated with the equation

$$\begin{aligned} \rho_n &= N(n)/(\pi L(n)^2/4) \\ &= (4A/\pi)\,[L(n)]^{D-2} \qquad \text{(s3.5)} \end{aligned}$$

where the area occupied by the cluster is $\pi L^2/4$ in terms of its diameter and $L(n) = 3^n d_0$ is the diameter of the nth cluster. Since $D < 2$, it follows from Eq. s3.5 that ρ_n decreases as n increases. In general, for a cluster fractal in E-dimensional Euclidean space, its number density can be calculated with the power-law expression:

$$\rho(L) = bA\, L^{D-E} \qquad \text{(s3.6)}$$

where b is an appropriate geometric factor ($b = 4/\pi$ in Eq. s3.5). Equation s3.6 epitomizes the important feature of cluster fractals, that *their porosity always increases as their size increases*. This feature can be used as an experimental criterion for evaluating whether a cluster is a fractal object. On the other hand, information as to cluster shape, or pore shape within a cluster, has no bearing on whether the cluster is fractal.[43]

Cluster fractals are defined by Eq. s3.1, and therefore any experimental technique that measures cluster size and dimension can be applied to determine the fractal dimension D. The most common techniques for doing this are electron microscopy and light scattering.[45] In the former method, N(L) and L are determined simultaneously by direct measurement on a set of representative micrographs.[7] If $D < 2$, the fact that micrographs are two-dimensional projections of three-dimensional objects does not affect the measured value of the fractal dimension. In the light-scattering method the intensity of radiation scattered by a suspension is determined over a range of scattering angles. If the light incident on the suspension is not absorbed by the colloids and is of long wavelength as compared to the spatial extent of a "primary particle," the intensity of the scattered light I_s can be modeled by[46]

$$I_s(Q,t) = I_0\bar{M}_M(t)\,[\,1 - \tfrac{1}{3}Q^2R^2(t)\,] \qquad (s3.7)$$

where $R^2(t) = \Sigma_q q^2[\rho_q(t)/\rho_0]R_q^{\,2}$ is a mean-square floccule radius, Q is a function of the angle of scattering and wavelength of the incident light, and I_0 is the intensity of light scattered by a suspension of "primary particles," assumed to be the initial state (cf. Eq. 6.58). For cluster fractals, the mass-average cluster size $\bar{M}_M$ and the root-mean-square cluster radius are connected by Eq. s3.1. Thus measurements of $I_s(Q,t)$ can be "inverted" to yield estimates of the fractal dimension.[46] Model forms of $\rho_q(t)$, such as those appearing in Eqs. 6.18 and 6.59, can also be inserted into Eq. s3.7 to predict the time dependence of the intensity of scattered light.

NOTES

1. Basic concepts of colloid chemistry that are relevant to soils are discussed in Chap. 10 of G. Sposito, *The Chemistry of Soils*, Oxford University Press, New York, 1989, and in the first three chapters of R. J. Hunter, *Foundations of Colloid Science*, Vol. I, Clarendon Press, Oxford, 1987. The definitions of *flocculation*, *floccule*, and *aggregate* given in the present chapter are those of its author.

2. See, for example, C. R. O'Melia, Particle–particle interactions in aquatic systems, *Colloids Surfaces* **39**:255 (1989) for a review of these phenomena from the perspective of colloid chemistry.

3. See, for example, M F. DeBoodt, M. H. B. Hayes, and A. Herbillon, *Soil Colloids and Their Associations in Aggregates*, Plenum Press, New York, 1990, for comprehensive surveys of these applications.

4. See, for example, Parts II and III of J. Buffle and H. P. van Leeuwen, *Environmental Particles*, Vol 1, Lewis Publishers, Boca Raton, FL, 1992, for surveys of these issues.

5. See Chapter V of R. Jullien and R. Botet, *Aggregation and Fractal Aggregates*, World Scientific, Singapore, 1987, for a description of cluster–cluster flocculation processes in the context of computer models.

6. P. Meakin, Fractal aggregates in geophysics, *Rev. Geophys.* **29**:317 (1991).

7. D. A. Weitz and M. Oliveria, Fractal structures formed by kinetic aggregation of aqueous gold colloids, *Phys. Rev. Lett.* **52**:1433 (1984). See also D. A. Weitz and J. S. Huang, Self-similar structures and the kinetics of aggregation of gold colloids, pp. 19–28 in *Kinetics of Aggregation and Gelation*, ed. by F. Family and D. P. Landau, Elsevier, Amsterdam, 1984.

8. See Chapter V (Table II) in Jullien and Botet, op. cit.,[5] and P. Meakin, op. cit.,[6] for discussions of these results.

9. Diffusionally mediated collisions between *two* floccules of equal size can be described by a *second-order* rate coefficient $K_D = 8\pi R\mathbf{D}$, where R is the radius and $\mathbf{D}$ is the diffusion coefficient of a floccule. Upon invoking the Stokes-Einstein relation, $\mathbf{D} = k_BT/6\pi\eta R$, one derives Eq. 6.2. For an introductory discussion of the second-order rate law for particle collisions, see, for example, Chap. 11 in P. C. Hiemenz, *Principles of Colloid and Surface Chemistry*, Marcel Dekker, New York, 1986.

10. D. A. Weitz, J. S. Huang, M. Y. Lin, and J. Sung, Dynamics of diffusion-

limited kinetic aggregation, *Phys. Rev. Lett.* **53**:1657 (1984).

11. C. Aubert and D. S. Cannell, Restructuring of colloidal silica aggregates, *Phys. Rev. Lett.* **56**:738 (1986); V. A. Hackley and M. A. Anderson, Effects of short-range forces on the long-range structure of hydrous iron oxide aggregates, *Langmuir* **5**:191 (1989).

12. D. A. Weitz, J. S. Huang, M. Y. Lin, and J. Sung, Limits of the fractal dimension for irreversible kinetic aggregation of gold colloids, *Phys. Rev. Lett.* **54**:1416 (1985).

13. An introductory—but thorough—discussion of the von Smoluchowski rate law is given in Chap. VII of H. Kruyt, *Colloid Science*, Vol. I, Elsevier, Amsterdam, 1952. See also Section 7.8 in R. J. Hunter, op. cit.[1]

14. R. M. Zipp, Aggregation kinetics via Smoluchowski's equation, pp. 191–199 in F. Family and D. P. Landau, op. cit.[7]

15. Technically, this result follows only if $k_{mn} < A(m + n)$, where A is a positive constant. If this condition is not met, cluster growth to a unit comprising all the "primary particles" can occur after a *finite* time interval (cf. Eq. 6.20b), corresponding to gel formation and a nonconstant M_1. For a discussion of this point, see F. Leyvraz, Critical exponents in the Smoluchowski equations of coagulation, pp. 201–204 in F. Family and D. P. Landau, op. cit.[7] and F. Leyvraz and H. R. Tschudi, Singularities in the kinetics of coagulation processes, *J. Phys. A* **14**:3389 (1981).

16. Floccule-size parameters and their measurement are discussed in Chap. 1, 3, and 5 of P. C. Hiemenz, op. cit.[9]

17. See, for example, pp. 636–639 in P. C. Hiemenz, op. cit.,[9] for an elementary discussion of diffusion-mediated particle collisions.

18. In general, the Brownian motions of two neighboring colloids will induce a correlation between their displacements that is not zero and **D** will exhibit a "cross-coupling" term in addition to the diffusion coefficients of the two colloids when widely separated. For a detailed discussion of **D**, see Section 4.1.2 in Th. G. M. van de Ven, *Colloidal Hydrodynamics*, Academic Press, London, 1989.

19. See, for example, Chap. 6 in H. J. V. Tyrrell and K. R. Harris, *Diffusion in Liquids*, Butterworths, London, 1984.

20. Corrections to Eq. 6.16d for clusters of different size are discussed in Chap. VII of H. Kruyt, op. cit.[13]

21. Equation 6.18 can be derived by introducing the "generating function"[14]

$$f(x,t) = \Sigma_q\,[\exp(qx) - 1]\,\rho_q(t)$$

into Eq. 6.17 to derive the differential equation

$$\frac{\partial f}{\partial t} = K_D[\,f(x,t)\,]^2$$

whose solution is

$$f(x,t) = \rho_0\,[\exp(x) - 1]\,/\,\{1 - [\exp(x) - 1]K_D\rho_0 t\}$$

for the initial condition $f(x,0) = \rho_0\,[\exp(x) - 1]$ that is consistent with that imposed in Eq. 6.18. With the transformation of variable $y \equiv \exp(x)$, it becomes apparent that $f(y, t)$ is a MacLaurin expansion in y with coefficients $\rho_q(t)$:

$$f(y, t) = \Sigma_q y^q \rho_q(t) - \Sigma_q \rho_q(t)$$
$$= \rho_0 (y - 1) / [1 - (y - 1) K_D \rho_0 t]$$

Therefore

$$\rho_q(t) = \frac{1}{q!} \left(\frac{\partial^q f}{\partial y^q} \right)_{y=0} \qquad (q = 1, 2, \ldots)$$

where the explicit solution for f(y, t) is used in calculating the partial derivatives. For example,

$$\rho_1(t) = \left(\frac{\partial f}{\partial y} \right)_{y=0} = \frac{\rho_0}{(1 + K_D \rho_0 t)^2}$$

$$\rho_2(t) = \left(\frac{\partial^2 f}{\partial y^2} \right)_{y=0} = \frac{\rho_0^2 K_D t}{(1 + K_D \rho_0 t)^3}$$

and so on, with the general result given by Eq. 6.18. Note also that f(0, t) = $M_0(t)$ and that

$$M_\alpha(t) = \left(\frac{\partial^\alpha f}{\partial y^\alpha} \right) \qquad (\alpha = 1, 2, \ldots)$$

An alternate derivation of Eq. 6.18 is given on p. 17f in R. Jullien and R. Botet, op. cit.[5]

22. See, for example, Section 2.7 of J. Feder, *Fractals*, Plenum Press, New York, 1988, for a discussion of homogeneity conditions. Homogeneous functions are described in detail in Section 11.1 of H. E. Stanley, *Introduction to Phase Transitions and Critical Phenomena*, Oxford University Press, New York, 1971.

23. R. M. Ziff, E. D. McGrady, and P. Meakin, On the validity of Smoluchowski's equation for cluster-cluster aggregation kinetics, *J. Chem. Phys.* **82**:5269 (1985).

24. See, for example, P. Meakin, Fractal aggregates, *Adv. Colloid Interface Sci.* 28:249 (1988) for a discussion of different model expressions for k_{mn}.

25. T. Vicsek and F. Family, Critical dynamics in cluster-cluster aggregation, pp. 111–115 in F. Family and D. P. Landau, op. cit.[7]

26. P. Meakin, The growth of fractal aggregates and their fractal measures, *Phase Transitions Crit. Phenomena* **12**:335 (1988).

27. P. Meakin, The effects of attractive and repulsive interaction on three-dimensional reaction-limited aggregation, *J. Colloid Interface Sci.* **134**:235 (1990).

28. See, for example, Chap. VII of H. Kruyt, op. cit.[13] or Chap. 11 of P. C. Hiemenz, op. cit.,[9] for an introductory discussion of Fuchsian kinetics.

29. J. N. Israelachvili, *Intermolecular and Surface Forces*, Academic Press, New York, 1992. A brief summary of research on the mathematical form of V(r) appears in Chap. 6 of G. Sposito, *The Surface Chemistry of Soils*, Oxford University Press, New

York, 1984. See also J. N. Israelachvili and P. M. McGuiggan, Forces between surfaces in liquids, *Science* **241**:795 (1988).

30. See, for example, p. 65 in E. T. Whittaker and G. N. Watson, *A Course of Modern Analysis*, Cambridge University Press, London, 1963.

31. D. A. Weitz and M. Y. Lin, Dynamic scaling of cluster-mass distributions in kinetic colloid aggregation, *Phys. Rev. Lett.* **57**:2037 (1986).

32. R. C. Ball, D. A. Weitz, T. A. Witten, and F. Leyvraz, Universal kinetics in reaction-limited aggregation, *Phys. Rev. Lett.* **58**:274 (1987).

33. See, for example, p. 251 in E. T. Whittaker and G. N. Watson, op. cit.[30]

34. The experimental stability ratio is defined generally as the rate of "fast" flocculation divided by the rate of "slow" coagulation (see, e.g., Section 7.8 in R. J. Hunter, op. cit.[1]). Equation 6.70 is a mathematical interpretation of this definition in terms of experimentally accessible quantities related to floccule size.

35. See Chap. XII in E. J. W. Verwey and J. Th. G. Overbeek, *Theory of the Stability of Lyophobic Colloids*, Elsevier, Amsterdam, 1948.

36. D. C. Prieve and E. Ruckenstein, Role of surface chemistry in primary and secondary coagulation and heterocoagulation, *J. Colloid Interface Sci.* **73**:539 (1980).

37. See Section 7.8.2 in R. J. Hunter, op. cit.[1]

38. The approximation in Eq. 6.72 requires that d be large compared to $1/\kappa$ ("diffuse double layer thickness") but small compared to the particle dimension. See Chaps. 4 and 6 in R. J. Hunter, op. cit.[1] It should be noted in passing that V(r) is, speaking strictly, not a potential energy but is instead a potential of mean force, a statistical thermodynamic quantity (hence the dependence of φ(d) on concentration, temperature, etc.). A complete discussion of this point is given by S. L. Carnie and G. M. Torrie, The statistical mechanics of the electrical double layer, *Adv. Chem. Phys.* **56**:141 (1984).

39. J. Th. G. Overbeek, Strong and weak points in the interpretation of colloid stability, *Adv. Colloid Interface Sci.* **16**:17 (1982); The rule of Schulze and Hardy, *Pure Appl. Chem.* **52**:1151 (1980).

40. See, for example, Chap. 12 in P. C. Hiemenz, op. cit.[9]

41. I. M. Metcalfe and T. W. Healy, Charge-regulation modeling of the Schulze-Hardy rule and related coagulation effects, *Faraday Discuss. Chem. Soc.* **90**:335 (1990).

42. L. Liang and J. J. Morgan, Chemical aspects of iron oxide coagulation in water: Laboratory studies and implications for natural systems, *Aquatic Sci.* **52**:32 (1990).

43. See Chaps. 2 and 3 in J. Feder, op. cit.[22] An introduction to fractal concepts with more mathematical detail is given by K. Falconer, *Fractal Geometry*, Wiley, Chichester, UK, 1990.

44. T. Vicsek, Fractal models for diffusion-controlled aggregation, *J. Phys. A* **16**:L647 (1983).

45. See, for example, Chap. 3.1.3 in D. Avnir, *The Fractal Approach to Heterogeneous Chemistry*, Wiley, Chichester, UK, 1989, for a review of these and other methods.

46. J. Feder and T. Jøssang, A reversible reaction limiting step in irreversible immunoglobulin aggregation, pp. 99–131 in *Scaling Phenomena in Disordered Systems*, ed. by R. Pynn and A. Skjeltorp, Plenum Press, New York, 1985; M. Y. Lin et al., Universality of fractal aggregates as probed by light scattering, *Proc. R. Soc. Lond.* *A***423**:71 (1989).

FOR FURTHER READING

Avnir, D., *The Fractal Approach to Heterogeneous Chemistry*, Wiley, Chichester, UK, 1989. The first three chapters of this edited treatise provide a comprehensive review of fractal concepts and experimental methodologies to measure the fractal dimension.

Family F., and D. P. Landau, *Kinetics of Aggregation and Gelation*, North-Holland, Amsterdam, 1984. An excellent compendium of articles on cluster fractals—both theory and experiment.

Feder, J., *Fractals*, Plenum Press, New York, 1988. Perhaps the most accessible introduction to fractal concepts, written with the *imprimatur* of Benoit Mandlebrot.

Hunter, R. J., *Foundations of Colloid Science*, Clarendon Press, Oxford, 1987 (Vol. I), 1989 (Vol. II). A comprehensive, edited treatise on all aspects of classical colloid chemistry—an excellent reference for any soil chemist.

Vicsek, T., *Fractal Growth Phenomena*, World Scientific, Singapore, 1992. A comprehensive review of cluster fractal growth processes. The experimental details of the use of light scattering to determine the fractal properties of floccules are given in three fine articles by M. Y. Lin, H. M. Lindsay, D. A. Weitz, R. C. Ball, R. Klein, and P. Meakin: Universality of fractal aggregates as probed by light scattering, *Proc. R. Soc. Lond. A***423**:71 (1989); The structure of fractal colloidal aggregates of finite extent, *J. Colloid Interface Sci.* **137**:263 (1990); Universal diffusion-limited colloid aggregation, *J. Phys. Condensed Matter* **2**:3093 (1990).

PROBLEMS

1. Light-scattering experiments on flocculating suspensions of silica colloids provided the data in the following table. (R is the average cluster radius.) Estimate the fractal dimension of the clusters formed and indicate whether the flocculation process is transport or reaction controlled. (*Hint*: Apply Eq. 6.1 and the concepts in Section 6.1.)

$\bar{M}_M$:	30	120	500	2,200	4,400
R(nm):	1.0	2.0	3.8	8.0	10.5

2. In the following table are data on the dependence of the average floccule radius, achieved after 500 s of flocculation, on the initial "primary particle" number density for a transport-controlled flocculation process. Estimate the fractal dimension of the floccules formed. (*Answer*: D $\approx$ 1.8, based on a log–log plot.)

$R(\mu m)$:	1.29	0.89	0.73	0.58
$\rho_0(10^{18} m^{-3})$:	1.5	0.76	0.51	0.39

3. Derive the zeroth-order kinetics model expression in Eq. 6.3 from the von Smoluchowski rate law. (*Hint*: Choose a suitable representation for N(t) and for the rate coefficient k_{mn} in Eq. 6.10.)

4. Measurements of the average floccule radius in a suspension of colloids, using light-scattering techniques, indicated the following time dependence: $R(t) = R_0(1 + \gamma t)^{\beta}$, where $R_0 = 5$ nm, $\gamma = 9.3\ s^{-1}$, and $\beta = 0.56$. Derive this equation, and estimate the initial number density of "primary particles" as well as the fractal dimension of the floccules formed. (*Answer*: $\rho_0 = 7.5 \times 10^{17}\ m^{-3}$ and $D = 1.78$.)

5. The average floccule radius in a flocculating colloidal suspension was measured as a function of time by light-scattering methods and the data were fitted to the equation:

$$R(t) = 8 \exp(t/2.24)$$

where R is in nanometers and t is in hours. Given that $\rho_0 = 1.6 \times 10^{18}\ m^{-3}$, estimate the rate coefficient k_F for the flocculation kinetics. (*Answer*: $k_F \approx 1.6 \times 10^{-22}\ m^3\ s^{-1}$.)

6. Direct particle counting of an initially monodisperse suspension was used to measure the time dependence of the q-moment M_0, as given in the following table. Examine these data for conformity to either transport- or reaction-controlled flocculation kinetics and estimate the characteristic time scale, $2/k_{11}\rho_0$, where $k_{11} = k_{mn}$ for $m = n = 1$. (*Answer*: $k^F_{mn} \approx 3.05 \times 10^{-22}\ m^3\ s^{-1} \equiv 2K_D/W_{mn}$, corresponding to $W_{mn} = 4.07 \times 10^4$ for all m, n.)

$M_0(t)\ (10^{16}\ m^{-3})$:	3.35	1.91	1.46	0.75	0.47
t (min):	0	2,550	4,050	11,250	20,100

7. Experimentally, it is observed that for sufficiently long flocculation times, semilog plots of $(\bar{M}^2_M \rho_q/\rho_0)$ versus $q/\bar{M}_M$ are approximately linear if flocculation is transport controlled, whereas log–log plots of $(\bar{M}^2_M \rho_q/\rho_0)$ versus $q/\bar{M}_M$ are approximately linear if flocculation is reaction controlled. Develop a theoretical basis for these two observations. (*Hint*: Review Sections 6.3 and 6.4.)

8. The asymptotic scaling expression for $\rho_q(t)$ in Eq. 6.49 appears to be accurate for models of both transport- and reaction-controlled flocculation

kinetics (with $B = \rho_0$). Examine the scaling property of this expression and deduce the requirements for a scale-invariant von Smoluchowski rate law. (*Hint*: Consider ρ_q as an explicit function of q and $\bar{M}_M$ only to infer $a = -2$ upon applying the homogeneity condition in Eq. 6.26. Then incorporate the scaling property in Eq. 6.35 into a scaled von Smoluchowski rate law like that in Eq. 6.47. If $\lambda^\delta t$ is chosen to be the scaling of the time variable, then scale invariance requires $\delta + \theta = 1$. Thus, for reaction-controlled kinetics modeled by Eq. 6.56, no scaling of the time variable is required; whereas for transport-controlled kinetics modeled by Eq. 6.16d, the time variable is scaled by λ to the first power.)

9. Often the dependence of W_{exp} on electrolyte concentration is linear in a log–log plot of data. Examine this possibility with the data in the following table. Develop an equation of the form $\ln W_{exp} = a + b \ln c$ to estimate ccc in mol m^{-3}. [*Answer*: ccc $\approx \exp(4.603/2.833) = 5$ mol m^{-3}.]

W_{exp}(light scattering):	489	203	28	5.8	2.4	1.7
$[CaCl_2]$ (mol m^{-3}):	0.4	0.8	1.6	2.8	3.1	3.8

10. Measurements of W_{exp} for a suspension of hematite (α-Fe_2O_3) colloids in the presence of octanoic acid [$CH_3(CH_2)_6COOH$] led to the results listed in the following table. Estimate graphically the point of zero charge of hematite with respect to octanoic acid. (*Answer*: 1 mol m^{-3} octanoic acid.)

W_{exp}(light scattering):	9.0	8.4	6.4	4.1	2.3	2.0	1.5	1.3
Acid concentration(mol m^{-3}):	0.0047	0.01	0.1	0.2	0.4	3.2	1.6	0.8

INDEX